Lecture Notes
in Control and Information Sciences 326

Editors: M. Thoma · M. Morari

He-Sheng Wang · Chee-Fai Yung · Fan-Ren Chang

H_∞ Control for Nonlinear Descriptor Systems

With 19 Figures

Springer

Authors

Dr. He-Sheng Wang
Department of Communications and Guidance Engineering
National Taiwan Ocean University
Keelung, Taiwan RoC, 20224

Professor Chee-Fai Yung
Department of Electrical Engineering
National Taiwan Ocean University
Keelung, Taiwan RoC, 202

Professor Fan-Ren Chang
Department of Electrical Engineering
National Taiwan University
Taipei, Taiwan RoC, 106

British Library Cataloguing in Publication Data
Wang, He-Sheng
H_∞ control for nonlinear descriptor systems.
(Lecture notes in control and information sciences ; 326)
1. H_∞ control
I. Title II. Yung, Chee-Fai III. Chang, Fan-Ren
629.8'36
ISBN-10: 1-84628-289-6

Library of Congress Control Number: 2005937006

Lecture Notes in Control and Information Sciences ISSN 0170-8643
ISBN-10: 1-84628-289-6 Printed on acid-free paper
ISBN-13: 978-1-84628-289-8

Typesetting: Data conversion by authors.
Final processing by PTP-Berlin Protago-TEX-Production GmbH, Germany
Cover-Design: design & production GmbH, Heidelberg

Printed in Germany

9 8 7 6 5 4 3 2 1

Springer Science+Business Media

springer.com

To my parents, my wife and daughter.

H. S. Wang

To my wife Joyce,
and our daughters Felicia and Sandia,
who may want to read this book some day.

C. F. Yung

To my wife and daughter.
In memory of my parents.

F. R. Chang

Preface

In this monograph, we study the $\mathcal{H}_\infty$ control problem, as well as some other related topics, for descriptor systems, which are described by a set of nonlinear differential-algebraic equations (DAEs) of the form

$$\begin{aligned} \boldsymbol{E}\dot{\boldsymbol{x}} &= \boldsymbol{F}(\boldsymbol{x},\, \boldsymbol{w},\, \boldsymbol{u}), \\ \boldsymbol{z} &= \boldsymbol{Z}(\boldsymbol{x},\, \boldsymbol{w},\, \boldsymbol{u}), \\ \boldsymbol{y} &= \boldsymbol{Y}(\boldsymbol{x},\, \boldsymbol{w},\, \boldsymbol{u}), \end{aligned}$$

where $\boldsymbol{E}$ is, in general, a singular matrix. Necessary and sufficient conditions are derived for the existence of a controller solving the standard nonlinear $\mathcal{H}_\infty$ control problem. We first give various sufficient conditions for the solvability of $\mathcal{H}_\infty$ control problem for nonlinear descriptor systems. Both state feedback and output feedback cases are considered. Then, a necessary condition for the output feedback control problem to be solvable are obtained in terms of two Hamilton–Jacobi inequalities plus a weak coupling condition. Moreover, a parameterization of a family of output feedback controllers solving the problem is also provided. All of the aforementioned results are then specialized to the linear case. For the linear case, the necessary and sufficient conditions for the corresponding problems to be solvable are expressed in terms of two hierarchically coupled generalized algebraic Riccati equations which may be considered to be the generalizations of the Riccati equations obtained by Doyle *et al.*(1989). When these conditions hold, state-space formulae for a controller solving the problem is also given. The approach used in this monograph is based on a generalized version of Bounded Real Lemma. Finally, the derivation of state-space formulae for all controllers solving the standard $\mathcal{H}_\infty$ control problem for descriptor systems are proposed. To establish the key formulae, a parameterization of all internally stabilizing controllers for descriptor systems is also given (both the linear and nonlinear cases are considered in this monograph). The results obtained may be considered as the generalization of the corresponding ones given in Doyle *et al.* (1989) and Doyle (1984).

Among other important relevent topics, we investigate the balanced realization, reduced-order controllers design and mixed $\mathcal{H}_2/\mathcal{H}_\infty$ control problems.

Balanced Realization The classical ideas are extended to the nonlinear descriptor systems circumstances. For a stable and index one nonlinear descriptor system, we propose an open-loop balancing method that use a past and a future energy functions to bring the descriptor system into a balanced form, which will give us a measure of importance of a state component; namely, a larger singular value function corresponds to a more important state component.

Reduced-Order Controller Design Based on the central controller built on the two Hamilton-Jacobi inequalities (or two generalized algebraic Riccati inequalities for linear cases), we develop a sufficient condition, under which a reduced-order normal controller (i.e., the $\boldsymbol{E}$-matrix is nonsingular) can be constructed to solve the standard $\mathcal{H}_\infty$ control problem for descriptor systems, provided that three auxiliary equations have solutions.

Mixed $\mathcal{H}_2$ and $\mathcal{H}_\infty$ Control A mixed $\mathcal{H}_2/\mathcal{H}_\infty$ control problem for descriptor systems is solved via a Nash game approach. We consider both linear and nonlinear cases, and give complete feedback control strategies for each case. The problems are solved via state feedback and static output feedback, respectively. It is worth noting that the control strategies are characterized by two cross-coupled generalized algebraic Riccati equations for linear descriptor systems, while for nonlinear descriptor systems they are built upon two cross-coupled Hamilton–Jacobi equations.

Taipei, TAIWAN
November 2005,

He-Sheng Wang
Chee-Fai Yung
Fan-Ren Chang

Contents

List of Figures

List of Notation

$\Vert \bullet \Vert_2$	$\mathcal{L}_2$ norm
$\Vert \boldsymbol{G} \Vert_\infty$	$\mathcal{H}_\infty$-norm of the transfer function $\boldsymbol{G}$
$\lambda(\boldsymbol{E}, \boldsymbol{A})$	the set of all generalized eigenvalues of the pair $(\boldsymbol{E}, \boldsymbol{A})$
$\rho(\bullet)$	spectral radius
$\sigma(\bullet)$	singular value
$\oplus$	direct sum
$\mathbb{B}_\delta$	open ball with radius δ
$\mathbb{C}$	filed of complex numbers
$\mathbb{C}^n$	n-dimensional complex plane
$\mathbb{C}^+$	open right-half plane
$\mathbb{C}^-$	open left-half plane
$\bar{\mathbb{C}}^+$	closed right-half plane
$\bar{\mathbb{C}}^-$	closed left-half plane
C^r	the set of all r times continuous differentiable functions
DAE	Differential-Algebraic Equation
ESPR	Extended Strictly Positive Real
FI	Full Information
GARE	Generalized Riccati Equation
GARI	Generalized Riccati Inequality
GLMI	Generalized Linear Matrix Inequality
$\Im m\boldsymbol{E}$	image space of the matrix $\boldsymbol{E}$
L_c	controllability gramian
L_o	observability gramian
LFT	Linear Fractional Transformation
$\boldsymbol{\mathcal{LFT}}(\boldsymbol{J}, \boldsymbol{Q})$	lower linear fractional transformation of the pair $(\boldsymbol{J}, \boldsymbol{Q})$
LHP	left-half plane
LMI	Linear Matrix Inequality

LTI	Linear Time Invariant
MIMO	Multi-Input Multi-Output
$\aleph(\boldsymbol{E})$	null space of the matrix $\boldsymbol{E}$
OF	Output Feedback
PR	Positive Real
$\mathbb{R}$	fields of real numbers
$\mathbb{R}^n$	n-dimensional real plane
$\mathbb{R}^+$	nonnegative real numbers
$\mathcal{RH}^2$	the set of all stable proper real rational transfer function matrices
$\mathcal{RH}^\infty$	the set of all stable strictly proper real rational transfer function matrices
RHP	right-half plane
SF	State Feedback
SISO	Single-Input Single Output
SPR	Strictly Positive Real
SVD	Singular Value Decomposition

1

Introduction

1.1 Why Differential-Algebraic Equations?

In system theory, a dynamical system is often considered as a set of ordinary differential or difference equations (ODE); these equations describe the relations between the system variables. As pointed out in [14], for the most general purpose of system analysis, one usually begins by defining the first order system

$$\boldsymbol{F}(\dot{\boldsymbol{x}}(t),\, \boldsymbol{x}(t)) = 0, \tag{1.1}$$

where $\boldsymbol{F}$ and $\boldsymbol{x}$ are vector-valued functions. The equation (1.1) is termed as *differential algebraic equations* (DAE), since it contains differential equations as well as a set of algebraic constraints. For control and systems engineers, it is usually assumed that (1.1) can be rewritten in an *explicit* form

$$\dot{\boldsymbol{x}}(t) = \boldsymbol{f}(\boldsymbol{x}(t)). \tag{1.2}$$

An ODE of the form (1.2) is called a state-variable (state-space) description in the systems and control society. Since then, theorems and design techniques being developed are largely based on (1.2). In fact, the state-variable descriptions have been the predominant tool in systems and control theory. While the representation (1.2) will continue to be very important, there has been an increasing interest in working directly with (1.1).

If (1.1) can, in principle, be rewritten as (1.2) with the same state variables $\boldsymbol{x}$, then it will be referred to as a system of implicit ODEs. In this monograph, we are especially interested in those problems for which this rewriting is impossible or less desirable. We consider the general nonlinear DAEs which are linear in the derivative

$$\boldsymbol{A}(\boldsymbol{x}(t))\dot{\boldsymbol{x}}(t) + \boldsymbol{f}(\boldsymbol{x}(t)) = 0. \tag{1.3}$$

Suppose that $\dfrac{\partial \boldsymbol{A}(\boldsymbol{x})}{\partial \boldsymbol{x}}$ has a constant rank. Then, in principle, locally the system (1.3) can be put in the *semi-explicit* form

$$\begin{aligned}\dot{\boldsymbol{x}}_1(t) &= \boldsymbol{f}_1(\boldsymbol{x}_1(t), \boldsymbol{x}_2(t)) \\ 0 &= \boldsymbol{f}_2(\boldsymbol{x}_1(t), \boldsymbol{x}_2(t)).\end{aligned} \tag{1.4}$$

A large class of physical systems can be modeled by this kind of DAEs. The paper of Newcomb *et al.* [91] gives many practical examples–including circuit and system design, robotics, neural network, etc.–and presents an excellent review on nonlinear DAEs. Many other applications of DAEs as well as numerical treatments can be found in [14]. An existence and uniqueness theory for nonlinear DAEs has been well developed in [97] by exploiting their underlying differential geometric structure. Recently, Venkatasubramanian *et al.* [113] have extensively studied the bifurcation phenomena of DAEs. They have also thoroughly investigated feasibility regions in differential-algebraic systems. The notion of feasibility regions provides a natural gateway to the stability theory of DAEs.

Depending on the area, system of the form (1.4). has different nomenclature in different fields. For example, control theorists and mathematicians have long been calling them *singular systems*[16][17][61][83], since the matrix on the derivative of the state-variables is generally singular, or sometimes they use the terminology *generalized state-space systems*[1][41][114], or at times *extended state-space systems*[55]. On the other hand, the name *descriptor systems* is most frequently used in the engineering economic systems community[78][79], since they give the natural description of the system, while numerical analysts call their descriptions *differential-algebraic equations*[14][81] [94], or differential equations with algebraic constraints[73]. In the circuits area the original name was *pseudostate*[59] but more recently these systems have been called *semistate systems*[26] because they are almost state described. This usage is now somewhat obsolete. In this monograph, we will use the terms descriptor systems and DAEs interchangably.

There are several reasons to consider systems of the form (1.4), rather than try to rewrite it as an ODE. Of great importance, we point out that, when physical problems are simulated, the model often takes the form of a DAE depicting a collection of relationships between variables of interest and some of their derivatives, namely the algebraic constraints. These relationships may even be generated automatically by a modeling or simulation program. In particular, the variables thus introduced usually have a physical significance. Changing the model to (1.2) may produce less meaningful state variables. If the original DAE can be solved directly, then it becomes easier for scientists or engineers to explore the effect of modeling changes and parameter variations. These advantages enable researchers to focus their attention on the physical problem of interest. On the other hand, although the state-space models are very useful, but the state variables thus introduced often do not provide a physical meaning[29][111]. Besides, some physical phenomena, like impulse, hysterisis which are important in circuit theory, cannot be treated properly in the state-space models[67][114]. Differential-algebraic equations representation provides a suitable way to handle such problems. It has been proven in the literature that DAE systems have higher capability in describing a physical system[66][91][114]. In fact, DAE system models appear more convenient and natural than state-space models in large scale systems, economics, networks, power, neural systems and elsewhere [66][78][91].

Example 1. Constrained Variational Problems [14][33]
The first example that can be well modeled by DAEs is a variational problem with constraints. Consider a constrained mechanical system with position $x(t)$, velocity $v = \dot{x}(t)$, kinetic energy $T(x(t), v(t))$, external force $f(x(t), v(t), t)$ and constraint $\phi(x(t)) = 0$. By the variational principle, the Euler-Lagrange formulation of the system can be put in the following form

$$\begin{aligned} \dot{x}(t) &= v(t) \\ \frac{\mathrm{d}}{\mathrm{d}t}\frac{\partial}{\partial v}T(x(t), v(t)) &= \frac{\partial T(x(t), v(t))}{\partial x} + f(x(t), v(t), t) + G^T\lambda \\ 0 &= \phi(x(t)), \end{aligned}$$

where $G = \frac{\partial \phi}{\partial x}$, and λ is the Lagrange multiplier. This system can be rewritten as

$$\frac{\partial^2 T}{\partial v^2}\dot{v}(t) = g(x(t), v(t), t) + G^T\lambda \tag{1.5a}$$

$$\dot{x}(t) = v(t) \tag{1.5b}$$

$$0 = \phi(x), \tag{1.5c}$$

where

$$g(x, v, t) \triangleq f(x, v, t) + \frac{\partial T}{\partial x} - \frac{\partial^2 T}{\partial x \partial v}\dot{x}.$$

In practical case, the matrix $\frac{\partial^2 T}{\partial v^2}$ is usually positive definite. Then multiplication of (1.5a) by $\left(\frac{\partial^2 T}{\partial v^2}\right)^{-1}$ converts the previous system into a semi-explicit DAE.

$$\dot{v}(t) = \left(\frac{\partial^2 T}{\partial v^2}\right)^{-1} g(x(t), v(t), t) + \left(\frac{\partial^2 T}{\partial v^2}\right)^{-1} G^T\lambda$$

$$\dot{x}(t) = v(t)$$
$$0 = \phi(x).$$

◇◇◇

Example 2. Singular Dynamic Leontieff Systems [62][77]
Consider an economic process that involves n interrelated production sectors. The relationships between the levels of production of the sectors can be described by a so-called *Leontieff Model*:

$$\boldsymbol{x}(k) = \boldsymbol{A}\boldsymbol{x}(k) + \boldsymbol{B}(x(k+1) - \boldsymbol{x}(k)) + \boldsymbol{w}(k). \tag{1.6}$$

Here the components of the n-dimensional vector $\boldsymbol{x}(k)$ are the levels of production of the sectors at time k. The vector $\boldsymbol{A}\boldsymbol{x}(k)$ should be interpreted as the capital that is required as direct input for production of $\boldsymbol{x}$; a coefficient a_{ij} of the *flow coefficient matrix* $\boldsymbol{A}$ indicates the amount of product i that is needed to produce one unit of product j. The vector $\boldsymbol{B}\boldsymbol{x}$ stands for the capital that is required to be in stock to be able to produce $\boldsymbol{x}$ in the next time period. A coefficient b_{ij} of the *stock coefficient matrix* $\boldsymbol{B}$ indicates the amount of product i that has to be in stock to be able to produce one unit of product j in the next time period. The vector $\boldsymbol{w}(k)$ represents the levels of production that are demanded. Econometric models of this type were considered by Leontieff in [64], in which both discrete time and continuous time cases were considered.

Usually, most of the elements in the stock coefficient matrix $\boldsymbol{B}$ are zero and $\boldsymbol{B}$ is often singular. This is because that productions in one sector does not require capital in stock from all the other sectors. Furthermore, in many cases, there are usually few sectors that offer capital in stock to other sectors. The representation (1.6) can be rewritten in the following form

$$\boldsymbol{B}\boldsymbol{x}(k+1) = (\boldsymbol{I} - \boldsymbol{A} + \boldsymbol{B})\boldsymbol{x}(k) - \boldsymbol{w}(k),$$

which is a descriptor form. This serves as a example that a descriptor form can arise naturally in modelling a practical dynamical system. ◇◇◇

Example 3. Electrical Circuit with Operational Amplifier[14][80]
Consider the electrical circuit of Fig. 1.1, which consists of a differential amplifier, a voltage source, and four resistors. To each node, the circuit equations are derived from Kirchoff's laws:

1. The algebraic sum of the currents into a node at any instant is zero.
2. The algebraic sum of the voltage drops around a loop at any instant is zero.

By Kirchoff's law, we can write down the circuit equations as

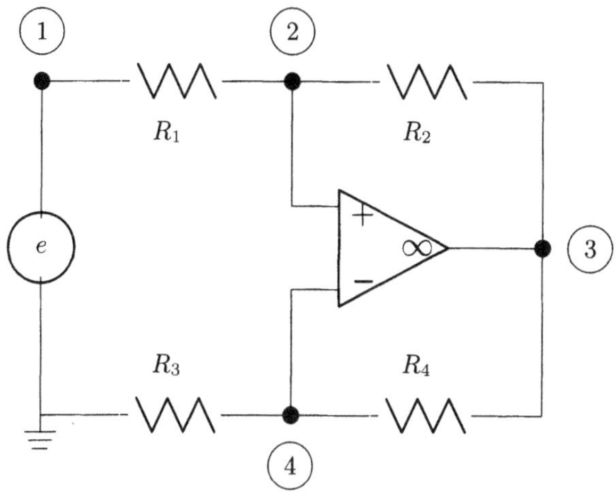

Fig. 1.1. Electrical circuit with Operational Amplifier(a)

$$\begin{bmatrix} \frac{1}{R_1} & -\frac{1}{R_1} & 0 & 0 & 1\,0 \\ -\frac{1}{R_1} & \frac{1}{R_1}+\frac{1}{R_2} & \frac{1}{R_2} & 0 & 0\,0 \\ 0 & -\frac{1}{R_2} & \frac{1}{R_2}+\frac{1}{R_4} & -\frac{1}{R_4} & 0\,1 \\ 0 & 0 & -\frac{1}{R_4} & \frac{1}{R_3}+\frac{1}{R_4} & 0\,0 \\ 1 & 0 & 0 & 0 & 0\,0 \\ 0 & 1 & 0 & -1 & 0\,0 \end{bmatrix} \begin{bmatrix} v_1 \\ v_2 \\ v_3 \\ v_4 \\ i_E \\ i_0 \end{bmatrix} = \begin{bmatrix} 0 \\ 0 \\ 0 \\ 0 \\ e \\ 0 \end{bmatrix} \tag{1.7}$$

where i_0 denotes the output current of the operational amplifier. Equation (1.7) is a purely algebraic linear system. It is solvable if and only if the coefficient matrix is nonsingular, or equivalently, its determinant

$$\mathbf{det} \begin{bmatrix} \frac{1}{R_1} & -\frac{1}{R_1} & 0 & 0 & 1\,0 \\ -\frac{1}{R_1} & \frac{1}{R_1}+\frac{1}{R_2} & \frac{1}{R_2} & 0 & 0\,0 \\ 0 & -\frac{1}{R_2} & \frac{1}{R_2}+\frac{1}{R_4} & -\frac{1}{R_4} & 0\,1 \\ 0 & 0 & -\frac{1}{R_4} & \frac{1}{R_3}+\frac{1}{R_4} & 0\,0 \\ 1 & 0 & 0 & 0 & 0\,0 \\ 0 & 1 & 0 & -1 & 0\,0 \end{bmatrix} = \frac{1}{R_2R_3} - \frac{1}{R_1R_4}$$

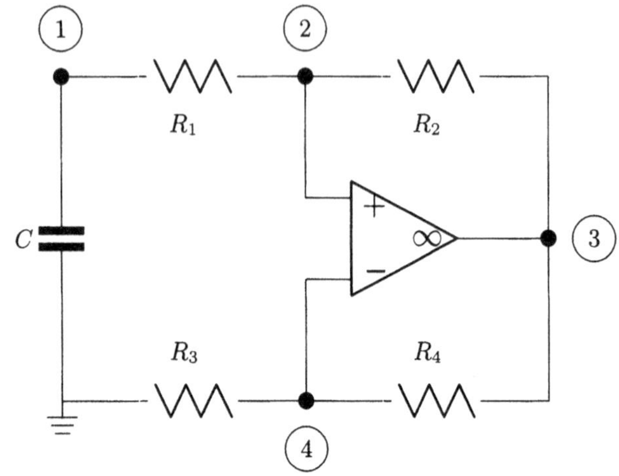

Fig. 1.2. Electrical circuit with Operational Amplifier (b)

is nonzero. This shows that the solvability of the system depends on the specific values of the resistances. In order to gain more insight of the differential-algebraic equations, let us replace the voltage source in the circuit given in Fig. 1.1 with a capacitor. The circuit equations now reads as follows (See Fig. 1.2)

$$\begin{bmatrix} C\dot{v}_1 \\ 0 \\ 0 \\ 0 \\ 0 \end{bmatrix} + \begin{bmatrix} \frac{1}{R_1} & -\frac{1}{R_1} & 0 & 0 & 0 \\ -\frac{1}{R_1} & \frac{1}{R_1}+\frac{1}{R_2} & \frac{1}{R_2} & 0 & 0 \\ 0 & -\frac{1}{R_2} & \frac{1}{R_2}+\frac{1}{R_4} & -\frac{1}{R_4} & 1 \\ 0 & 0 & -\frac{1}{R_4} & \frac{1}{R_3}+\frac{1}{R_4} & 0 \\ 0 & 1 & 0 & -1 & 0 \end{bmatrix} \begin{bmatrix} v_1 \\ v_2 \\ v_3 \\ v_4 \\ i_0 \end{bmatrix} = \begin{bmatrix} 0 \\ 0 \\ 0 \\ 0 \\ 0 \end{bmatrix}. \quad (1.8)$$

The determinant of the matrix pencil of system (1.8) is

$$sC\left(\frac{1}{R_1}\frac{1}{R_4} - \frac{1}{R_2}\frac{1}{R_3}\right) - \frac{1}{R_1}\frac{1}{R_2}\frac{1}{R_3} \triangleq sG + H,$$

which is never identically zero for nonzero resistances. Now the number of state variables is equal to the degree of the polynomial $\mathbf{det}[sG + H]$. Equivalently, this is the number of independent initial conditions which must be specified. Thus, if $\frac{1}{R_1}\frac{1}{R_4} - \frac{1}{R_2}\frac{1}{R_3} = 0$, the only solution is the zero solution. If $\frac{1}{R_1}\frac{1}{R_4} -$

$\frac{1}{R_2}\frac{1}{R_3} \neq 0$, there is one parameter family of solutions to (1.8). This shows that the number of state variables are determined not only by the topology of the circuit, but also by the specific resistance values. ◊◊◊

1.2 Control Problems Based on DAEs

The desirability of working directly with DAEs has been recognized for twenty years by scientists and engineers in several areas. For the purpose of control, the to-be-controlled plant considered in this monograph is usually described in the following descriptor form:

$$\dot{\boldsymbol{x}}_1 = \boldsymbol{F}_1(\boldsymbol{x}_1, \boldsymbol{x}_2, \boldsymbol{w}, \boldsymbol{u}), \tag{1.9}$$

$$0 = \boldsymbol{F}_2(\boldsymbol{x}_1, \boldsymbol{x}_2, \boldsymbol{u}), \tag{1.10}$$

$$\boldsymbol{z} = \boldsymbol{Z}(\boldsymbol{x}, \boldsymbol{w}, \boldsymbol{u}), \tag{1.11}$$

$$\boldsymbol{y} = \boldsymbol{Y}(\boldsymbol{x}, \boldsymbol{w}, \boldsymbol{u}), \tag{1.12}$$

Here $\boldsymbol{u}$ is the control input, $\boldsymbol{w}$ is the exogenous input (disturbances to-be-rejected or signals to-be-tracked), $\boldsymbol{y}$ is the measured output, and $\boldsymbol{z}$ denotes a set of the penalty variables (tracking errors, cost variables). In the state-space $\mathcal{X}$, dynamic state variables $\boldsymbol{x}_1$ and instantaneous state variables $\boldsymbol{x}_2$ are distinguished.

The control theory based on descriptor system models has been well estabilished for many years with the practical outcome that the shortcomings of state-variable theory are often overcome [91]. For the linear descriptor systems, Cobb first gave a necessary and sufficient condition for the existence of an optimal solution to linear quadratic optimization problem[19] and also extensively studied the notions of controllability, observability and duality in descriptor systems[20]. Lewis[66], Bender *et al.*[10] and Takaba *et al.*[111] constructed different kinds of Riccati equations for solving linear quadratic regulator problems based on certain assumptions. Some excellent results on pole placement[92] and robust control[29][119], to name only a few, were also obtained.

Recently, Copeland and Safonov used the descriptor-system-like models to solve the singular $\mathcal{H}_2$ and $\mathcal{H}_\infty$ control problems in which the plants have pure imaginary (including infinity) poles or zeros [21]. Solutions to $\mathcal{H}_\infty$ control problem for descriptor systems were given in Takaba *et al.* [111]. They dealt with the problem using a J-spectral factorization, thus their proofs were involved. Moreover, only sufficient conditions for solutions to exist were given. Most recently, Masubuchi *et al.*[82] have considered a similar problem by using a matrix inequalities approach. They treated a more general problem with less assumptions. Their solutions were obtained by use of a version of Bounded Real Lemma and given in terms of linear matrix inequalities(LMI) which may be solved by existing numerical tools. However, they gave a necessary and sufficient condition in terms of two generalized algebraic Riccati inequalities(GARI) involving two unknown parameters plus two to-be-determined variables. On the other hand, compared to the existing results for linear case, less efforts have been made to investigate the problems of robust control for nonlinear descriptor systems. The problem of feedback stabilization of nonlinear descriptor systems have been investigated in [106], while the paper of Wu and Mizukami[123] elaborated the stability and robust stabilization of nonlinear descriptor systems with uncertainty. A Hamilton–Jacobi like theorem was given in Xu and Mizukami[125]. Recently, Boutayeb and Darouach[13], and Shields[107] have developed some results concerning the observers for nonlinear descriptor systems.

The monograph continues this line of research to study the $\mathcal{H}_\infty$ control problem for descriptor systems. We give a comprehensive investigation on the contraction property of descriptor systems. The most general motivation of the $\mathcal{H}_\infty$ control problem stems from the small gain theorem, which is originally addressed in the two celebrating papers by Professor Zames[137][138]. If the system uncertainties (linear and/or nonlinear) can be characterized by bounded real property, then the classical results in stability theory can be used to guarantee robust stability provided an appropriate closed-loop system has an $\mathcal{L}_2$ gain strictly less than (or equal to) a prescribed attenuation level γ. Consider the interconnected system shown in Fig. 1.3. Suppose that $M(s) \in R^{p\times q}(s)$ is a stable transfer matrix and the interconnection is well defined. Then the classical *Small Gain Theorem* tells us that the loop $M - \Delta$ is well defined and internally stable, for all $\Delta(s) \in \mathcal{RH}^\infty$ with $\|\Delta\|_\infty \leq \frac{1}{\gamma}$, if and only if $\|M(s)\|_\infty < \gamma$.

It is well known that, to solve the $\mathcal{H}_\infty$ control problem for conventional state-variable systems, we always need certain materials: Lyapunov stability theorem to verify closed-loop internal stability, Hamilton–Jacobi inequalities (equations) for obtaining $\mathcal{H}_\infty$ controllers for nonlinear systems, algebraic Riccati equation and bounded real lemma in deriving the central controller for linear systems; Youla parameterization in characterizing all stabilizing controllers [128]. (One often gets the impression that those materials in fact constitute the bottleneck of this problem.) This monograph can be regarded as an extension of those above mentioned results to the descriptor systems circumstances and we indeed use this extension to solve the $\mathcal{H}_\infty$ control problem for descriptor systems. Perhaps this is the first attempt in the literature to comprehensively study the $\mathcal{H}_\infty$ control problems for both linear and nonlinear descriptor systems. Most of the previous work for conventional state-variable systems can be generalized, *mutatis mutandis*, to descriptor systems. In particular, we will derive the necessary and sufficient conditions for the existence

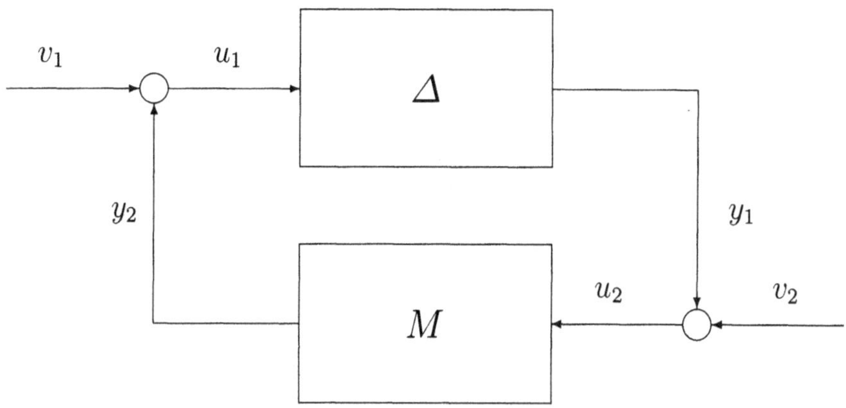

Fig. 1.3. Small Gain Theorem

of a controller solving the standard nonlinear $\mathcal{H}_\infty$ control problem. We first give various sufficient conditions for the solvability of $\mathcal{H}_\infty$ control problem for nonlinear descriptor systems. Both state feedback and output feedback cases are considered. Then, a necessary condition for the output feedback control problem to be solvable are obtained in terms of two Hamilton–Jacobi inequalities plus a weak coupling condition. Moreover, a parameterization of a family of output feedback controllers solving the problem is also provided. All of the aforementioned results are then specialized to the linear case. For the linear case, the necessary and sufficient conditions for the corresponding problems to be solvable are expressed in terms of two hierarchically coupled generalized algebraic Riccati equations which may be considered to be the generalizations of the Riccati equations obtained by Doyle *et al.*[24]. When these conditions hold, state-space formulae for a controller solving the problem is also given. The approach used in this monograph is based on a generalized version of Bounded Real Lemma. Finally, the derivation of state-space formulae for all controllers solving the standard $\mathcal{H}_\infty$ control problem for descriptor systems are proposed. To establish the key formulae, a parameterization of all internally stabilizing controllers for descriptor systems is also given (both the linear and nonlinear cases are considered in this monograph). The results obtained may be considered as the generalization of the corresponding ones given in Doyle *et al.*[24] and Doyle[23].

1.3 Highlights of the Monograph

The remainder of the monograph is organized in the following and the key results are highlighted accordingly.

Chapter 2 reviews some elementary results concering descriptor systems, including controllability and obsrvability, existence and uniqueness of solution of DAEs, and the Lyapunov stability theorems. Furthermore, we explore certain theorems and properties–namely the LaSalle's Invariance Principle, dissipativeness, Bounded Real Lemma, and LQ optimization–in the descriptor systems case.

In Chapter 3, we characterize all controllers that internally stabilize a given dynamical system $\boldsymbol{\Sigma}$, which is described by a set of differential-algebraic equations. The construction of the controller parameterization is done by a simple change of variables and some direct algebraic calculations provided that a state feedback law and an observer gain are available. Moreover, to serve as an illustration, we give a complete parameterization of all solutions to the extended positive real control problems (ESPR) for linear descriptor systems by using a Youla parameterization approach. In this case, the state feedback law and the observer gain are constructed properly from the *control* generalized algebraic Riccati equation and the *filter* generalized algebraic Riccati equation, respectively.

Chapter 4 is devoted to the $\mathcal{H}_\infty$ control problems for descriptor systems. Both the nonlinear and linear cases are studied. We first solve the $\mathcal{H}_\infty$ control problems via static state feedback. Based on the state feedback law thus obtained, the output feedback controllers are then obtained by a simple idea beginning with a change of variables and some algebraic calculations. It is shown that the solutions to the nonlinear $\mathcal{H}_\infty$ control problems are characterized by two *Hamilton–Jacobi inequalities.* In the linear case, the solutions are given in terms of two hierarchical-coupled generalized algebraic Riccati equations. Finally, a family of output feedback controllers for nonlinear descriptor systems as well as a complete parameterization of output feedback controllers for linear descriptor systems are given.

Chapter 5 considers the problem of reducing the order of a descriptor system by the balanced truncation method. The balanced realization has the property that mode i is equally controllable and observable, with these properties being measured in terms of a number $\sigma_i \geq 0$. As σ_i decreases, the corresponding level of controllability and observability becomes less important. The model reduction method that applies the truncation operation to a balanced realization is known as balanced truncation. For this algorithm, the absolute error of the reduced system with respect to the full order system is guaranteed to satisfy the twice-the-sum-of-the-tail bound

$$\|G - \hat{G}\|_\infty \leq 2(\sigma_{k+1} \ldots \sigma_r).$$

The precise definition of the values k and r will be given in Chapter 5.

In Chapter 6, two further topics concerning the $\mathcal{H}_\infty$ control problem for descriptor systems are investigated. The first is the reduced-order controller design. The plant considered in Chapter 4 is, in fact, a generalized plant, which may contain some weighting functions. It is shown from Chapter 4 that full-order controllers constructed from two Hamilton–Jacobi inequalities for nonlinear systems or two generalized algebraic Riccati equations for linear systems have state dimensions no less than that of the generalized plant. Since the generalized plant is built from the physical plant and some weighting functions that are used to reflect performance and robustness requirements, the order of the generalized plant may be very high. In this case, design of the low-order controllers is of importance. Accordingly, we propose an algebraic method to the reduced-order $\mathcal{H}_\infty$ controllers design. It is shown that the reduced-order nonlinear controllers, respectively linear controllers, can be constructed from the solutions of two Hamilton–Jacobi inequalities, respectively two generalized algebraic Riccati inequalities, and three auxiliary equations.

The next topic considered is a mixed $\mathcal{H}_2$ and $\mathcal{H}_\infty$ control problem for descriptor systems. We solve the problem by using a two-player, non-zero sum differential game approach, namely the Nash game approach. The key idea behind the proposed methodology is to use two performance indices associated with a two-player Nash game to represent the $\mathcal{H}_2$ and $\mathcal{H}_\infty$ criteria separately. We consider the infinite horizon case, and give complete feedback control strategies. The control strategies are characterized by two cross-coupled Hamilton–Jacobi equations for nonlinear descriptor systems. For the linear case, the solutions are then given in terms of two cross-coupled generalized algebraic Riccati equations. The results obtained here generalize some recent results in the literature.

2

Elements of Descriptor Systems Theory

In this chapter we summarize some basic definitions and preliminary results that will be used throughout this book. Most of the treatments are purely algebraic and the definitions are fairly standard. In particular, we give a generalized version of the LaSalle's Invariance Principle, which is useful in the proof of internal stability in the subsequent chapters. Furthermore, Bounded Real Lemma, which is a basic tool for solving the $\mathcal{H}_\infty$ control problems, is extended to the descriptor systems in a natural way.

2.1 Differential Algebraic Equations

Consider a descriptor system described by the following set of differential-algebraic equations

$$\dot{\boldsymbol{x}}_1 = \boldsymbol{F}_1(\boldsymbol{x}_1,\, \boldsymbol{x}_2,\, \boldsymbol{u}), \qquad \boldsymbol{x}(0) \in \Im m \boldsymbol{E} \text{ given} \tag{2.1a}$$

$$0 = \boldsymbol{F}_2(\boldsymbol{x}_1,\, \boldsymbol{x}_2,\, \boldsymbol{u}), \qquad \boldsymbol{u} \in \mathbb{U} \subset \mathbb{R}^m \tag{2.1b}$$

$$\boldsymbol{y} = \boldsymbol{H}(\boldsymbol{x}_1,\, \boldsymbol{x}_2,\, \boldsymbol{u}), \qquad \boldsymbol{y} \in \mathbb{Y} \subset \mathbb{R}^p \tag{2.1c}$$

or in a more compact form

$$\begin{aligned} \boldsymbol{E}\dot{\boldsymbol{x}} &= \boldsymbol{F}(\boldsymbol{x},\, \boldsymbol{u}), \\ \boldsymbol{y} &= \boldsymbol{H}(\boldsymbol{x},\, \boldsymbol{u}), \end{aligned} \tag{2.2}$$

where $\boldsymbol{E} = \begin{bmatrix} \boldsymbol{I}_r & 0 \\ 0 & 0 \end{bmatrix}$ ($\boldsymbol{I}_r$ is an $r \times r$ identity matrix) and $\boldsymbol{x} = [\boldsymbol{x}_1^T \;\; \boldsymbol{x}_2^T]^T \triangleq (x_1,\, \ldots, x_n)^T$ are local coordinates for an n-dimensional state-space manifold $\mathbb{X}$. In the state-space $\mathbb{X}$, state variables are divided into two parts, namely the dynamic state variables $\boldsymbol{x}_1$ and instantaneous state variables $\boldsymbol{x}_2$. The dynamics of the states $\boldsymbol{x}_1$ is directly defined by (2.1a), while the dynamics of $\boldsymbol{x}_2$ is such that the system satisfies the constraint (2.1b). The parameter $\boldsymbol{u}$, usually termed as "control input", defines a specific system configuration and the operating condition. In many cases, the algebraic constraint (2.1b) of the full DAEs can be eliminated (usually due to the consistence of initial conditions). As a consequence, the DAEs reduce to a well-known state-variable system. In fact, the state-variable descriptions have been the predominant tool in systems and control theory. Nevertheless, in some cases this kind of elimination is not possible (often due to inconsistent initial conditions), since it may result in loss of accuracy or loss of necessary information. In this monograph, we consider only the case that DAE (2.1a)-(2.1b) admits a unique solution provided that the function $\boldsymbol{F}(\bullet, \bullet)$ is smooth enough. Note that we do not assume the smoothness property of the solution $\boldsymbol{x}$. The concept of solvability is given in the following definition which is taken from [14]. Denote by $\mathcal{T}$ an open subinterval of $\mathbb{R}$, Ω a connected open subset of $\mathbb{R}^n$. Let $\tilde{\Omega}$ be a connected open subset of a local r-dimensional manifold $\mathbb{M}$ of $\mathbb{R}^n$ (a trivial selection of $\mathbb{M}$ is $\Im m\boldsymbol{E}$).

Definition 2.1. *Let $\mathcal{T}$, Ω and $\tilde{\Omega}$ be given. Suppose that $\boldsymbol{F}_1$ and $\boldsymbol{F}_2$ are differentiable functions from $\Omega \times \mathbb{U}$ to $\mathbb{R}^n$. Then the DAE (2.1a)-(2.1b) is said to be solvable in Ω for all $t \in \mathcal{T}$ if there exists an r-dimensional family of solutions $\phi(t, c)$ defined on $\mathcal{T} \times \tilde{\Omega}$, such that*

1. *$\phi(t, c)$ is defined on all $\mathcal{T}$ for each $c \in \tilde{\Omega}$.*
2. *$(\phi(t, c), \dot{\phi}(t, c)) \in \Omega \times \Omega$ for $(t, c) \in \mathcal{T} \times \Omega$.*
3. *If $\varphi(t)$ is any other solution with $(\varphi(t), \dot{\varphi}(t)) \in \Omega \times \Omega$, then $\varphi(t) = \phi(t, c)$ for some $c \in \tilde{\Omega}$.*
4. *The graph of ϕ as a function of (t, c) is an $r + 1$-dimensional manifold.* □

In view of this definition, if the DAE (2.1a)-(2.1b) is solvable, the dimension r is independent of $t_0 \in \mathcal{T}$. As a matter of fact, this r-dimensional manifold can be identified with $\Im m\, \boldsymbol{E}$, i.e., the range space of $\boldsymbol{E}$. An initial condition given in this manifold is called consistent, in the sense that the solutions will continuously depend on it. However, it is more convenient to allow the initial conditions being *inconsistent*, i.e. those solutions admit jump behavior, especially in the circuit theory. See, for example, [66][91][114].

A property known as *index* plays a crucial role in the classification and behavior of DAEs. The following definition is also taken from [14].

Definition 2.2. [14] *The index of the DAE (2.2) is the the minimum number of times that all or part of (2.2) must be differentiated with respect to t in order to determine $(\dot{\boldsymbol{x}}_{r+1}, \ldots, \dot{\boldsymbol{x}}_n)$ as a continuous function of $(\boldsymbol{x}_{r+1}, \ldots, \boldsymbol{x}_n)$.* □

In accordance with this definition, an implicit ordinary differential equation (ODE)(i.e., $\boldsymbol{E}$ is nonsingular) has index zero. We should stress that the index of nonlinear DAE systems is, in general, a local property. Hence, it is more correct to say *local index* rather than just *index*. Nevertheless, we only consider the local phenomena of DAEs in this monograph so that we shall simply use *index* to mean *local index* unless otherwise stated. Suppose that DAE (2.1a)-(2.1b) has index κ. Then, by Definition 2.2, its solutions also satisfy an ODE

$$\dot{\boldsymbol{x}} \quad = \quad \boldsymbol{G}(\boldsymbol{x},\, \boldsymbol{u}), \tag{2.3}$$

where $\boldsymbol{u} = \begin{bmatrix} \boldsymbol{u} \ \dot{\boldsymbol{u}} \ \ldots \ \boldsymbol{u}^{(i)} \end{bmatrix}$, for some $i \leq \kappa$. In (2.3), $\boldsymbol{G}$ is a function of partial derivatives of $\boldsymbol{F}$. Equation (2.3) has a partucular interest in its own right, although not all trajectories $\boldsymbol{x}(\bullet)$ satisfying (2.3) should be solutions of the original DAE. Theoretically, solutions of (2.3) contain higher derivatives of both input functions and system coefficients. This characteristic suggests that, in general, solutions of a DAE will behave in a similar way. It also introduces several ramifications for numerical procedures. See [14] for details.

In order to motivate the next concept, at first we shall consider a much simpler class of DAEs, namely the linear constant coefficient (time-invariant) DAEs, which are now well understood and have particular importance in the electric circuits and control theory. Consider the following linear time-invariant descriptor system

$$\boldsymbol{\Sigma} : \begin{cases} \boldsymbol{E}\dot{\boldsymbol{x}} = \boldsymbol{A}\boldsymbol{x} + \boldsymbol{B}\boldsymbol{u}, \\ \quad \boldsymbol{y} = \boldsymbol{C}\boldsymbol{x}, \end{cases} \quad \boldsymbol{E}\boldsymbol{x}(0) \in \Im m\boldsymbol{E} \text{ given}, \tag{2.4}$$

where $\boldsymbol{A}, \boldsymbol{B} \in \mathbb{R}^{n \times n}$ are constant matrices. From a frequency-domain point of view, (2.4) has a unique solution, for a given initial condition $\boldsymbol{x}(0)$ and a continuous forcing function $\boldsymbol{u}$, if and only if the matrix polynominal $s\boldsymbol{E} - \boldsymbol{A}$, where s is a complex variable, is invertible. Henceforth the nature of the solutions to (2.4) is characterized by this polynominal matrix. A polynominal matrix of this form is called a *matrix pencil*, which is originally taken from geometry where a one-parameter family of curves is termed as a *pencil* of curves. Obviously, the pencil $s\boldsymbol{E}-\boldsymbol{A}$ is invertible if and only if the determinant of $s\boldsymbol{E} - \boldsymbol{A}$, denoted by $\det(s\boldsymbol{E} - \boldsymbol{A})$, is not identically zero as a function of s. A matrix pencil having this property is then called *regular*. We have the following theorem concerning the solvability of (2.4).

Theorem 2.3. *The linear time-invariant DAE (2.4) is solvable if and only if the matrix pencil $s\boldsymbol{E} - \boldsymbol{A}$ is regular.* □

Perhaps the most appealing structural form to better perceive the intrinic of DAE (2.4) is the *Weierstrass form*. It is a specified Kronecker canonical form for a regular matrix pencil. Let $\boldsymbol{x} = \boldsymbol{Q}\hat{\boldsymbol{x}}$, and premultiply by $\boldsymbol{P}$ in (2.4), where $\boldsymbol{P}$ and $\boldsymbol{Q}$ are $n \times n$ nonsingular matrices. Then, if the pair $\{\boldsymbol{E},\, \boldsymbol{A}\}$ is regular, we can always find a suitable pair of nonsingular matrices $\{\boldsymbol{P},\, \boldsymbol{Q}\}$ such that the DAE (2.4) has the following Weierstrass form[32]:

$$\begin{bmatrix} \boldsymbol{I} & 0 \\ 0 & \boldsymbol{N} \end{bmatrix} \begin{bmatrix} \dot{\hat{\boldsymbol{x}}}_1 \\ \dot{\hat{\boldsymbol{x}}}_2 \end{bmatrix} = \begin{bmatrix} \boldsymbol{A}_1 & 0 \\ 0 & \boldsymbol{I} \end{bmatrix} \begin{bmatrix} \hat{\boldsymbol{x}}_1 \\ \hat{\boldsymbol{x}}_2 \end{bmatrix} + \begin{bmatrix} \boldsymbol{B}_1 \\ \boldsymbol{B}_2 \end{bmatrix} \begin{bmatrix} \boldsymbol{u}_1 \\ \boldsymbol{u}_2 \end{bmatrix}, \quad \boldsymbol{y} = \begin{bmatrix} \boldsymbol{C}_1 & \boldsymbol{C}_2 \end{bmatrix} \begin{bmatrix} \hat{\boldsymbol{x}}_1 \\ \hat{\boldsymbol{x}}_2 \end{bmatrix}, \tag{2.5}$$

where $\boldsymbol{N}$ is a nilpotent matrix (Note that we use the same symbol $\boldsymbol{I}$ to denote different dimensional identity matrices). Recall that a matrix $\boldsymbol{N}$ is said to be a nilpotent matrix with index (nilpotency) κ if $\boldsymbol{N}^{\kappa} = 0$ but $\boldsymbol{N}^{\kappa-1} \neq 0$. We should stress that $\boldsymbol{x}_1$ and $\hat{\boldsymbol{x}}_1$, respectively $\boldsymbol{x}_2$ and $\hat{\boldsymbol{x}}_2$, may not have the same dimensions, in general. The portion of (2.5) relevent to $\hat{\boldsymbol{x}}_1$ is an ODE, thus has a solution for any initial value of $\boldsymbol{x}_1$ and any continuous input function $\boldsymbol{u}_1$. On the other hand, the other portion of (2.5) relevent to $\hat{\boldsymbol{x}}_2$ has only one solution, which also completely determined the initial value of $\hat{\boldsymbol{x}}_2$. To see this, observe that the equation concerning $\boldsymbol{x}_2$ can be rewritten as

$$(\boldsymbol{I} + \boldsymbol{N}\frac{\mathrm{d}}{\mathrm{d}t})\hat{\boldsymbol{x}}_2 = \boldsymbol{u}_2 \tag{2.6}$$

Under the assumption that $s\boldsymbol{E} - \boldsymbol{A}$ is regular, we can solve the previous equation for $\hat{\boldsymbol{x}}_2$ to get

$$\hat{\boldsymbol{x}}_2 = \left(\boldsymbol{N}\frac{\mathrm{d}}{\mathrm{d}t} + \boldsymbol{I}\right)^{-1} \boldsymbol{u}_2 = \sum_{i=0}^{\kappa-1} (-1)^i \boldsymbol{N} \boldsymbol{u}_2^{(i)}. \tag{2.7}$$

The last equality follows from $(\boldsymbol{N}\frac{\mathrm{d}}{\mathrm{d}t})^{\kappa} = 0$. Equation (2.7) shows that the solution of (2.5) can involve derivatives of order up to $\kappa - 1$ of the forcing function if the matrix pencil $s\boldsymbol{E} - \boldsymbol{A}$ has a nilpotent index κ. From the above observation, one can easily recognize the similarity between the index of DAEs and nilpotency of $\boldsymbol{N}$. In fact, it can be shown that the nilpotent index coincides with the index of the DAE (2.4). See [14]. Furthermore, it can also be clearly seen from (2.7) that not all initial conditions of (2.5) admit a smooth solution for higher index DAEs. Those initial conditions that do admit smooth solutions are called *consistent*. Although one might feel uncomfortable to have a DAE with inconsistent initial conditions, they are useful in constructing particular physical phenomena (e.g., hysteresis), however. A further observation about (2.5) can be made: In the case that the $\boldsymbol{N}$-matrix has nilpotency one (i.e., $\boldsymbol{N} = 0$), the solution of (2.6) becomes pure algebraic; in other words, $\hat{\boldsymbol{x}}_2$ will not contain any impulsive behavior. It is preferable for practical control engineers since the impulsive dynamics may cause system damage, in the sense that they produce infinite instantaneous power. Such DAEs are said to be of *index one*. The following definition is made for general nonlinear descriptor systems.

Definition 2.4. [14] *The DAE (2.2) is said to be of (uniform) index one if the index of the constant coefficient system*

$$\boldsymbol{E}\dot{\boldsymbol{w}}(t) = \boldsymbol{F}_{\boldsymbol{x}}(\hat{\boldsymbol{x}}(t), 0)\boldsymbol{w}(t) = \boldsymbol{g}(t)$$

is one for all $\hat{\boldsymbol{x}}$ in a neighborhood of the graph of the solution, where $\boldsymbol{F}_{\boldsymbol{x}}$ denotes the Jacobian matrix $\dfrac{\partial \boldsymbol{F}}{\partial \boldsymbol{x}}$. □

The notion of *index one* provides an easy way to guarantee the solvability of a given DAE. Rewrite the DAE (2.2) in the following form:

$$\begin{bmatrix} \boldsymbol{I}_r & 0 \\ 0 & 0 \end{bmatrix} \begin{bmatrix} \dot{\boldsymbol{x}}_1 \\ \dot{\boldsymbol{x}}_2 \end{bmatrix} = \begin{bmatrix} \boldsymbol{F}_1(\boldsymbol{x}_1, \boldsymbol{x}_2, \boldsymbol{u}) \\ \boldsymbol{F}_2(\boldsymbol{x}_1, \boldsymbol{x}_2, \boldsymbol{u}) \end{bmatrix}$$

Suppose that the above DAE is of index one. Then, from Definition 2.4, it is necessary that $\dfrac{\partial}{\partial \boldsymbol{x}_2}\boldsymbol{F}_2(\boldsymbol{x}_1, \boldsymbol{x}_2, 0)$ is nonsingular around the equilibrium point $\boldsymbol{x} = 0$. Consequently, by the Implicit Function Theorem, there exists a function $\boldsymbol{h}(\bullet)$ satisfying

$$[\boldsymbol{F}_2(\boldsymbol{x}_1, \boldsymbol{x}_2, 0)]_{\boldsymbol{x}_2=\boldsymbol{h}(\boldsymbol{x}_1)} = 0, \quad \boldsymbol{h}(0) = 0$$

such that the DAE reduces to an ODE

$$\dot{\boldsymbol{x}}_1 = \boldsymbol{F}_1(\boldsymbol{x}_1, \boldsymbol{h}(\boldsymbol{x}_1), \boldsymbol{u}),$$

which is always solvable provided that $\boldsymbol{F}_1$ is smooth enough. In view of Definition 2.1, this implies that DAE (2.2) is solvable.

2.2 Stability, Controllability and Observability

In this section, we study the fundamentals in system theory, such as stability, controllability and observability. The concept explained here are essential for later chapters. Let's first introduce the notion of feasibility regions of DAEs. Consider the following differential-algebraic equations.

$$\boldsymbol{E}\dot{\boldsymbol{x}}(t) = \boldsymbol{F}(\boldsymbol{x}(t)), = \begin{bmatrix} \boldsymbol{F}_1(\boldsymbol{x}_1(t), \boldsymbol{x}_2(t)) \\ \boldsymbol{F}_2(\boldsymbol{x}_1(t), \boldsymbol{x}_2(t)) \end{bmatrix} \tag{2.8}$$

where $\boldsymbol{E} = \begin{bmatrix} \boldsymbol{I}_r & 0 \\ 0 & 0 \end{bmatrix}$ and $\boldsymbol{x} \in \mathbb{X}$. Without loss of generality, we shall always assume that $\boldsymbol{F}(\boldsymbol{x})$ satisfies $\boldsymbol{F}(0) = 0$. The following definition is readily obtained.

Definition 2.5. *For the constrained system (2.8), define the set* **EQ** *of all equilibrium points and the set* **OP** *of all stable equilibrium points, respevtively, as follows*

$$\mathbf{EQ} \triangleq \{\boldsymbol{x} \in \mathbb{X}, \boldsymbol{F}(\boldsymbol{x}) = 0\},$$

and

$$\mathbf{OP} \triangleq \{(\boldsymbol{x}_1, \boldsymbol{x}_2) \in \mathbf{EQ} : \frac{\partial \boldsymbol{F}_2}{\partial \boldsymbol{x}_2} \text{ is nonsingular and all eigenvalues of}$$

$$\mathbf{J} = \frac{\partial}{\partial \boldsymbol{x}_1}\boldsymbol{F}_1 - \frac{\partial}{\partial \boldsymbol{x}_2}\boldsymbol{F}_1(\frac{\partial}{\partial \boldsymbol{x}_2}\boldsymbol{F}_2)^{-1}\frac{\partial}{\partial \boldsymbol{x}_1}\boldsymbol{F}_2 \textit{ have negative real parts. } \}.$$

Then, given an equilibrium point $\boldsymbol{x}(t_0)$, the connected component **FR** *of* **OP** *containing $\boldsymbol{x}(t_0)$ is called the local feasibility region of $\boldsymbol{x}(t_0)$.* □

Clearly, if the system (2.8) has a local feasibility region in a neighborhood of the equilibrium $\boldsymbol{x}(0) = 0$, then the DAE is of index one. Moreover, the ODE $\dot{\boldsymbol{x}}_1 = \boldsymbol{F}_1(\boldsymbol{x}_1, \boldsymbol{x}_2)$ is locally asymptotically stable (in the sense of Lyapunov) over the constrained surface $\boldsymbol{F}_2(\boldsymbol{x}_1, \boldsymbol{x}_2) = 0$. That is to say: Given a DAE with a local feasibility region, the solution trajectory $\boldsymbol{x} = \boldsymbol{x}(t;\, t_0, \boldsymbol{E}\boldsymbol{x}(0))$ is locally an r-dimensional stable manifold. Next, since the DAE is of index one, there exists a unique solution $\boldsymbol{x}_2 = \boldsymbol{h}(\boldsymbol{x}_1)$ satisfying $\boldsymbol{F}_2(\boldsymbol{x}_1, \boldsymbol{h}(\boldsymbol{x}_1)) = 0$ with $\boldsymbol{h}(0) = 0$. Hence, if $\boldsymbol{x}_1(t) \to 0$, then $\boldsymbol{x}_2(t) \to 0$. This motivates the next few definitions about the stability of DAEs, which is relevant to the previous subject. Denote by $\mathbb{B}_\delta$ the open ball with radius δ, i.e., $\mathbb{B} \triangleq \{\boldsymbol{x} \in \mathbb{R}^n,\ \|\boldsymbol{x}\| < \delta\}$, and $\boldsymbol{x} = \boldsymbol{x}(t;\, t_0, \boldsymbol{E}\boldsymbol{x}(0))$ the solution trajectory. $\mathbb{R}^+ \triangleq [0, \infty)$.

Definition 2.6. *The equilibrium point $\boldsymbol{x} = 0$ is called stable if for every $\epsilon > 0$, and any $t_0 \in \mathbb{R}^+$, there exists a $\delta = \delta(\epsilon,\, t_0) > 0$ such that if $\boldsymbol{x}(0) \in \mathbb{B}_\delta$, then $\|\boldsymbol{x}(t;\, t_0, \boldsymbol{E}\boldsymbol{x}(0))\| < \epsilon$ for all $t \geq t_0$.* □

Note. The initial condition is given in the form $\boldsymbol{E}\boldsymbol{x}(0) \in \Im m\boldsymbol{E}$, i.e., it is a consistent initial condition.

Similarly, the definitions of uniform stability, asymptotical stability uniformly asymptotical stability for conventional systems can be generalized in a natural way to the DAE systems. The paper of Wu *et al.*[123] has discussed this subject in a comprehensive manner. Hence, we shall not reproduce here. Instead, we give a modified definition of locally asymptotical stability for DAEs.

Definition 2.7. *Consider (2.8). The equilibrium point $\boldsymbol{x} = 0$ is said to be asymptotically stable if it is stable and, in addition, there exists a $\delta_0(t_0) > 0$ such that*

$$\boldsymbol{x}(0) \in \mathbb{B}_{\delta_0} \Longrightarrow \lim_{t\to\infty} \|\boldsymbol{x}(t;\, t_0, \boldsymbol{E}\boldsymbol{x}(0))\| = 0.$$

Moreover, if $\left(\dfrac{\partial \boldsymbol{F}_2}{\partial \boldsymbol{x}_2}\right)_{(\boldsymbol{x}_1,\boldsymbol{x}_2)=0}$ is nonsingular, we call the equilibrium asymptotically stable with index one. □

Remark. In view of Definitions 2.5 and 2.7, it is clear that locally there exists a neighborhood $\mathbb{M}$ of $\boldsymbol{x} = 0$ such that $\mathbb{M}$ is identified with the local feasibility region containing $\boldsymbol{x} = 0$. Henceforth, we shall use both notions interchangably.

The main difference between Definition 2.4 of [123] and the previous definition can be declared as follows. In [123], it is assumed the DAE is of index one in order to provide the stability property. However, in our opinion, state variables in the state-space $\mathbb{X}$ should be distinguished, namely the dynamic state variable $\boldsymbol{x}_1$ and the instantaneous variable $\boldsymbol{x}_2$. Furthermore, as mentioned before, the dynamics of $\boldsymbol{x}_1$ are governed by the ODE $\dot{\boldsymbol{x}}_1 = \boldsymbol{F}_1(\boldsymbol{x}_1, \boldsymbol{x}_2)$ which is an integrated behavior defined for all $t \in \mathbb{R}^+$. Meanwhile, $\boldsymbol{x}_2$ is immediately decided by the initial condition at $t = 0$; otherwise, it is such that the DAE satisfies the constraint $\boldsymbol{F}_2(\boldsymbol{x}_1, \boldsymbol{x}_2) = 0$. This shows that the index of DAEs affects only the transient behavior and has no influence on the system stability for consistent initial conditions. Nevertheless, we should emphasize that, for higher index DAEs, the systems would possess impulsive behavior in certain time transiences which may result in system damage. But we shall distinguish such a phenomenon from the instability of DAEs in this monograph. In this way, the aforementioned definition will be consistent for both the nonlinear and linear DAEs. To see this, consider the linear constant descriptor system

$$\begin{bmatrix} \boldsymbol{I} & 0 \\ 0 & \boldsymbol{N} \end{bmatrix} \begin{bmatrix} \dot{\hat{\boldsymbol{x}}}_1 \\ \dot{\hat{\boldsymbol{x}}}_2 \end{bmatrix} = \begin{bmatrix} \boldsymbol{A}_1 & 0 \\ 0 & \boldsymbol{I} \end{bmatrix} \begin{bmatrix} \hat{\boldsymbol{x}}_1 \\ \hat{\boldsymbol{x}}_2 \end{bmatrix}. \tag{2.9}$$

Again, we assume the descriptor system is given in the Weierstrass form. The following definition is the linear system counterpart of Definition 2.7. In the linear case, "stabl" means "asymptotically stable". Denote by $\mathbb{C}^-$ the open left plane in the frequency domain.

Definition 2.8. *Consider (2.4). Suppose that the matrix pencil $s\boldsymbol{E} - \boldsymbol{A}$ is regular, then its Weierstrass form decomposition (2.9) is well-defined.*

1. *the descriptor system* (2.4) *is said to be stable if all the eigenvalues of $\boldsymbol{A}_1$ lie in the $\mathbb{C}^-$.*
2. *the descriptor system* (2.4) *is said to be* impulse-free *if it is of index one, i.e., $\boldsymbol{N} = 0$.*
3. *the descriptor system* (2.4) *is* admissible *if it is stable and impulse-free.*

□

In the literature, it is well known that a linear descriptor system contains three different modes: finite dynamic modes, impulsive modes and nondynamic modes. For a detailed definition of this terminology, see [10]. However, for the sake of completeness, we shall briefly review these notions as follows. Suppose that $\{\boldsymbol{E}, \boldsymbol{A}\}$ is regular with $\text{rank}\boldsymbol{E} = r < n$ and $q \overset{\triangle}{=} \deg\det(s\boldsymbol{E} - \boldsymbol{A})$. Then $\{\boldsymbol{E}, \boldsymbol{A}\}$ have q finite dynamic modes, $r - q$ impulsive modes and $n - r$ nondynamic modes. Furthermore, if $r = q$, then there exist no impulsive modes and the system is impulse-free. While connecting with Definition 2.8, it is not difficult to see that the finite dynamic modes are relative to the eigenvalues of $\boldsymbol{A}_1$ or, equivalently, the eigenvalues of the pair $\{\boldsymbol{E}, \boldsymbol{A}\}$ defined as follows.

$$\boldsymbol{\lambda}(\boldsymbol{E},\, \boldsymbol{A}) \triangleq \{\lambda \in \mathbb{C} \mid \lambda \boldsymbol{E}\boldsymbol{x} = \boldsymbol{A}\boldsymbol{x}, \quad \boldsymbol{x} \in \mathbb{R}^n,\, \boldsymbol{x} \neq 0\}$$

Moreover, it has as many impulsive modes as rank $\boldsymbol{N}$. The nondynamic modes correspond to the zero eigenvalues of $\boldsymbol{E}$ with their associated eigenvectors form a minimal basis of $\aleph(\boldsymbol{E})$, the null space of $\boldsymbol{E}$.

The concepts of controllability and observability for descriptor systems are introduced next. To be clear, we first contemplate the linear case.

Definition 2.9. *Consider the linear constant coefficient descriptor system (2.4). Suppose that matrix pencil $s\boldsymbol{E} - \boldsymbol{A}$ is regular. Then*

1. *System (2.4) is finite dynamic stabilizable if there exists a constant matrix $\boldsymbol{K}$ such that the pair $\{\boldsymbol{E},\, \boldsymbol{A} + \boldsymbol{B}\boldsymbol{K}\}$ is stable.*
2. *System (2.4) is finite dynamic detectable if there exists a constant matrix $\boldsymbol{L}$ such that the pair $\{\boldsymbol{E},\, \boldsymbol{A} + \boldsymbol{L}\boldsymbol{C}\}$ is stable.*
3. *System (2.4) is impulse controllable if there exists a constant matrix $\boldsymbol{K}$ such that the pair $\{\boldsymbol{E},\, \boldsymbol{A} + \boldsymbol{B}\boldsymbol{K}\}$ is impulse-free.*
4. *System (2.4) is impulse observable if there exists a constant matrix $\boldsymbol{L}$ such that the pair $\{\boldsymbol{E},\, \boldsymbol{A} + \boldsymbol{L}\boldsymbol{C}\}$ is impulse-free.*
5. *System (2.4) is finite dynamic stabilizable and impulse controllable if there exists a constant matrix $\boldsymbol{K}$ such that the pair $\{\boldsymbol{E},\, \boldsymbol{A}+\boldsymbol{B}\boldsymbol{K}\}$ is admissible.*
6. *System (2.4) is finite dynamic detectable and impulse observable if there exists a constant matrix $\boldsymbol{K}$ such that the pair $\{\boldsymbol{E},\, \boldsymbol{A}+\boldsymbol{L}\boldsymbol{C}\}$ is admissible.*

□

Similar to the conventional linear systems, the stabilizability and detectability of a descriptor system as well as the impulse controllability and impulse observability can be verified through some algebraic or geometric criteria. Denote by $\bar{\mathbb{C}}^+$ the closed right half plane in the frequency domain. Set $\mathbb{X} = \mathbb{S} \oplus \mathbb{F}$ with $\hat{\boldsymbol{x}}_1 \in \mathbb{S}$ and $\hat{\boldsymbol{x}}_2 \in \mathbb{F}$, respectively.

Theorem 2.10. *Consider (2.4) and its Weierstrass form decomposition (2.5). The following statements are equivalent.*

(i) *The triple $\{\boldsymbol{E},\, \boldsymbol{A},\, \boldsymbol{B}\}$ is finite dynamic stabilizable.*
(ii) *The matrix $[s\boldsymbol{E} - \boldsymbol{A} \quad \boldsymbol{B}]$ has full row rank for all $s \in \bar{\mathbb{C}}^+$.*
(iii) *For all scalars $\lambda \in \bar{\mathbb{C}}^+$ and vector $x \in \mathbb{C}^n$ satisfying $x^*\boldsymbol{A} = x^*\boldsymbol{E}\lambda$, it implies that $x^*\boldsymbol{B} \neq 0$.*
(iv) *There exists a matrix $\boldsymbol{K}_1$ such that $\boldsymbol{A}_1 + \boldsymbol{B}_1\boldsymbol{K}_1$ is Hurwitz.*
(v) *There exists a matrix $\boldsymbol{K}$ such that $\{\boldsymbol{E},\, \boldsymbol{A} + \boldsymbol{B}\boldsymbol{K}\}$ is stable.*

□

The next theorem is devoted to the impulse controllability.

Theorem 2.11. *Consider (2.4) and (2.5). The following statements are equivalent.*

(i) *The system (2.4) is impulse controllable.*
(ii) *The fast subsystem*

$$\begin{aligned} N\dot{\hat{x}}_2 &= \hat{x}_2 + B_2 u \\ y_2 &= C_2 \hat{x}_2 \end{aligned} \tag{2.10}$$

is impulse controllable.
(iii) $\Im m N + \Im m B_2 + \aleph(N) = \mathbb{F}$.
(iv) $\Im m E + A\aleph(E) + \Im m B = \mathbb{X}$.
(v) *There exists a constant matrix* K *such that* $\deg\det(sE - (A + BK)) = \operatorname{rank} E$.

□

The results concerning the detectability and impulse observability of descriptor systems can be made in a dual manner.

Theorem 2.12. *Consider (2.4) and its Weierstrass form decomposition (2.5). The following statements are equivalent.*

(i) *The triple* $\{E, A, C\}$ *is finite dynamic detectable.*
(ii) *The matrix* $\begin{bmatrix} sE - A \\ C \end{bmatrix}$ *has full column rank for all* $s \in \bar{\mathbb{C}}^+$.
(iii) *For all scalars* $\lambda \in \bar{\mathbb{C}}^+$ *and vector* $x \in \mathbb{C}^n$ *satisfying* $Ax = \lambda Ex$, *it implies that* $Cx \neq 0$.
(iv) *There exists a matrix* L_1 *such that* $A_1 + L_1 C_1$ *is Hurwitz.*
(v) *There exists a matrix* C *such that* $\{E, A + LC\}$ *is stable.*

□

The next theorem contains some necessary and sufficient conditions for the observability of System (2.4) at infinity.

Theorem 2.13. *Consider (2.4) and (2.5). The following statements are equivalent.*

(i) *The system (2.4) is impulse observable.*
(ii) *The fast system* (2.10) *is impulse observable.*
(iii) $\aleph(N) \cap \aleph(C_2) \cap \Im m N = 0$.
(iv) $\aleph(E) \cap A^{-1}(\Im m E) \cap \aleph(C) = 0$.
(v) *There exists a constant matrix* L *such that* $\deg\det(sE - (A + LC)) = \operatorname{rank} E$.

□

2.3 Lyapunov Stability Theorems

Stability theory plays an important role in systems and control engineering and practice. Many different kinds of control problems, arising in the study of dynamical systems, have direct connection with stability problems. We are concerned with stability of equilibrium points in this section. For conventional state-space systems, the stability of equilibrium points is usually characterized in the sense of Lyapunov. We will extend Lyapunov stability theory for conventional system to descriptor systems in a natural way with emphasis on asymptotical stability and LaSalle Invariance Principle. Denote by C^n the set of all n times continuously differentiable function.

Theorem 2.14. *Consider (2.8) with $\boldsymbol{F}(0) = 0$. Let $\boldsymbol{x}(0) \in \Im m\boldsymbol{E}$ be given. Suppose that there exists a C^2 function $\mathsf{V} : \mathbb{X} \longrightarrow \mathbb{R}^+$ vanishing at $\boldsymbol{E}\boldsymbol{x} = 0$ and positive elsewhere which satisfies the following properties*

(i) $\dfrac{\partial}{\partial \boldsymbol{x}}\mathsf{V} = \boldsymbol{V}^T(\boldsymbol{x})\boldsymbol{E}$ *for some* C^1 *function* $\boldsymbol{V} : \mathbb{X} \longrightarrow \mathbb{R}^n$, *and*
(ii) $\dot{\mathsf{V}} = \boldsymbol{V}^T(\boldsymbol{x})\boldsymbol{F}(\boldsymbol{x}) \leq 0.$

Then the equilibrium point $\boldsymbol{x} = 0$ of System (2.8) is (locally) stable. If, furthermore, the following condition also holds

(iii) $\dot{\mathsf{V}} = \boldsymbol{V}^T(\boldsymbol{x})\boldsymbol{F}(\boldsymbol{x}) < 0.$

Then the equilibrium point $\boldsymbol{x} = 0$ is (locally) asymptotically stable. □

Proof. Given $\epsilon > 0$, choose $\mathrm{r} \in (0, \epsilon]$ such that

$$\mathbb{B}_{\mathrm{r}} = \{\boldsymbol{x} \in \mathbb{X} \mid \|\boldsymbol{x}\| \leq \mathrm{r}\} \subset \mathbb{X}.$$

Let $\alpha \triangleq \min_{\|\boldsymbol{x}\|=\mathrm{r}} \mathsf{V}(\boldsymbol{x})$. Then $\alpha > 0$ by hypothesis on $\mathsf{V}(\bullet)$. Take $\beta \in (0, \alpha)$, and let

$$\boldsymbol{\Omega}_\beta = \{\boldsymbol{x} \in \mathbb{B}_{\mathrm{r}} \mid \mathsf{V}(\boldsymbol{x}) \leq \beta\}.$$

We first claim that $\boldsymbol{\Omega}_\beta$ is in the interior of $\mathbb{B}_{\mathrm{r}}$. This fact can be shown by contradiction. Suppose $\boldsymbol{\Omega}_\beta$ is not in the interior of $\mathbb{B}_{\mathrm{r}}$, then there exists a point $\boldsymbol{P} \in \boldsymbol{\Omega}_\beta$ lying on the boundary of $\mathbb{B}_{\mathrm{r}}$ such that $\mathsf{V} \geq \alpha > \beta$. But, for all $\boldsymbol{x} \in \boldsymbol{\Omega}_\beta$, we have $\mathsf{V}(\boldsymbol{x}) \leq \beta$, which is a contradiction. The set $\boldsymbol{\Omega}_\beta$ has the property that any trajectory starting in $\boldsymbol{\Omega}_\beta$ at $t = 0$ stays in $\boldsymbol{\Omega}_\beta$ for all $t \geq 0$. This follows from (iii), because

$$\dot{\mathsf{V}}(\boldsymbol{x}(t)) \leq 0 \Rightarrow \mathsf{V}(\boldsymbol{x}(t)) \leq \mathsf{V}(\boldsymbol{x}(0)) \leq \beta, \quad \forall t \geq 0.$$

Choose a $\delta > 0$ such that

$$\|\boldsymbol{x}\| \leq \delta \Rightarrow \mathsf{V}(\boldsymbol{x}(t)) < \beta.$$

This can always be done since $\mathsf{V}(\boldsymbol{x}(t))$ is continuous and $\mathsf{V}(0) = 0$. Then, by construction, the three sets $\mathbb{B}_\delta$, $\boldsymbol{\Omega}_\beta$, and $\mathbb{B}_{\mathrm{r}}$ satisfy

$$\mathbb{B}_\delta \subset \boldsymbol{\Omega}_\beta \subset \mathbb{B}_{\mathrm{r}},$$

and

$$\boldsymbol{x}(0) \in \mathbb{B}_\delta \Rightarrow \boldsymbol{x}(0) \in \boldsymbol{\Omega}_\beta \Rightarrow \boldsymbol{x}(t) \in \boldsymbol{\Omega}_\beta \Rightarrow \boldsymbol{x}(t) \in \mathbb{B}_{\mathtt{r}}.$$

Therefore,

$$\|\boldsymbol{x}(0)\| < \delta \Rightarrow \|\boldsymbol{x}(t;\, t_0, \boldsymbol{E}\boldsymbol{x}(0))\| < \mathtt{r} \leq \epsilon, \quad \forall t \geq 0.$$

This shows that the equilibrium point $\boldsymbol{x} = 0$ is stable. Now assume (iii) holds as well. To show asymptotical stability. we need to show that $\boldsymbol{x}(t) \to 0$ as $t \to \infty$; that is, for every $\mathsf{a} > 0$, there is a $T > 0$ such that $\|\boldsymbol{x}(t)\| < \mathsf{a}$, for all $t > T$. By repeating the previous arguments, we then arrive at: for every $\mathsf{a} > 0$, there exists a $\mathsf{b} > 0$ such that $\boldsymbol{\Omega}_{\mathsf{b}} \subset \mathbb{B}_{\mathsf{a}}$. Hence, it is sufficient to show that $\mathsf{V}(\boldsymbol{x}(t)) \to 0$ as $t \to \infty$. Observe that

$$\mathsf{V}(\boldsymbol{x}(t)) \to \mathsf{c} \geq 0, \ \text{ as } t \to \infty.$$

This is because $\mathsf{V}(\boldsymbol{x}(t))$ is monotonically decreasing and bounded form below by zero. We need to show that c is eventually zero. To see this, suppose, by contradiction, $\mathsf{c} > 0$. Then by continuity of $\mathsf{V}(\boldsymbol{x}(t))$, there is a $\mathsf{d} > 0$ such that $\mathbb{B}_{\mathsf{d}} \subset \boldsymbol{\Omega}_{\mathsf{c}}$. The limit $\mathsf{V}(\boldsymbol{x}(t)) \to \mathsf{c} > 0$ implies that the trajectory $\boldsymbol{x}(t;\, t_0, \boldsymbol{E}\boldsymbol{x}(0))$ lies outside the ball $\mathbb{B}_{\mathsf{a}}$ for all $t \geq 0$. Let $-\gamma = \max_{\mathsf{d} \leq \|\boldsymbol{x}\| \leq \mathtt{r}} \dot{\mathsf{V}}(\boldsymbol{x}(t))$, which exists because the continuous function $\dot{\mathsf{V}}(\boldsymbol{x}(t))$ has a maximum over the compact set $\{\mathsf{d} \leq \|\boldsymbol{x}\| \leq \mathtt{r}\}$ (This is a basic fact of mathematical analysis). By (iii), $-\gamma < 0$. It follows that

$$\mathsf{V}(\boldsymbol{x}(t)) = \mathsf{V}(\boldsymbol{x}(0)) + \int_0^t \dot{\mathsf{V}}(\boldsymbol{x}(\tau))\mathrm{d}\tau \leq \mathsf{V}(\boldsymbol{x}(0)) - \gamma t. \tag{2.11}$$

Since the right-hand side of Equation (2.11) will eventually become negative, the inequality contradicts the assumptions that $\mathsf{c} > 0$. **Q.E.D.**

Theorem 2.14 has nothing to do with the index of Descriptor System (2.8). In the following, we will derive an improved version of the previous theorem. It provides a sufficient condition for the existence of local feasibility regions.

Theorem 2.15. *Consider DAE (2.8) with $\boldsymbol{F}(0) = 0$. Let $\boldsymbol{x}(0) \in \Im m\boldsymbol{E}$ be given. Suppose that there exists a C^3 function $\mathsf{V} : \mathbb{X} \to \mathbb{R}$ vanishing at $\boldsymbol{E}\boldsymbol{x} = 0$ and positive elsewhere which satisfies the following properties.*

i) $\dfrac{\partial}{\partial \boldsymbol{x}}\mathsf{V}(\boldsymbol{x}) = \boldsymbol{V}^T(\boldsymbol{x})\boldsymbol{E}$ *for some C^2 function $\boldsymbol{V} : \mathbb{X} \to \mathbb{R}^n$, and*
ii) $\boldsymbol{V}^T(\boldsymbol{x})\boldsymbol{F}(\boldsymbol{x}) < 0$, *and*
iii) $\boldsymbol{E}^T\boldsymbol{V}_{\boldsymbol{x}} = \boldsymbol{V}_{\boldsymbol{x}}^T\boldsymbol{E} \geq 0$, *where $\boldsymbol{V}_{\boldsymbol{x}}$ denotes the* Jacobian *of $\boldsymbol{V}(\boldsymbol{x})$, i.e., $\boldsymbol{V}_{\boldsymbol{x}} = \dfrac{\partial \boldsymbol{V}(\boldsymbol{x})}{\partial \boldsymbol{x}}$.*

Then the equilibrium point $\boldsymbol{x} = \boldsymbol{0}$ is locally asymptotically stable with index one. □

Proof. We first show that DAE (2.8) is of index one. Set

$$\boldsymbol{V}(\boldsymbol{x}(t)) = \begin{bmatrix} \boldsymbol{V}_1(\boldsymbol{x}_1(t),\, \boldsymbol{x}_2(t)) \\ \boldsymbol{V}_2(\boldsymbol{x}_1(t),\, \boldsymbol{x}_2(t)) \end{bmatrix}.$$

It is easy to see that

$$\frac{\partial^2}{\partial \boldsymbol{x}^2}\boldsymbol{V}^T(\boldsymbol{x})\boldsymbol{F}(\boldsymbol{x})\bigg|_{\boldsymbol{x}=0} = \boldsymbol{V}^T(0)_{\boldsymbol{x}}\boldsymbol{F}_{\boldsymbol{x}}(0) + \boldsymbol{F}_{\boldsymbol{x}}^T(0)\boldsymbol{V}_{\boldsymbol{x}} < 0. \tag{2.12}$$

Condition iii) implies that $\frac{\partial}{\partial \boldsymbol{x}_2}\boldsymbol{V}_1(\boldsymbol{x}) = 0$. In view of Equation (2.12), this in turn implies that $\frac{\partial}{\partial \boldsymbol{x}_2}\boldsymbol{F}_2(0)$ is nonsingular. Consequently, by the continuity of $\boldsymbol{F}$, DAE (2.8) is of index one.

Now consider the following constant coefficient DAE

$$\boldsymbol{E}\dot{\boldsymbol{x}} = \boldsymbol{F}_{\boldsymbol{x}}(0)\boldsymbol{x}, \tag{2.13}$$

which is a linearization of the nonlinear DAE (2.8) around the equilibrium point $\boldsymbol{x} = 0$. Upon applying an equivalent transformation of the form

$$\{\boldsymbol{P},\, \boldsymbol{Q}\} = \left\{ \begin{bmatrix} I & -\frac{\partial \boldsymbol{F}_1}{\partial \boldsymbol{x}_2}(0)\left(\frac{\partial \boldsymbol{F}_2}{\partial \boldsymbol{x}_2}(0)\right)^{-1} \\ 0 & I \end{bmatrix},\, \begin{bmatrix} I & 0 \\ -\left(\frac{\partial \boldsymbol{F}_2}{\partial \boldsymbol{x}_2}(0)\right)^{-1}\frac{\partial \boldsymbol{F}_2}{\partial \boldsymbol{x}_1}(0) & 0 \end{bmatrix} \right\}$$

on the constant pair $\{\boldsymbol{E},\, \boldsymbol{F}_{\boldsymbol{x}}(0)\}$, we then arrive at a new set of algebraic inequality.

$$\begin{cases} \begin{bmatrix} \boldsymbol{V}_{11} & 0 \\ \boldsymbol{V}_{21} & \boldsymbol{V}_{22} \end{bmatrix}^T \begin{bmatrix} \frac{\partial \boldsymbol{F}_1}{\partial \boldsymbol{x}_1}(0) - \frac{\partial \boldsymbol{F}_1}{\partial \boldsymbol{x}_2}(0)\left(\frac{\partial \boldsymbol{F}_2}{\partial \boldsymbol{x}_2}(0)\right)^{-1}\frac{\partial \boldsymbol{F}_2}{\partial \boldsymbol{x}_1}(0) & 0 \\ 0 & \frac{\partial \boldsymbol{F}_2}{\partial \boldsymbol{x}_2}(0) \end{bmatrix} \\ \quad + \begin{bmatrix} \frac{\partial \boldsymbol{F}_1}{\partial \boldsymbol{x}_1}(0) - \frac{\partial \boldsymbol{F}_1}{\partial \boldsymbol{x}_2}(0)\left(\frac{\partial \boldsymbol{F}_2}{\partial \boldsymbol{x}_2}(0)\right)^{-1}\frac{\partial \boldsymbol{F}_2}{\partial \boldsymbol{x}_1}(0) & 0 \\ 0 & \frac{\partial \boldsymbol{F}_2}{\partial \boldsymbol{x}_2}(0) \end{bmatrix}^T \begin{bmatrix} \boldsymbol{V}_{11} & 0 \\ \boldsymbol{V}_{21} & \boldsymbol{V}_{22} \end{bmatrix} < 0. \\ \boldsymbol{E}^T \begin{bmatrix} \boldsymbol{V}_{11} & 0 \\ \boldsymbol{V}_{21} & \boldsymbol{V}_{22} \end{bmatrix} = \begin{bmatrix} \boldsymbol{V}_{11} & 0 \\ \boldsymbol{V}_{21} & \boldsymbol{V}_{22} \end{bmatrix}^T \boldsymbol{E} \geq 0. \end{cases} \tag{2.14}$$

The (1, 1)-block of equation (2.14) implies that

$$\begin{aligned} &\boldsymbol{V}_{11}[(\boldsymbol{F}_1)_{\boldsymbol{x}_1} - (\boldsymbol{F}_1)_{\boldsymbol{x}_2}\left((\boldsymbol{F}_2)_{\boldsymbol{x}_2}\right)^{-1}(\boldsymbol{F}_2)_{\boldsymbol{x}_1}] \\ &+[(\boldsymbol{F}_1)_{\boldsymbol{x}_1} - (\boldsymbol{F}_1)_{\boldsymbol{x}_2}\left((\boldsymbol{F}_2)_{\boldsymbol{x}_2}\right)^{-1}(\boldsymbol{F}_2)_{\boldsymbol{x}_1}]^T\boldsymbol{V}_{11} < 0 \end{aligned}$$

with $\boldsymbol{V}_{11} > 0$. This shows that the equilibrium point $\boldsymbol{x} = 0$ belongs to the local feasibility region. Hence the system is asymptotically stable with index one. **Q.E.D.**

The next theorem is an extension of the well known LaSalle Invariance Principle.

Theorem 2.16. *Consider the DAE*

$$\dot{\boldsymbol{x}}(t) = \boldsymbol{f}_1(\boldsymbol{x}(t),\, \boldsymbol{y}(t)), \tag{2.15a}$$
$$0 = \boldsymbol{f}_2(\boldsymbol{x}(t),\, \boldsymbol{y}(t)), \tag{2.15b}$$

where $\boldsymbol{f}_1$, $\boldsymbol{f}_2$ are continuous differentiable functions. Suppose the DAE is of index one. Let $(\boldsymbol{x}, \boldsymbol{y}) = (0,\, 0)$ be an equilibrium point for DAE (2.15a) (2.15b). Let $\mathrm{V}(\boldsymbol{x},\, \boldsymbol{y}) : D \longrightarrow \mathbb{R}^+ = [0,\, \infty)$ be a smooth positive definite function on a neighborhood D of $(\boldsymbol{x}, \boldsymbol{y}) = (\mathbf{0},\, \mathbf{0})$, such that $\dot{\mathrm{V}}(\boldsymbol{x},\, \boldsymbol{y}) \leq 0$. Let $\boldsymbol{S} = \{(\boldsymbol{x},\, y) \in D \,|\, \dot{\mathrm{V}} = 0\}$, and suppose that no solution can stay forever in $\boldsymbol{S}$, other than the trivial solution $\lim_{t\to\infty} \boldsymbol{x}(t) = 0$. Then, the origin is locally asymptotically stable. □

Proof. The proof is straightforward. Since the DAE is of index one, as far as the equation (2.15b) is concerned, there exists a unique solution $\boldsymbol{y} = \boldsymbol{G}(\boldsymbol{x})$ such that

$$\boldsymbol{f}_2(\boldsymbol{x}, \boldsymbol{G}(\boldsymbol{x})) = 0, \qquad \text{with} \quad \boldsymbol{G}(0) = 0,$$

provided by the Implicit Function Theorem. In this case, if $\lim_{t\to\infty} \boldsymbol{x}(t) = 0$, then $\lim_{t\to\infty} \boldsymbol{y}(t) = 0$. The condition $\lim_{t\to\infty} \boldsymbol{x}(t) = 0$ is a direct consequence of the usual LaSalle Invariance Principle. **Q.E.D.**

The aforementioned results in this section can be applied to the linear constant coefficient system (2.4) without any difficulty.

Theorem 2.17. *Consider the descriptor system (2.4). The following statements are true.*

i) *Suppose that the triple $\{\boldsymbol{E},\, \boldsymbol{A},\, \boldsymbol{C}\}$ is finite dynamic detectable and impulse observable. Then $\{\boldsymbol{E},\, \boldsymbol{A}\}$ is stable and impulse-free if and only if there exists a solution $\boldsymbol{X}_E$ to the generalized Lyapunov equation*

$$\begin{cases} \boldsymbol{A}^T\boldsymbol{X} + \boldsymbol{X}^T\boldsymbol{A} + \boldsymbol{C}^T\boldsymbol{C} = 0, \\ \boldsymbol{E}^T\boldsymbol{X} = \boldsymbol{X}^T\boldsymbol{E}, \end{cases} \tag{2.16}$$

with $\boldsymbol{E}^T\boldsymbol{X}_E = \boldsymbol{X}_E^T\boldsymbol{E} \geq 0$.

ii) *Suppose that $\{\boldsymbol{E},\, \boldsymbol{A},\, \boldsymbol{B}\}$ is finite dynamic stabilizable and impulse controllable. Then $\{\boldsymbol{E},\, \boldsymbol{A}\}$ is stable and impulse-free if and only if there exists a solution $\boldsymbol{Y}_E$ to the generalized Lyapunov equation*

$$\begin{cases} \boldsymbol{A}\boldsymbol{Y} + \boldsymbol{Y}^T\boldsymbol{A}^T + \boldsymbol{B}\boldsymbol{B}^T = 0, \\ \boldsymbol{E}\boldsymbol{Y} = \boldsymbol{Y}^T\boldsymbol{E}^T, \end{cases} \tag{2.17}$$

with $\boldsymbol{E}\boldsymbol{Y}_E = \boldsymbol{Y}_E^T\boldsymbol{E}^T \geq 0$.

iii) *The pair $\{\boldsymbol{E},\, \boldsymbol{A}\}$ is stable and impulse-free if and only if there exists a solution $\boldsymbol{P}$ to the following generalized Lyapunov inequality*

$$\begin{cases} \boldsymbol{A}^T\boldsymbol{P} + \boldsymbol{P}^T\boldsymbol{A} < 0, \\ \boldsymbol{E}^T\boldsymbol{P} = \boldsymbol{P}^T\boldsymbol{E}, \end{cases} \tag{2.18}$$

with $\boldsymbol{E}^T\boldsymbol{P} = \boldsymbol{P}^T\boldsymbol{E} \geq 0$. *Furthermore,* $\boldsymbol{x}^T\boldsymbol{E}^T\boldsymbol{P}\boldsymbol{x} = 0$ *if and only if* $\boldsymbol{E}\boldsymbol{x} = 0$.

iv) *Suppose that the triple* $\{\boldsymbol{E}, \boldsymbol{A}, \boldsymbol{C}\}$ *is finite dynamic detectable and impulse observable. Then* $\{\boldsymbol{E}, \boldsymbol{A}\}$ *is stable and impulse-free if and only if there exists a solution* $\boldsymbol{X}$ *to the generalized Lyapunov inequality*

$$\begin{cases} \boldsymbol{A}^T\boldsymbol{X} + \boldsymbol{X}^T\boldsymbol{A} + \boldsymbol{C}^T\boldsymbol{C} \leq 0, \\ \boldsymbol{E}^T\boldsymbol{X} = \boldsymbol{X}^T\boldsymbol{E}, \end{cases} \tag{2.19}$$

with $\boldsymbol{E}^T\boldsymbol{X} = \boldsymbol{X}^T\boldsymbol{E} \geq 0$.

v) *Suppose that* $\{\boldsymbol{E}, \boldsymbol{A}, \boldsymbol{B}\}$ *is finite dynamic stabilizable and impulse controllable. Then* $\{\boldsymbol{E}, \boldsymbol{A}\}$ *is stable and impulse-free if and only if there exists a solution* $\boldsymbol{Y}$ *to the generalized Lyapunov inequality*

$$\begin{cases} \boldsymbol{A}\boldsymbol{Y} + \boldsymbol{Y}^T\boldsymbol{A}^T + \boldsymbol{B}\boldsymbol{B}^T \leq 0, \\ \boldsymbol{E}\boldsymbol{Y} = \boldsymbol{Y}^T\boldsymbol{E}^T, \end{cases} \tag{2.20}$$

with $\boldsymbol{E}\boldsymbol{Y} = \boldsymbol{Y}^T\boldsymbol{E}^T \geq 0$.

□

Proof. We only prove i). The others are similar. To prove i), we need the following preliminary result.

Proposition 2.18. *Consider the descriptor system (2.10). Assume that it is impulse observable. Suppose that there exists a solution* $\boldsymbol{X}_{22}$ *to the following set of algebraic equations.*

$$\boldsymbol{X}_{22} + \boldsymbol{X}_{22}^T + \boldsymbol{C}_2^T\boldsymbol{C}_2 = 0, \tag{2.21}$$

$$\boldsymbol{N}^T\boldsymbol{X}_{22} = \boldsymbol{X}_{22}^T\boldsymbol{N}. \tag{2.22}$$

The the system is impulse-free, i.e., $\boldsymbol{N} = 0$. *Conversely, if* $\boldsymbol{N} = 0$, *then System (2.10) is impulse-free.* □

Proof of Proposition 2.18. Since, by hypothesis, the triple $\{\boldsymbol{N}, \boldsymbol{I}, \boldsymbol{C}_2\}$ is impulse observable, then, by duality, the triple $\{\boldsymbol{N}^T, \boldsymbol{I}, \boldsymbol{C}_2^T\}$ is impulse controllable. Hence $\Im m\boldsymbol{N}^T + \Im m\boldsymbol{C}_2^T + \aleph(\boldsymbol{N}^T) = \mathbb{F}$. Let $\hat{\kappa}$ be the nilpotency of $\boldsymbol{N}^T$. Then

$$\Im m(\boldsymbol{N}^T)^{\hat{\kappa}-1} = \Im m(\boldsymbol{N}^T)^{\hat{\kappa}-1}\boldsymbol{C}_2^T + \Im m(\boldsymbol{N}^T)^{\hat{\kappa}} = \Im m(\boldsymbol{N}^T)^{\hat{\kappa}-1}\boldsymbol{C}_2^T. \tag{2.23}$$

Recall that $\hat{\kappa}$ is the minimum number such that $\boldsymbol{N}^{\hat{\kappa}} = 0$. Premultiplying (2.21) by $(\boldsymbol{N}^T)^{\hat{\kappa}-1}$ and postmultiplying by $(\boldsymbol{N})^{\hat{\kappa}-1}$ yields

$$\begin{aligned} &(\boldsymbol{N}^T)^{\hat{\kappa}-1}\boldsymbol{X}_{22}^T(\boldsymbol{N})^{\hat{\kappa}-1} + (\boldsymbol{N}^T)^{\hat{\kappa}-1}\boldsymbol{X}_{22}(\boldsymbol{N})^{\hat{\kappa}-1} \\ &\quad = -(\boldsymbol{N}^T)^{\hat{\kappa}-1}\boldsymbol{C}_2^T\boldsymbol{C}_2(\boldsymbol{N})^{\hat{\kappa}-1} \\ \Rightarrow\ &(\boldsymbol{N}^T)^{\hat{\kappa}-1}\boldsymbol{X}_{22}^T\boldsymbol{N}(\boldsymbol{N})^{\hat{\kappa}-2} + (\boldsymbol{N}^T)^{\hat{\kappa}-2}\boldsymbol{N}^T\boldsymbol{X}_{22}(\boldsymbol{N})^{\hat{\kappa}-1} \\ &\quad = -(\boldsymbol{N}^T)^{\hat{\kappa}-1}\boldsymbol{C}_2^T\boldsymbol{C}_2(\boldsymbol{N})^{\hat{\kappa}-1} \\ \Rightarrow\ &(\boldsymbol{N}^T)^{\hat{\kappa}}\boldsymbol{X}_{22}(\boldsymbol{N})^{\hat{\kappa}-2} + (\boldsymbol{N}^T)^{\hat{\kappa}-2}\boldsymbol{X}_{22}^T(\boldsymbol{N})^{\hat{\kappa}} \\ &\quad = -(\boldsymbol{N}^T)^{\hat{\kappa}-1}\boldsymbol{C}_2^T\boldsymbol{C}_2(\boldsymbol{N})^{\hat{\kappa}-1} \end{aligned}$$

It is obvious that both terms in the left-hand side of the previous equation are zero, so that $(\boldsymbol{N}^T)^{\hat{\kappa}-1}\boldsymbol{C}_2^T = 0$. Hence, from (2.23), we get $\Im m(\boldsymbol{N}^T)^{\hat{\kappa}-1} = 0$, which is a contradiction. This implies that $\boldsymbol{N} = 0$, so that $\{\boldsymbol{N},\, \boldsymbol{I}\}$ is impulse-free. The rest of the proof is plain. ◊◊◊

We can now prove Theorem 2.17.

Sufficiency. Without loss of generality, assume that $\{\boldsymbol{E},\, \boldsymbol{A},\, \boldsymbol{C}\}$ are given in the Weierstrass form

$$\boldsymbol{E} = \begin{bmatrix} \boldsymbol{I} & 0 \\ 0 & \boldsymbol{N} \end{bmatrix}, \quad \boldsymbol{A} = \begin{bmatrix} \boldsymbol{A}_1 & 0 \\ 0 & \boldsymbol{I} \end{bmatrix}, \quad \boldsymbol{C} = \begin{bmatrix} \boldsymbol{C}_1 & \boldsymbol{C}_2 \end{bmatrix}.$$

Partition $\boldsymbol{X} = \begin{bmatrix} \boldsymbol{X}_{11} & \boldsymbol{X}_{12} \\ \boldsymbol{X}_{21} & \boldsymbol{X}_{22} \end{bmatrix}$ accordingly. Then the (2, 2)-block of GLE (2.16) shows that

$$\begin{aligned} \boldsymbol{X}_{22} + \boldsymbol{X}_{22}^T + \boldsymbol{C}_2^T\boldsymbol{C}_2 &= 0, \\ \boldsymbol{N}^T\boldsymbol{X}_{22} &= \boldsymbol{X}_{22}^T\boldsymbol{N}. \end{aligned}$$

This implies that $\{\boldsymbol{E},\, \boldsymbol{A}\}$ is impulse-free. Consider $\mathsf{V}(\boldsymbol{x}(t)) = \frac{1}{2}\boldsymbol{x}^T\boldsymbol{E}^T\boldsymbol{X}\boldsymbol{x}$ a candidate Lyapunov function. Along any trajectory $\boldsymbol{x}(\bullet)$ of the system, we have

$$\frac{\mathrm{d}\mathsf{V}(\boldsymbol{x}(t))}{\mathrm{d}t} = -\|\boldsymbol{C}\boldsymbol{x}\|^2 \le 0.$$

This proves that the equilibrium $\boldsymbol{x} = \boldsymbol{0}$ of the system is stable in the sense of Lyapunov, i.e., the pair $\{\boldsymbol{E},\, \boldsymbol{A}\}$ does not have eigenvalues with $\Re e[s] > 0$. In order to complete the proof, we have to show that it does not have eigenvalues on the imaginary axis. Observe that any trajectory $\boldsymbol{x}(\bullet)$ such that $\dot{\mathsf{V}}(\boldsymbol{x}(t)) = 0$ for all $t \ge 0$ is necessary a trajectory of $\boldsymbol{E}\dot{\boldsymbol{x}} = \boldsymbol{A}\boldsymbol{x}$ such that $\boldsymbol{x}(t)$ is bounded and $\boldsymbol{C}\boldsymbol{x} = 0$. Suppose, by contradiction, that $\boldsymbol{x}(\bullet)$ is bounded and does not converge to zero as $t \to \infty$, then necessarily there exists an eigenvalue of $\{\boldsymbol{E},\, \boldsymbol{A}\}$ on the imaginary axis such that the corresponding eigenvector $\boldsymbol{v}$ satisfies $\boldsymbol{C}\boldsymbol{v} = 0$. That is, there exists a real number $\omega \in \mathbb{R}$ and a vector $\boldsymbol{v}$ such that

$$\begin{bmatrix} \boldsymbol{A} - \mathrm{j}\omega\boldsymbol{E} \\ \boldsymbol{C} \end{bmatrix} \boldsymbol{v} = 0.$$

This contradicts the fact that $\{\boldsymbol{E},\, \boldsymbol{A},\, \boldsymbol{C}\}$ is finite dynamic detectable. Hence the set $\boldsymbol{S} = \{\boldsymbol{x} \in \mathbb{X} |\, \dot{\mathsf{V}}(\boldsymbol{x}(t)) = 0\}$ contains only the trajectory with $\lim_{t\to\infty} \boldsymbol{x}(t) = 0$. By Theorem 2.16, it is concluded that the equilibrium point $\boldsymbol{x} = 0$ is asymptotically stable (or simply *stable*).

Necessity. Suppose that $\{\boldsymbol{E},\, \boldsymbol{A}\}$ is stable and impulse-free. Then, without loss of generality, we can assume that $\{\boldsymbol{E},\, \boldsymbol{A},\, \boldsymbol{C}\}$ has a Weierstrass form given as

$$\{\boldsymbol{E},\, \boldsymbol{A},\, \boldsymbol{C}\} = \left\{ \begin{bmatrix} \boldsymbol{I} & 0 \\ 0 & 0 \end{bmatrix}, \begin{bmatrix} \boldsymbol{A}_1 & 0 \\ 0 & \boldsymbol{I} \end{bmatrix}, \begin{bmatrix} \boldsymbol{C}_1 & \boldsymbol{C}_2 \end{bmatrix} \right\}$$

with $\boldsymbol{A}_1$ Hurwitz. By the classical Lyapunov stability theorem, there exists a matrix $\boldsymbol{X}_{11} \geq 0$ satisfying the Lyapunov equation

$$\boldsymbol{A}_1^T \boldsymbol{X}_{11} + \boldsymbol{X}_{11} \boldsymbol{A}_1 + \boldsymbol{C}_1^T \boldsymbol{C}_1 = 0.$$

Setting

$$\boldsymbol{X} = \begin{bmatrix} \boldsymbol{X}_{11} & 0 \\ -\boldsymbol{C}_2^T \boldsymbol{C}_1 & -\frac{1}{2} \boldsymbol{C}_2^T \boldsymbol{C}_2 \end{bmatrix}$$

yields

$$\begin{cases} \boldsymbol{A}^T \boldsymbol{X} + \boldsymbol{X}^T \boldsymbol{A} + \boldsymbol{C}^T \boldsymbol{C} = 0, \\ \boldsymbol{E}^T \boldsymbol{X} = \boldsymbol{X}^T \boldsymbol{E} = \begin{bmatrix} \boldsymbol{X}_{11} & \boldsymbol{0} \\ \boldsymbol{0} & \boldsymbol{0} \end{bmatrix} \geq 0. \end{cases}$$

This completes the proof. **Q.E.D.**

2.4 Dissipative Descriptor Systems

Assume that a descriptor system

$$\boldsymbol{\Sigma} \triangleq \begin{cases} \boldsymbol{E}\dot{\boldsymbol{x}} = \boldsymbol{F}(\boldsymbol{x}, \boldsymbol{u}), \\ \boldsymbol{y} = \boldsymbol{H}(\boldsymbol{x}, \boldsymbol{u}), \end{cases} \quad \boldsymbol{E}\boldsymbol{x}(t_0) \in \Im m \boldsymbol{E}, \tag{2.24}$$

is given together with a real-valued function $s(\bullet, \bullet)$ defined on $\mathbb{U} \times \mathbb{Y}$, which will be called the *supply rate*. Suppose that for any $t_1, t_0 \in \mathbb{R}^+$, $\boldsymbol{u} \in \mathbb{U}$ and $\boldsymbol{y} \in \mathbb{Y}$, the function $s = s(\boldsymbol{u}(t), \boldsymbol{y}(t))$ is locally integrable. The following definition is an extension of the notion of dissipativity first introduced by Willems[121] for systems in state-space model.

Definition 2.19. *A descriptor system $\boldsymbol{\Sigma}$ is said to be dissipative with respective to the supply rate $s : \mathbb{U} \times \mathbb{Y} \rightarrow \mathbb{R}$ if there exists a nonnegative function $S : \mathbb{X} \rightarrow \mathbb{R}^+$, called the storage function, such that for all $t_1 \geq t_0$, $\boldsymbol{x}(t_0) \in \Im m \boldsymbol{E} \subset \mathbb{X}$ and all input function $\boldsymbol{u} \in \mathbb{U}$,*

$$S(\boldsymbol{x}(t_0)) + \int_{t_0}^{t_1} s(\boldsymbol{u}(t), \boldsymbol{y}(t)) \mathrm{d}t \geq S(\boldsymbol{x}_1), \tag{2.25}$$

where $\boldsymbol{x}_1 = \boldsymbol{x}(t_1; t_0, \boldsymbol{E}\boldsymbol{x}(t_0), \boldsymbol{u})$ (we denote $\boldsymbol{x}(t_1; t_0, \boldsymbol{E}\boldsymbol{x}(t_0), \boldsymbol{u})$ the state at time t_1 reached from the initial state $\boldsymbol{x}(t_0)$ at time t_0 by applying the input $\boldsymbol{u}(t)$ to the descriptor system $\boldsymbol{\Sigma}$). □

The inequality (2.25) is called the dissipation inequality. The definition of dissipative systems postulates the existence of a storage function which satisfies a dissipation inequality involving a given function called the supply rate. It expresses the fact that the energy stored in system $\boldsymbol{\Sigma}$ at any time t_1 cannot exceed the amount of energy that is stored in the system plus the external energy that is supplied to the system during the time period $t_0 \to t_1$. One important choice of supply rate is

$$s(\boldsymbol{u}(t), \boldsymbol{y}(t)) := \gamma^2\|\boldsymbol{u}(t)\|^2 - \|\boldsymbol{y}\|^2, \quad \gamma > 0. \tag{2.26}$$

Suppose that $\boldsymbol{x}(t_0) = 0$ and $\boldsymbol{\Sigma}$ is dissipative with respect to the supply rate (2.26). Then there exists $S \geq 0$ satisfying

$$\int_{t_0}^{t_1} (\gamma^2\|\boldsymbol{u}(t)\|^2 - \|\boldsymbol{y}\|^2 \mathrm{d}t \geq S(\boldsymbol{x}(t_1)) \geq 0$$

for all $t_1 \geq t_0$ and $\boldsymbol{u}(t)$. The previous inequality implies that, for any $T \geq 0$, we have

$$\int_0^T \|\boldsymbol{y}\|^2 \mathrm{d}t \leq \int_0^T \|\boldsymbol{u}(t)\|^2 \mathrm{d}t + S(\boldsymbol{x}(0)).$$

If $\boldsymbol{x}(0) = 0$, then the inequality reduces to

$$\int_0^T \|\boldsymbol{y}\|^2 \mathrm{d}t \leq \int_0^T \|\boldsymbol{u}(t)\|^2 \mathrm{d}t. \tag{2.27}$$

Inequality (2.27) can be rewritten as

$$\frac{\|\boldsymbol{y}(t)\|_2}{\|\boldsymbol{u}(t)\|_2} \leq \gamma, \tag{2.28}$$

where $\| \bullet \|_2$ denotes the usual $\mathcal{L}_2$-norm, namely $\| \bullet \|_2 \triangleq \sqrt{\int_0^T \| \bullet \|^2 \mathrm{d}t}$. Inequality (2.28) implies that the input-output maps $\boldsymbol{G}_{\boldsymbol{E}\boldsymbol{x}(t_0)}$ of the system $\boldsymbol{\Sigma}$ for every initial condition $\boldsymbol{E}\boldsymbol{x}(0)$ have $\mathcal{L}_2$ – gain $\leq \gamma$. The $\mathcal{L}_2$ – gain of $\boldsymbol{\Sigma}$ is defined as

$$\gamma(\boldsymbol{\Sigma}) \triangleq \inf \{\gamma \,|\, \|\boldsymbol{G}_{\boldsymbol{E}\boldsymbol{x}_0}(\boldsymbol{u}(t))\|_2 \leq \gamma \|\boldsymbol{u}(t)\|_2 \}.$$

This leads to the following definition.

Definition 2.20. *Given a descriptor system* $\boldsymbol{\Sigma}$. $\boldsymbol{\Sigma}$ *is said to have* $\mathcal{L}_2$ – *gain* $\leq \gamma$ *if it is dissipative with respect to the supply rate* $s(\boldsymbol{u}, \boldsymbol{y}) : \gamma^2\|\boldsymbol{u}\|^2 - \|\boldsymbol{y}\|^2$. *It is said to have* $\mathcal{L}_2$ – *gain* $< \gamma$ *if there exists* $\tilde{\gamma} < \gamma$ *such that* $\boldsymbol{\Sigma}$ *has* $\mathcal{L}_2$ – *gain* $\leq \tilde{\gamma}$. □

A rather theoretical method to determine whether a descriptor system $\boldsymbol{\Sigma}$ is dissipative is given in the next theorem.

Theorem 2.21. *Consider (2.24). Let the supply rate s be given. Then system (2.24) is dissipative with respect to* $s(\boldsymbol{u}, \boldsymbol{y})$ *if and only if*

$$S_a(\boldsymbol{x}) = \sup_{\substack{\boldsymbol{u}(\bullet) \\ T \geq 0}} - \int_0^T s(\boldsymbol{u}(t), \boldsymbol{y}(t)) \mathrm{d}t, \quad \boldsymbol{E}\boldsymbol{x}(0) \text{ given}$$

is finite for all $\boldsymbol{x} \in \mathbb{X}$. Moreover, under this condition, $S_a(\boldsymbol{x})$ is itself a storage function, and all other possible storage function $S(\boldsymbol{x})$ satisfy

$$S_a(\boldsymbol{x}) \leq S(\boldsymbol{x}), \quad \forall \boldsymbol{x} \in \mathbb{X}.$$

□

Theorem 2.21 gives a method which, in theory, can be used to verify whether or not a dynamical system is dissipative. However, it is somehow a difficult task since we have to check all elements $\boldsymbol{x}$ of $\mathbb{X}$. Fortunately, if the system is *reachable* from some initial point $\boldsymbol{x}^*$, then we need only to demonstrate this property for $\boldsymbol{x}^*$. This is summarized in the next theorem. Before giving this result, we should at first investigate the concept of reachability of the nonlinear descriptor system $\boldsymbol{\Sigma}$.

Definition 2.22. *The descriptor system $\boldsymbol{\Sigma}$ is said to be reachable from $\boldsymbol{x}^* \in \Im m\boldsymbol{E}$ if, for any $\boldsymbol{x} \in \Im m\boldsymbol{E} \subset \mathbb{X}$, there exists a t and $\boldsymbol{u} \in \mathbb{U}$ such that*

$$\boldsymbol{x} = \boldsymbol{x}(0;\ -t, \boldsymbol{x}^*, \boldsymbol{u})$$

□

Theorem 2.23. *Consider (2.24). Suppose that $\boldsymbol{\Sigma}$ is reachable from $\boldsymbol{x}^* \in \Im m\boldsymbol{E} \subset \mathbb{X}$. Then $\boldsymbol{\Sigma}$ is dissipative if and only if $S_a(\boldsymbol{x}^*)$ is finite.* □

Assume that $\boldsymbol{\Sigma}$ is reachable from $\boldsymbol{x}^* \in \Im m\boldsymbol{E}$. Consider the function $S_r : \mathbb{X} \to \mathbb{R}$ defined as

$$S_r(\boldsymbol{x}(t)) = \inf_{\substack{\boldsymbol{u}(\bullet) \\ T \geq 0}} \int_{-T}^{0} s(\boldsymbol{u}(t), \boldsymbol{y}(t)) \mathrm{d}t, \quad \boldsymbol{E}\boldsymbol{x}(-T) = \boldsymbol{x}^*, \text{ and } \boldsymbol{E}\boldsymbol{x}(0) \text{ are fixed.}$$

This function is called the *required supply* associated with the supply rate s. Note that possibly $S_r(\boldsymbol{x}) = -\infty$ for some $\boldsymbol{x}$. Clearly, the function thus defined satisfies the dissipation inequality (2.25). Moreover, we can see that $\boldsymbol{\Sigma}$ is dissipative with respect to $s(\boldsymbol{u}, \boldsymbol{y})$ if and only if $S_r(\boldsymbol{x}(t))$ is finite. With this observation in hand, we then arrive at the following theorem.

Theorem 2.24. *Consider (2.24). Suppose that $\boldsymbol{\Sigma}$ is reachable from $\boldsymbol{x}^* \in \Im m\boldsymbol{E} \subset \mathbb{X}$. Then $S_r(\boldsymbol{x}(t))$ satisfies the dissipation inequality (2.25). Furthermore, $\boldsymbol{\Sigma}$ is dissipative if and only if $S_r(\boldsymbol{x}(t))$ is bounded from below for all $\boldsymbol{x} \in \Im m\boldsymbol{E}$ (i.e., $S_r(\boldsymbol{x}(t)) > -\infty$). Under this condition, if $S(\boldsymbol{x})$ is any other storage function, then $S(\boldsymbol{x})$ satisfies*

$$S(\boldsymbol{x}(t)) \leq S_r(\boldsymbol{x}(t)) + S(\boldsymbol{x}^*), \quad \text{for all} \boldsymbol{x} \in \Im m\boldsymbol{E} \subset \mathbb{X},$$

and $S_r(\boldsymbol{x}) + S(\boldsymbol{x}^*)$ is itself a storage function. □

Suppose that system $\boldsymbol{\Sigma}$ is dissipative with C^1 storage function $\mathsf{V}(\boldsymbol{x})$ which satisfies $\frac{\partial \mathsf{V}}{\partial \boldsymbol{x}} = \boldsymbol{V}^T\boldsymbol{E}$. Then the dissipation inequality (2.25) can be conveniently replaced by its equivalent *infinitesimal* version

$$\boldsymbol{V}^T\boldsymbol{F}(\boldsymbol{x}(t), \boldsymbol{u}(t)) \leq s(\boldsymbol{u}(t), \boldsymbol{y}(t)), \tag{2.29}$$

which is called *differential dissipation inequality* [105]. Inequality (2.29) has advantage over the dissipation inequality (2.25) because it does not involve solution trajectories of the system $\boldsymbol{\Sigma}$. Moreover, as will be seen later, it can be easily connected with the Lyapunov stability criterion. From now on, we shall only consider the particular supply rate $s:\ \gamma^2\|\boldsymbol{u}(t)\|^2 - \|\boldsymbol{y}\|^2$, with $\gamma > 0$.

Theorem 2.25. *Consider System (2.24) with $\boldsymbol{x}(0) \in \Im m\boldsymbol{E}$ given. Suppose that the matrix $\boldsymbol{D}^T\mathrm{d} - \gamma^2 I$ is negative definite and $\{\boldsymbol{E},\ \boldsymbol{A},\ \boldsymbol{B}\}$ is impulse observable, where*

$$\boldsymbol{D} = \left(\frac{\partial}{\partial \boldsymbol{u}}\boldsymbol{h}\right)_{(\boldsymbol{x},\boldsymbol{u})=(0,0)}, \quad \boldsymbol{A} = \left(\frac{\partial}{\partial \boldsymbol{x}}\boldsymbol{F}\right)_{(\boldsymbol{x},\boldsymbol{u})=(0,0)}$$

and

$$\boldsymbol{B} = \left(\frac{\partial}{\partial \boldsymbol{x}}\boldsymbol{h}\right)_{(\boldsymbol{x},\boldsymbol{u})=(0,0)}$$

Suppose that any bounded trajectory $\boldsymbol{x}(t)$ of the system $\boldsymbol{E}\dot{\boldsymbol{x}} = \boldsymbol{F}(\boldsymbol{x}(t),\ 0)$ satisfying $\boldsymbol{h}(\boldsymbol{x}(t),\ 0) = 0$ for all $t \geq 0$ is such that $\lim_{t\to\infty}\boldsymbol{x}(t) = 0$. Suppose also that there exists a C^3 function $\mathsf{V}:\ \mathbb{X} \longrightarrow \mathbb{R}^+$ vanishing at $\boldsymbol{E}\boldsymbol{x} = 0$ and positive elsewhere which satisfies the following properties.

i) $\frac{\partial}{\partial \boldsymbol{x}}\mathsf{V} = \boldsymbol{V}^T(\boldsymbol{x})\boldsymbol{E}$ *for some* C^2 *function* $\boldsymbol{V}:\mathbb{X} \longrightarrow \mathbb{R}^n$

ii) $Y_0(\boldsymbol{x},\boldsymbol{u},\boldsymbol{y}) \triangleq \boldsymbol{V}^T(\boldsymbol{x})\boldsymbol{F}(\boldsymbol{x},\ \boldsymbol{u}) + \|\boldsymbol{y}\|^2 - \gamma^2\|\boldsymbol{u}\|^2 \leq 0.$

iii) $\boldsymbol{E}^T\boldsymbol{V}_{\boldsymbol{x}} = \boldsymbol{V}_{\boldsymbol{x}}^T\boldsymbol{E}.$

Then $\boldsymbol{\Sigma}$ has an $\mathcal{L}_2$ gain less than or equal to γ and the equilibrium point $\boldsymbol{x} = 0$ is locally asymptotically stable with index one. □

Proof. Set $S(\boldsymbol{x}(t)) = \mathsf{V}(\boldsymbol{x}(t))$. Then, as $t_1 \to t_0$, Inequality (2.25) reduce to Condition ii). We now prove asymptotical stability and index-one property. We first show that $\boldsymbol{\Sigma}$ is of index one. Setting $\boldsymbol{u} \equiv \boldsymbol{0}$ in Y_0 and taking the second order partial derivative of Y_0 with respect to $\boldsymbol{x}$ around $(\boldsymbol{x},\ \boldsymbol{u}) = (\boldsymbol{0},\boldsymbol{0})$ yields

$$\boldsymbol{A}^T\boldsymbol{V}_{\boldsymbol{x}}(0) + \boldsymbol{V}_{\boldsymbol{x}}^T(0)A + \boldsymbol{B}^T\boldsymbol{B} \leq 0. \tag{2.30}$$

Since $\{\boldsymbol{E},\ \boldsymbol{A},\ \boldsymbol{B}\}$ is impulse observable, inequality (2.30) along with condition iii) implies that DAE (2.24) is of index one. To prove asymptotical stability, observe that along any trajectory $\boldsymbol{x}(\bullet)$ of $\boldsymbol{\Sigma}$ with $\boldsymbol{u} \equiv 0$ we have

$$\frac{\mathrm{d}\mathsf{V}(\boldsymbol{x}(t))}{\mathrm{d}t} \leq -\|\boldsymbol{y}\|^2 \leq 0.$$

This shows that the equilibrium point $\boldsymbol{x} = \mathbf{0}$ of $\boldsymbol{\Sigma}$ is stable. In addition, observe that any trajectory $\boldsymbol{x}(\bullet)$ such that $\dot{\mathsf{V}}(\boldsymbol{x}(t)) = 0$ for all $t \geq 0$ is necessarily a trajectory of $\boldsymbol{E}\dot{\boldsymbol{x}} = \boldsymbol{F}(\boldsymbol{x}, \mathbf{0})$ such that $\boldsymbol{x}(t)$ is bounded and $\boldsymbol{h}(\boldsymbol{x}, \mathbf{0}) = \mathbf{0}$ for all $t \geq 0$. Hence, by hypothesis, it is concluded that $\lim_{t\to\infty} \boldsymbol{x}(t) = 0$ by using Theorem 2.16. **Q.E.D.**

Let us get back to the linear constant coeffieicent descriptor system (2.4). Define a function $H : \mathbb{X} \times \mathbb{R}^n \times \mathbb{R}^m \longrightarrow \mathbb{R}$, which is called a *Hamiltonian function*, as follows

$$H(\boldsymbol{x}, \boldsymbol{p}, \boldsymbol{u}) = \boldsymbol{p}^T(\boldsymbol{A}\boldsymbol{x} + \boldsymbol{B}\boldsymbol{u}) + \boldsymbol{x}^T\boldsymbol{C}^T\boldsymbol{C}\boldsymbol{x} - \gamma^2\boldsymbol{u}^T\boldsymbol{u}. \tag{2.31}$$

The function thus defined is a polynominal of degree two in $\boldsymbol{u}$. It has a unique maximum at $\boldsymbol{u}_*(\boldsymbol{x},\boldsymbol{p}) = \frac{1}{2\gamma^2}\boldsymbol{B}^T\boldsymbol{p}$. Substituting $\boldsymbol{u}_*(\boldsymbol{x},\boldsymbol{p})$ in (2.31) yields

$$H_* \triangleq \boldsymbol{p}^T\boldsymbol{A}\boldsymbol{x} + \frac{1}{4\gamma^2}\boldsymbol{p}^T\boldsymbol{B}\boldsymbol{B}^T\boldsymbol{p} + \boldsymbol{x}^T\boldsymbol{C}^T\boldsymbol{C}\boldsymbol{x}.$$

Therefore, the Hamiltonian function can be expressed in a Taylor series of order two given by

$$H(\boldsymbol{x},\boldsymbol{p},\boldsymbol{u}) = H_*(\boldsymbol{x},\boldsymbol{p}) - \gamma^2(\boldsymbol{u} - \boldsymbol{u}_*(\boldsymbol{x},\boldsymbol{p}))^T(\boldsymbol{u} - \boldsymbol{u}_*(\boldsymbol{x},\boldsymbol{p})). \tag{2.32}$$

Now, assume that there exists a C^1 function $\mathsf{V} : \mathbb{X} \longrightarrow \mathbb{R}$, vanishing at $\boldsymbol{E}\boldsymbol{x} = \mathbf{0}$, which satisfies $\dfrac{\partial \mathsf{V}(\boldsymbol{x})}{\partial \boldsymbol{x}} = \boldsymbol{V}^T(\boldsymbol{x})\boldsymbol{E}$ such that the identity $H_*(\boldsymbol{x}, \boldsymbol{u}) = 0$ holds. Under this condition, it is readily seen that Equation (2.32) implies

$$\boldsymbol{V}^T(\boldsymbol{x})(\boldsymbol{A}\boldsymbol{x} + \boldsymbol{B}\boldsymbol{u}) + \boldsymbol{x}^T\boldsymbol{C}^T\boldsymbol{C}\boldsymbol{x} - \gamma^2\boldsymbol{u}^T\boldsymbol{u} \leq 0. \tag{2.33}$$

Inequality (2.33) is simply a restatement of the fact that, for any $t \in \mathbb{R}^+$,

$$\frac{\mathrm{d}\mathsf{V}(\boldsymbol{x}(t))}{\mathrm{d}t} + \|\boldsymbol{y}(t)\|^2 - \gamma^2\|\boldsymbol{u}(t)\|^2 \leq 0.$$

This shows that the $\mathcal{L}_2$ – gain of the system $\boldsymbol{\Sigma}$ is less than or equal to γ. Suppose now that the function $\mathsf{V}(\boldsymbol{x})$ is given by a quadratic form: $\mathsf{V}(\boldsymbol{x}) = \boldsymbol{x}^T\boldsymbol{E}^T\boldsymbol{X}\boldsymbol{x}$ with $\boldsymbol{E}^T\boldsymbol{X} = \boldsymbol{X}^T\boldsymbol{E}$. Then $\dfrac{\partial \mathsf{V}(\boldsymbol{x})}{\partial \boldsymbol{x}} = 2\boldsymbol{x}^T\boldsymbol{E}^T\boldsymbol{X}$. Substitue $\boldsymbol{V}(\boldsymbol{x}) = \boldsymbol{X}\boldsymbol{x}$ into $H_*(\boldsymbol{x}, \boldsymbol{V}(\boldsymbol{x})) = 0$ to get

$$2\boldsymbol{x}^T\boldsymbol{X}^T\boldsymbol{A}\boldsymbol{x} + \frac{1}{\gamma^2}\boldsymbol{x}^T\boldsymbol{X}^T\boldsymbol{B}\boldsymbol{B}^T\boldsymbol{X}\boldsymbol{x} + \boldsymbol{x}^T\boldsymbol{C}^T\boldsymbol{C}\boldsymbol{x} = 0,$$

which holds if and only if $\boldsymbol{X}$ is a solution to the *generalized algebraic Riccati equation* (GARE)

$$\begin{cases} \boldsymbol{A}^T\boldsymbol{X} + \boldsymbol{X}^T\boldsymbol{A} + \boldsymbol{C}^T\boldsymbol{C} + \frac{1}{\gamma^2}\boldsymbol{X}^T\boldsymbol{B}\boldsymbol{B}^T\boldsymbol{X} = 0, \\ \boldsymbol{E}^T\boldsymbol{X} = \boldsymbol{X}^T\boldsymbol{E}. \end{cases}$$

That is to say, if there exists a solution to the GARE, then $\boldsymbol{\Sigma}$ has an $\mathcal{L}_2$-gain which is less than or equal to γ. As a matter of fact, the property thus shown is also a necessary condition for a linear constant coefficient descriptor system to be dissipative with respect to the supply rate $s: \gamma^2\|\boldsymbol{u}(t)\|^2 - \|\boldsymbol{y}\|^2$. This very important result is the main subject of the following lemma. It is a generalization of the well-known *Bounded Real Lemma.*

Lemma 2.26. (Bounded Real Lemma) *Consider (2.4). Suppose the pair $\{\boldsymbol{E}, \boldsymbol{A}\}$ is regular so that its Weierstrass form decomposition (2.5) is well-defined. The following statements are equivalent.*
(1) *$\{\boldsymbol{E}, \boldsymbol{A}\}$ is stable, impulse-free and $\|\boldsymbol{G}(s)\|_\infty < \gamma$, where*

$$\boldsymbol{G}(s) \triangleq \boldsymbol{C}(s\boldsymbol{E} - \boldsymbol{A})^{-1}\boldsymbol{B}.$$

(2) *$\{\boldsymbol{E}, \boldsymbol{A}\}$ is stable, impulse-free and $\gamma^2\boldsymbol{I} - \boldsymbol{B}_2^T\boldsymbol{C}_2^T\boldsymbol{C}_2\boldsymbol{B}_2 > 0$. Furthermore, the Hamiltonian system*

$$\begin{bmatrix} \boldsymbol{E} & 0 \\ 0 & \boldsymbol{E}^T \end{bmatrix} \begin{bmatrix} \dot{\boldsymbol{x}} \\ \dot{\boldsymbol{\lambda}} \end{bmatrix} = \begin{bmatrix} \boldsymbol{A} & \frac{1}{\gamma^2}\boldsymbol{B}\boldsymbol{B}^T \\ -\boldsymbol{C}^T\boldsymbol{C} & -\boldsymbol{A}^T \end{bmatrix} \begin{bmatrix} \boldsymbol{x} \\ \boldsymbol{\lambda} \end{bmatrix}$$

is regular, impulse-free and has no finite dynamic modes on the imaginary axis.

(3) *There exists an admissible solution to the GARE*

$$\begin{cases} \boldsymbol{A}^T\boldsymbol{X} + \boldsymbol{X}^T\boldsymbol{A} + \boldsymbol{C}^T\boldsymbol{C} + \frac{1}{\gamma^2}\boldsymbol{X}^T\boldsymbol{B}\boldsymbol{B}^T\boldsymbol{X} = 0, \\ \boldsymbol{E}^T\boldsymbol{X} = \boldsymbol{X}^T\boldsymbol{E} \geq 0. \end{cases} \tag{2.34}$$

(4) *There exists a nonsingular matrix $\boldsymbol{P}$ satisfying the generalized algebraic Riccati inequality*

$$\begin{cases} \boldsymbol{P}^T\boldsymbol{A} + \boldsymbol{A}^T\boldsymbol{P} + \boldsymbol{C}^T\boldsymbol{C} + \frac{1}{\gamma^2}\boldsymbol{P}^T\boldsymbol{B}\boldsymbol{B}^T\boldsymbol{P} < 0 \\ \boldsymbol{E}^T\boldsymbol{P} = \boldsymbol{P}^T\boldsymbol{E} \geq 0 \end{cases}$$

□

Proof. We prove the equivalence between (1), (2), and (3). The proof of (1) ⇔ (4) can be found in Masubuchi *et al.* [82].

(1) ⇒ (2).
Without loss of generality, we can assume that the system (2.4) in the Weierstrass form (2.5). Since $\{\boldsymbol{E}, \boldsymbol{A}\}$ is impulse-free, $\boldsymbol{N} = 0$. Writing $\boldsymbol{x} = \begin{bmatrix} \hat{\boldsymbol{x}}_1 \\ \hat{\boldsymbol{x}}_2 \end{bmatrix}$ and $\boldsymbol{\lambda} = \begin{bmatrix} \boldsymbol{\lambda}_1 \\ \boldsymbol{\lambda}_2 \end{bmatrix}$ compatible with (2.5), the Hamiltonian system can be put in the form

$$\begin{bmatrix} I & 0 & 0 & 0 \\ 0 & I & 0 & 0 \\ 0 & 0 & 0 & 0 \\ 0 & 0 & 0 & 0 \end{bmatrix} \begin{bmatrix} \dot{x}_1 \\ \dot{\lambda}_1 \\ \dot{\lambda}_2 \\ \dot{x}_2 \end{bmatrix} = \begin{bmatrix} A_1 & \frac{1}{\gamma^2} B_1 B_1^T & 0 & \frac{1}{\gamma^2} B_1 B_2^T \\ -C_1^T C_1 & -A_1^T & -C_1^T C_2 & 0 \\ 0 & \frac{1}{\gamma^2} B_2 B_1^T & I & \frac{1}{\gamma^2} B_2 B_2^T \\ -C_2^T C_1 & 0 & -C_2^T C_2 & -I \end{bmatrix} \begin{bmatrix} x_1 \\ \lambda_1 \\ \lambda_2 \\ x_2 \end{bmatrix}.$$

This system is clearly regular and impulse-free since $\|G(s)\|_\infty < \gamma$ implies $\gamma^2 I - B_2^T C_2^T C_2 B_2 > 0$. This implies that

$$\begin{bmatrix} I & \frac{1}{\gamma^2} B_2 B_2^T \\ -C_2^T C_2 & -I \end{bmatrix}$$

is nonsingular. Furthermore, it is easy to see that

$$\det\left(s \begin{bmatrix} E & 0 \\ 0 & E^T \end{bmatrix} - \begin{bmatrix} A & \frac{1}{\gamma^2} BB^T \\ -C^T C & -A^T \end{bmatrix} \right)$$

$$= \det(sE - A) \det(sE^T + A^T) \det(\gamma^2 I - G^*(s) G(s))$$

It follows that the characteristic polynomial has no pure imaginary roots.

(2) $\Rightarrow$ (3).
This part follows immediately by Proposition A.4.

(3) $\Rightarrow$ (1).
Suppose, by hypothesis, that there exists an admissible solution to the GARE:

$$\begin{cases} A^T X + X^T A + C^T C + \frac{1}{\gamma^2} X^T BB^T X = 0, \\ E^T X = X^T E \geq 0. \end{cases} \tag{2.35}$$

Set $\tilde{C} = \frac{1}{\gamma} B^T X$. Since $\{E, A + \frac{1}{\gamma^2} BB^T X\}$ is admissible, $\{E, A, \tilde{C}\}$ is finite dynamics detectable and impulse observable. Rearrange (2.35) as:

$$\begin{cases} A^T X + X^T A + \tilde{C}^T \tilde{C} = -C^T C \leq 0, \\ E^T X = X^T E \geq 0, \end{cases}$$

which is a generalized Lyapunov inequality. Then, by Proposition 2, $\{E, A\}$ is stable and impulse-free. Next, we show that $\|C(sE - A)^{-1} B\|_\infty < \gamma$. Observe that Equation (2.34) implies that

$$\begin{aligned} &\frac{1}{\gamma^2} B^T (-j\omega E - A^T)^{-1} C^T C (j\omega E - A)^{-1} B \\ &= I - [I - \frac{1}{\gamma^2} B^T X (-j\omega E - A)^{-1} B]^T [I - \frac{1}{\gamma^2} B^T X (j\omega E - A)^{-1} B] \end{aligned} \tag{2.36}$$

for all $\omega \geq 0$. It follows that $\|\boldsymbol{C}(s\boldsymbol{E}-\boldsymbol{A})^{-1}\boldsymbol{B}\|_\infty \leq \gamma$. To complete the proof, we need to show that the strict inequality holds. Suppose, by contradiction, that there exists an $\bar{\omega} \geq 0$ such that $\|\boldsymbol{C}(\mathrm{j}\bar{\omega}\boldsymbol{E}-\boldsymbol{A})^{-1}\boldsymbol{B}\| = \gamma$. Then equation (2.36) implies that there exists a vector p_0 such that $[\boldsymbol{I}-\boldsymbol{B}^T\boldsymbol{X}(\mathrm{j}\bar{\omega}\boldsymbol{E}-\boldsymbol{A})^{-1}\boldsymbol{B}]p_0 = 0$. This gives that $\det[\boldsymbol{I}-\boldsymbol{B}^T\boldsymbol{X}(\mathrm{j}\bar{\omega}\boldsymbol{E}-\boldsymbol{A})^{-1}\boldsymbol{B}] = 0$. Now, by a standard result on determinants, we have

$$\det[\mathrm{j}\bar{\omega}\boldsymbol{E}-\boldsymbol{A}-\frac{1}{\gamma^2}\boldsymbol{B}\boldsymbol{B}^T\boldsymbol{X}] = \det[\mathrm{j}\bar{\omega}\boldsymbol{E}-\boldsymbol{A}]\det[\boldsymbol{I}-\frac{1}{\gamma^2}\boldsymbol{B}^T\boldsymbol{X}(\mathrm{j}\bar{\omega}\boldsymbol{E}-\boldsymbol{A})^{-1}\boldsymbol{B}].$$

Thus $\det[\mathrm{j}\bar{\omega}\boldsymbol{E}-\boldsymbol{A}-\frac{1}{\gamma^2}\boldsymbol{B}\boldsymbol{B}^TX] = 0$. This contradicts the fact that $\boldsymbol{X}$ is an admissible solution to GARE (2.34). Hence, we can conclude that $\|\boldsymbol{C}(s\boldsymbol{E}-\boldsymbol{A})^{-1}\boldsymbol{B}\|_\infty < \gamma$. This completes the proof. **Q.E.D.**

Example. Bounded Real Systems
Consider a descriptor system $\boldsymbol{G}(s)$, which has a generalized state-space representation given by

$$\boldsymbol{G}(s) := \begin{cases} \boldsymbol{E}\dot{x}(t) = \boldsymbol{A}\boldsymbol{x}(t) + \boldsymbol{B}\boldsymbol{u}(t), \\ \boldsymbol{y}(t) = \boldsymbol{C}\boldsymbol{x}(t) \end{cases}$$
$$\triangleq \begin{cases} \begin{bmatrix} 1 & 0 \\ -2 & 0 \end{bmatrix}\dot{\boldsymbol{x}}(t) = \begin{bmatrix} -1 & 0 \\ 4 & 1 \end{bmatrix}\boldsymbol{x}(t) + \begin{bmatrix} 1 \\ -1 \end{bmatrix}\boldsymbol{u}(t) \\ \boldsymbol{y}(t) = \begin{bmatrix} 1 & 0 \end{bmatrix}\boldsymbol{x}(t), \end{cases}$$

It is easy to verify that

$$\boldsymbol{X}_B := \begin{bmatrix} 1 & 0 \\ 0 & 0 \end{bmatrix}$$

is a solution that satisfies GARE (2.34) with $\gamma = 1$. This shows that $\|\boldsymbol{G}(s)\|_\infty \leq 1$ (Note that the solution $\boldsymbol{X}_B$ is not an admissible one; hence, in general, we have to include the equality sign.). As a matter of fact, by direct mathematical analysis, it can be shown that $\|\boldsymbol{G}(s)\|_\infty = 1$. With the matrix $\boldsymbol{X}_B$ in hand, we can construct a *spectral factor* $\Phi(s)$ for $1-\boldsymbol{G}^*(s)\boldsymbol{G}(s)$, which is given by

$$\Phi(s) \triangleq 1 + (-1)\boldsymbol{B}^T\boldsymbol{X}_B(s\boldsymbol{E}-\boldsymbol{A})^{-1}\boldsymbol{B} = \frac{s}{s+1},$$

such that

$$1-\boldsymbol{G}^*(s)\boldsymbol{G}(s) = \Phi^*(s)\Phi(s).$$

Setting $s=0$ in the above identity yields $1-\boldsymbol{G}^*(0)\boldsymbol{G}(0) = 0$. This shows that $\|\boldsymbol{G}(s)\|_\infty = 1$. ◇◇◇

Notes and References

1. Existence and uniqueness property of solutins to the differential-algebraic equations investigated in this chapter is only a special case of the general

existence and uniqueness problem for fully implicit DAEs. An existence and uniqueness theory for general fully implicit DAEs has been well developed in [96][97] by exploiting their underlying differential geometric structure. See also Brenan *et al.*[14], Hairer and Wanner[40].

2. Higher index nonlinear DAEs can exhibit different physical phenomena. See [113]. Although the high-index DAE might induce damage to the control systems, it has attracted many attentions in the field of numerical analysis. For a great many results and ramifications on numerical treatment, the readers may consult [4] [14] and the references quoted there.
3. The theory of dissipative systems, while laid down by Willems in the fundamental paper[121], has now been well developed [42][43][105]. Our treatment in dissipative descriptor sytems is largely based on Isidori[47] and van der Schaft[105]. The generalized version of the Bounded Real Lemma (Lemma 2.26) is taken from [116].

3

Youla Parameterization for Descriptor Systems

For control engineers, the stability of the given control systems has always been the main concern. We often encounter controller synthesis problems of the following form: Given a plant $\boldsymbol{G}$, design a controller $\boldsymbol{K}$ such that the closed-loop system is stable and satisfies certain given performance criteria. It is thus convenient to have a parameterization of the class of controllers which stabilize the plant and to optimize the performance criteria within this class. This approach has been quite successful in the modern control systems theory, the result being the Youla–Kucera parameterization. The objective of this chapter is to extend the results which gave the Youla parameterization for nonlinear systems, as obtained via a state-space approach, of [46] and [76] to the nonlinear descriptor systems case. By providing such a parameterization for descriptor systems, it is hoped that controller synthesis problems for descriptor systems may also become tractable. In particular, one would like to tackle the $\mathcal{H}_\infty$ control problems in this way.

3.1 Parameterization of All Stabilizing Controllers - Nonlinear Case

Consider the following nonlinear descriptor system

$$\boldsymbol{\Sigma} : \begin{cases} \boldsymbol{E}\dot{\boldsymbol{x}} = \boldsymbol{f}(\boldsymbol{x}, \boldsymbol{u}), \\ \quad \boldsymbol{y} = \boldsymbol{h}(\boldsymbol{x}), \boldsymbol{E}\boldsymbol{x}(0) \text{ given} \end{cases} \tag{3.1}$$

where $\boldsymbol{x} = [\boldsymbol{x}_1^T \;\; \boldsymbol{x}_2^T]^T \triangleq (x_1, \ldots, x_n)^T$ are local coordinates for an n-dimensional state-space manifold $\mathcal{X}$, $\boldsymbol{u} \in \mathbb{U} \subset \mathbb{R}^m$, and $\boldsymbol{y} \in \mathbb{Y} \subset \mathbb{R}^p$. The measured output $\boldsymbol{y}$ is fed back to $\boldsymbol{u}$ by $\boldsymbol{u} \triangleq \boldsymbol{K}\boldsymbol{y}$. Suppose the controller $\boldsymbol{K}$ is, in general, also a descriptor system described by

$$\boldsymbol{K}_{\boldsymbol{y}} : \begin{cases} \hat{\boldsymbol{E}}_0\dot{\hat{\boldsymbol{x}}}_0 = \boldsymbol{f}_0(\hat{\boldsymbol{x}}_0, \boldsymbol{y}), \\ \quad \boldsymbol{u} = \boldsymbol{h}_0(\hat{\boldsymbol{x}}_0, \boldsymbol{y}), \qquad \boldsymbol{E}\hat{\boldsymbol{x}}(0) \quad \text{given.} \end{cases} \tag{3.2}$$

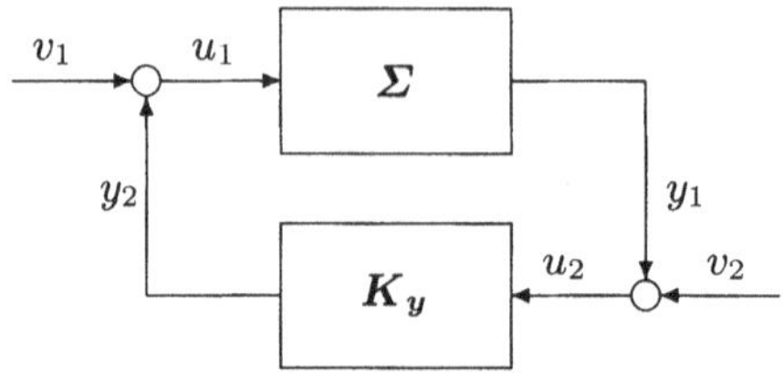

Fig. 3.1. Well-posedness analysis diagram for nonlinear systems

In this section, we will investigate first the state feedback case, namely $y = x$. We'll give a complete state-space parameterization of all stabilizing state feedback controllers for a given nonlinear descriptor system. The problem to be worked out here is formulated as follows.

Problem 3.1. *Consider (3.1) with $y = x$. The state feedback stabilization problem is to find all state feedback controllers $K_{y=x}$ such that the closed-loop system is asymptotically stable with index one.* ◊◊◊

The following assumptions are made.

(**A1**) The feedback interconnection shown in Fig. 3.1 is well-posed. The well-posedness corresponds to the existence and uniqueness of solutions for u_1 and u_2 for each choice of v_1 and v_2.

(**A2**) There exists a function $\alpha(x)$ such that

$$E\dot{x} = f(x, \alpha(x))$$

is asymptotically stable with index one.

Theorem 3.2. *Consider (3.1) with $y = x$. Suppose that Assumptions (A1) and (A2) hold. Then a parameterization of all internally stabilizing controllers via state feedback is given by*

$$K_{SF} := \mathcal{LFT}(J_{SF}, Q_{SF}) \tag{3.3}$$

(See Fig. 3.2) with

$$J_{SF}: \begin{cases} E\dot{\xi} = f(x, u) - \pi(x - \xi), \\ u = \alpha(\xi) + v, \\ r = \xi, \end{cases} \qquad Q_{SF}: \begin{cases} E_Q\dot{q} = f_Q(x, \xi, q), \\ v = h_Q(x, \xi, q), \end{cases}$$

where the systems

$$E\dot{\xi} = \pi(\xi),$$

and

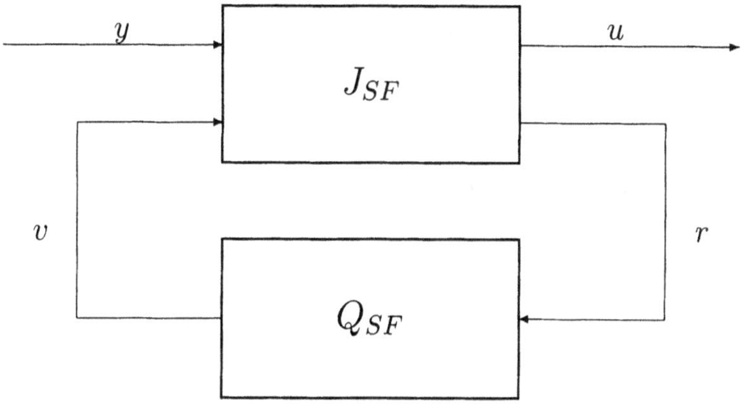

Fig. 3.2. Parameterization of all state feedback controllers

$$\boldsymbol{E}_Q\dot{\boldsymbol{q}} = \boldsymbol{f}_Q(0, 0, \boldsymbol{q})$$

are asymptotically stable with index one in a neighborhood of the origin with $\boldsymbol{f}_Q(\boldsymbol{x}, \boldsymbol{x}, \boldsymbol{q}) = \boldsymbol{f}_Q(0, 0, \boldsymbol{q})$, $\boldsymbol{h}_Q(\boldsymbol{x}, \boldsymbol{x}, \boldsymbol{q}) = \boldsymbol{h}_Q(0, 0, \boldsymbol{q})$ *and* , $\boldsymbol{h}_Q(0, 0, 0) = 0$. □

Proof.
Sufficiency. By setting $\boldsymbol{e} = \boldsymbol{x} - \boldsymbol{\xi}$, the closed-loop system can be put in the following form.

$$\begin{aligned} \boldsymbol{E}\dot{\boldsymbol{x}} &= \boldsymbol{f}(\boldsymbol{x}, \boldsymbol{\alpha}(\boldsymbol{x} - \boldsymbol{e}) + \boldsymbol{h}_Q(\boldsymbol{x}, \boldsymbol{x} - \boldsymbol{e}, \boldsymbol{q})), \\ \boldsymbol{E}_Q\dot{\boldsymbol{q}} &= \boldsymbol{f}_Q(\boldsymbol{x}, \boldsymbol{x} - \boldsymbol{e}, \boldsymbol{q}), \\ \boldsymbol{E}\dot{\boldsymbol{e}} &= \boldsymbol{\pi}(\boldsymbol{e}). \end{aligned} \tag{3.4}$$

Thereafter, by Lemma B.3, System (3.4) is asymptotically stable with index one at the equilibrium point.

Necessity. Suppose that the state feedback stabilization problem is solved by the controller given as follows.

$$\boldsymbol{K}_c : \begin{cases} \boldsymbol{E}_c\dot{\boldsymbol{x}}_c = \boldsymbol{f}_c(\boldsymbol{x}, \boldsymbol{x}_c), \\ \quad\;\; \boldsymbol{u} = \boldsymbol{h}_c(\boldsymbol{x}, \boldsymbol{x}_c), \end{cases} \tag{3.5}$$

where $\boldsymbol{f}_c(0, 0) = 0$ and $\boldsymbol{h}_c(0, 0) = 0$. Then, by hypothesis, it implies that the closed-loop system

$$\begin{aligned} \boldsymbol{E}\dot{\boldsymbol{x}} &= \boldsymbol{f}(\boldsymbol{x}, \boldsymbol{h}_c(\boldsymbol{x}, \boldsymbol{x}_c)) \\ \boldsymbol{E}_c\dot{\boldsymbol{x}}_c &= \boldsymbol{f}_c(\boldsymbol{x}, \boldsymbol{x}_c) \end{aligned}$$

is asymptotically stable with index one at the origin. We need to show that $\boldsymbol{K}_c = \boldsymbol{K}_{SF}$ by an appropriate selection of $\boldsymbol{Q}$. To this end, set $\boldsymbol{y} \triangleq \boldsymbol{q} + \boldsymbol{x} - \boldsymbol{\xi}$ and

$$E_Q\dot{q} = f_Q(x,\, \xi,\, q) = \begin{bmatrix} f(\eta,\, h_c(q_2, \eta)) - \pi(x - \xi) \\ f_c(q_2,\, \eta) \end{bmatrix} \tag{3.6}$$

$$h_Q(x,\, \xi, q) = -\alpha(q_1) + h_c(q_2,\, \eta). \tag{3.7}$$

It is easily shown that (3.6) and (3.7) satisfy the hypothesis of the theorem. Observe that the controller K_{SF} can be written in the following form.

$$\begin{aligned} E\dot{\xi} &= f(x,\, u) - \pi(x - \xi), \\ E\dot{q}_1 &= f(\eta,\, h_c(q_2, \eta)) - \pi(x - \xi), \\ E_c\dot{q}_2 &= f_c(q_2,\, \eta), \\ u &= \alpha(\xi) + h_Q(x,\, \xi, q) = \alpha(\xi) - \alpha(q_1) + h_c(q_2,\, \eta). \end{aligned} \tag{3.8}$$

Now set $q_1(0) = \xi(0)$. Then, in view of (3.8), we have $q_1 \equiv \xi(t)$. Therefore (3.8) can be rewritten as

$$\begin{aligned} E\dot{\xi} &= f(x,\, h_c(q_2, x)) - \pi(x - \xi), \\ E_c\dot{q}_2 &= f_c(q_2,\, x), \\ u &= h_c(q_2,\, x). \end{aligned}$$

Obviously, the above system is equivalent to

$$\begin{aligned} E_c\dot{q}_2 &= f_c(q_2,\, x), \\ u &= h_c(q_2,\, x), \end{aligned}$$

which is exactly the controller K_c. This completes the proof. **Q.E.D.**

Let us now get back to the output feedback control case. The problem to be solved at this time is stated as follows.

Problem 3.3. *Consider (3.1) and the output feedback controller K_y. The output feedback stabilization problem is to find all output feedback controllers $K_{y=h(x)}$ such that the resulting closed-loop system is asymptotically stable with index one.* ◊◊◊

An additional assumption is needed.

(A2) There exists a function $\ell(\bullet,\, \bullet)$ such that the system

$$\begin{aligned} E\dot{x} &= f(x,\, u) \\ E\dot{\xi} &= f(\xi,\, u) + \ell(\xi,\, h(x)) \end{aligned} \tag{3.9}$$

is asymptotically stable with index one around the origin.

Assumption (A2) can be viewed as a version of weak detectability[115] for general nonlinear descriptor systems. The main result for the output feedback case is summarized as follows.

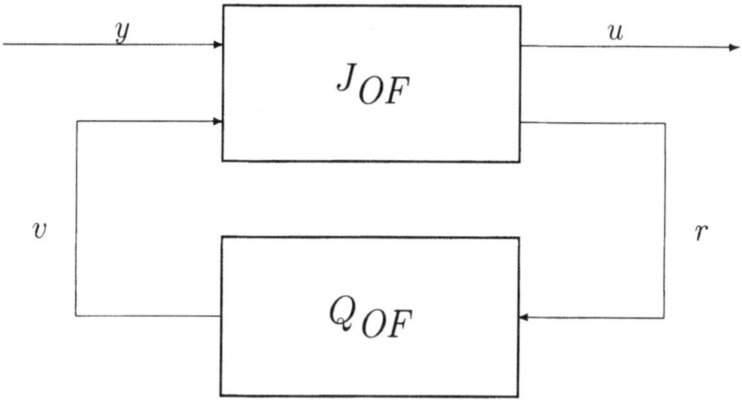

Fig. 3.3. Parameterization of all output feedback controllers

Theorem 3.4. *Consider (3.1). Suppose that Assumptions (A1)-(A3) hold. Then a parameterization of all internally stabilizing controllers via output feedback is given by*

$$\boldsymbol{K}_{OF} := \boldsymbol{\mathcal{LFT}}(\boldsymbol{J}_{OF}, \boldsymbol{Q}_{OF}) \tag{3.10}$$

(See Fig. 3.3) with

$$\boldsymbol{J}_{OF}: \begin{cases} \boldsymbol{E}\dot{\boldsymbol{\xi}} = \boldsymbol{f}(\boldsymbol{x}, \boldsymbol{u}) - \boldsymbol{\ell}(\boldsymbol{\xi}, \boldsymbol{y}), \\ \boldsymbol{u} = \boldsymbol{\alpha}(\boldsymbol{\xi}) + \boldsymbol{v}, \\ \boldsymbol{r} = \boldsymbol{\xi}, \end{cases} \qquad \boldsymbol{Q}_{OF}: \begin{cases} \boldsymbol{E}_Q\dot{\boldsymbol{q}} = \boldsymbol{f}_Q(\boldsymbol{x}, \boldsymbol{\xi}, \boldsymbol{q}), \\ \boldsymbol{v} = \boldsymbol{h}_Q(\boldsymbol{x}, \boldsymbol{\xi}, \boldsymbol{q}), \end{cases}$$

where $\boldsymbol{\ell}(\bullet, \bullet)$ is such that (3.9) is asymptotically stable with index one. The parameter

$$\boldsymbol{Q}_{OF}: \begin{cases} \boldsymbol{E}_Q\dot{\boldsymbol{q}} = \boldsymbol{f}_Q(\boldsymbol{x}, \boldsymbol{\xi}, \boldsymbol{q}), \\ \boldsymbol{v} = \boldsymbol{h}_Q(\boldsymbol{x}, \boldsymbol{\xi}, \boldsymbol{q}), \end{cases}$$

can be chosen almost arbitrarily but have to possess locally asymptotical stablility with index one in a neighborhood of the origin with $\boldsymbol{f}_Q(\boldsymbol{x}, \boldsymbol{x}, \boldsymbol{q}) = \boldsymbol{f}_Q(0, 0, \boldsymbol{q})$, $\boldsymbol{h}_Q(\boldsymbol{x}, \boldsymbol{x}, \boldsymbol{q}) = \boldsymbol{h}_Q(0, 0, \boldsymbol{q})$ and , $\boldsymbol{h}_Q(0, 0, 0) = 0$. □

Proof. The proof is pretty much similiar to the proof of Theorem 3.2. The sufficiency follows directly from Lemma B.3 for it is easily deduced that the closed-loop system is asymptotically stable with index one. For the proof of necessity part, we construct the following functions

$$\begin{aligned} \boldsymbol{f}_Q(\boldsymbol{\xi}, \boldsymbol{q}, \boldsymbol{y}) &\triangleq \begin{bmatrix} \boldsymbol{f}(\boldsymbol{q}_1, \boldsymbol{h}_c(\boldsymbol{q}_2, \boldsymbol{h}(\boldsymbol{q}_1) + \boldsymbol{y} - \boldsymbol{h}(\boldsymbol{\xi}))) + \boldsymbol{\ell}(\boldsymbol{\xi}, \boldsymbol{y}) \\ \boldsymbol{f}_c(\boldsymbol{q}_2, \boldsymbol{h}(\boldsymbol{q}_1) + \boldsymbol{y} - \boldsymbol{h}(\boldsymbol{\xi})) \end{bmatrix}, \\ \boldsymbol{h}_Q(\boldsymbol{\xi}, \boldsymbol{q}, \boldsymbol{y}) &\triangleq -\boldsymbol{\alpha}(\boldsymbol{q}_1) + \boldsymbol{h}_c(\boldsymbol{q}_2, \boldsymbol{h}(\boldsymbol{q}_1) + \boldsymbol{y} - \boldsymbol{h}(\boldsymbol{\xi})). \end{aligned}$$

The rest of the proof can be made in an almost identical argument used in the proof of Theorem 3.2. **Q.E.D.**

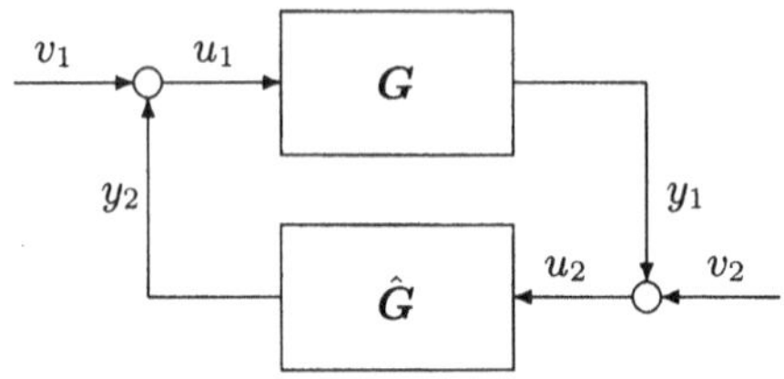

Fig. 3.4. Well-posedness analysis diagram for linear systems

3.2 Parameterization of All Stabilizing Controllers - Linear Case

We begin this section by studying the well-posedness problem arising from the interconnection of two descriptor systems. Assume that both $\boldsymbol{G}$ and $\hat{\boldsymbol{G}}$ in Fig. 3.4 are real rational transfer matrices (may be improper) with the following state equations:

$$\boldsymbol{G} \triangleq \begin{cases} \boldsymbol{E}\dot{\boldsymbol{x}} = \boldsymbol{A}\boldsymbol{x} + \boldsymbol{B}\boldsymbol{u}_1, \\ \boldsymbol{y}_1 = \boldsymbol{C}\boldsymbol{x} + \boldsymbol{D}\boldsymbol{u}_1, \end{cases}$$

and

$$\hat{\boldsymbol{G}} \triangleq \begin{cases} \hat{\boldsymbol{E}}\dot{\hat{\boldsymbol{x}}} = \hat{\boldsymbol{A}}\hat{\boldsymbol{x}} + \hat{\boldsymbol{B}}\boldsymbol{u}_2, \\ \boldsymbol{y}_2 = \hat{\boldsymbol{C}}\hat{\boldsymbol{x}} + \hat{\boldsymbol{D}}\boldsymbol{u}_2, \end{cases}$$

respectively. We say that the feedback loop is *well-posed* if the transfer matrix from $\begin{bmatrix} \boldsymbol{v}_1 \\ \boldsymbol{v}_2 \end{bmatrix}$ to $\begin{bmatrix} \boldsymbol{u}_1 \\ \boldsymbol{u}_2 \end{bmatrix}$ is well-defined. Note that we do not make the implication that the closed-loop system is necessarily proper in order to be well-posed. As a matter of fact, we have the following result.

Lemma 3.5. *Consider the feedback interconnection shown in Fig. 3.4. Then the feedback loop is well-posed if and only if $\boldsymbol{I} - \hat{\boldsymbol{D}}\boldsymbol{D}$ is invertible.* □

Proof. Observe that the closed-loop system can be put in the following descriptor form:

$$\begin{aligned} \begin{bmatrix} \boldsymbol{E} & 0 \\ 0 & \hat{\boldsymbol{E}} \end{bmatrix} \begin{bmatrix} \dot{\boldsymbol{x}} \\ \dot{\hat{\boldsymbol{x}}} \end{bmatrix} &= \begin{bmatrix} \boldsymbol{A} & 0 \\ 0 & \hat{\boldsymbol{A}} \end{bmatrix} \begin{bmatrix} \boldsymbol{x} \\ \hat{\boldsymbol{x}} \end{bmatrix} + \begin{bmatrix} \boldsymbol{B} & 0 \\ 0 & \hat{\boldsymbol{B}} \end{bmatrix} \begin{bmatrix} \boldsymbol{u}_1 \\ \boldsymbol{u}_2 \end{bmatrix} \\ \begin{bmatrix} \boldsymbol{y}_1 \\ \boldsymbol{y}_2 \end{bmatrix} &= \begin{bmatrix} \boldsymbol{C} & 0 \\ 0 & \hat{\boldsymbol{C}} \end{bmatrix} \begin{bmatrix} \boldsymbol{x} \\ \hat{\boldsymbol{x}} \end{bmatrix} + \begin{bmatrix} \boldsymbol{D} & 0 \\ 0 & \hat{\boldsymbol{D}} \end{bmatrix} \begin{bmatrix} \boldsymbol{u}_1 \\ \boldsymbol{u}_2 \end{bmatrix} \\ \begin{bmatrix} \boldsymbol{u}_1 \\ \boldsymbol{u}_2 \end{bmatrix} &= \begin{bmatrix} \boldsymbol{v}_1 \\ \boldsymbol{v}_2 \end{bmatrix} + \begin{bmatrix} 0 & \boldsymbol{I} \\ \boldsymbol{I} & 0 \end{bmatrix} \begin{bmatrix} \boldsymbol{y}_1 \\ \boldsymbol{y}_2 \end{bmatrix} \end{aligned}$$

The last two equations can be rewritten as

$$\begin{bmatrix} \boldsymbol{I} & -\hat{\boldsymbol{D}} \\ -\boldsymbol{D} & \boldsymbol{I} \end{bmatrix} \begin{bmatrix} \boldsymbol{u}_1 \\ \boldsymbol{u}_2 \end{bmatrix} = \begin{bmatrix} 0 & \hat{\boldsymbol{C}} \\ \boldsymbol{C} & 0 \end{bmatrix} \begin{bmatrix} \boldsymbol{x} \\ \hat{\boldsymbol{x}} \end{bmatrix} + \begin{bmatrix} \boldsymbol{v}_1 \\ \boldsymbol{v}_2 \end{bmatrix}$$

It is clear that the transfer matrix from $\begin{bmatrix} \boldsymbol{v}_1 \\ \boldsymbol{v}_2 \end{bmatrix}$ to $\begin{bmatrix} \boldsymbol{u}_1 \\ \boldsymbol{u}_2 \end{bmatrix}$ can be uniquely defined if and only if the inverse

$$\begin{bmatrix} \boldsymbol{I} & -\hat{\boldsymbol{D}} \\ -\boldsymbol{D} & \boldsymbol{I} \end{bmatrix}^{-1}$$

exists, or equivalently, $\boldsymbol{I} - \hat{\boldsymbol{D}}\boldsymbol{D}$ is invertible. **Q.E.D.**

Let us see a simple example to illustrate the idea.

Example.[1] Consider two conventional linear time-invariant systems.

$$G_1 : \begin{cases} \begin{bmatrix} \dot{x}_1 \\ \dot{x}_2 \end{bmatrix} = \begin{bmatrix} -1 & 3 \\ 0 & -2 \end{bmatrix} \begin{bmatrix} x_1 \\ x_2 \end{bmatrix} + \begin{bmatrix} 1.5 \\ 4 \end{bmatrix} u_1, \\ y_1 = \begin{bmatrix} 0.5 \; 3.2 \end{bmatrix} \begin{bmatrix} x_1 \\ x_2 \end{bmatrix}, \end{cases}$$

$$G_2 : \begin{cases} \begin{bmatrix} \dot{\hat{x}}_1 \\ \dot{\hat{x}}_2 \end{bmatrix} = \begin{bmatrix} -3 & 0 \\ 2 & -2 \end{bmatrix} \begin{bmatrix} \hat{x}_1 \\ \hat{x}_2 \end{bmatrix} + \begin{bmatrix} 3.3 \\ 5 \end{bmatrix} u_2, \\ y_2 = \begin{bmatrix} 1.7 \; 2 \end{bmatrix} \begin{bmatrix} \hat{x}_1 \\ \hat{x}_2 \end{bmatrix}, \end{cases}$$

The two systems are both strict proper. Hence, by a classical linear systems result (See, e.g., Chen[18]), the interconnection of G_1-G_2 is well-posed. Henceforth, the conventional Youla parameterization [23][127] can be applied to obtain all stabilizing controllers. However, systems G_1 and G_2 can also be expressed in descriptor forms, say

$$G_1 := \begin{cases} \begin{bmatrix} 1\,0\,0 \\ 0\,1\,0 \\ 0\,0\,0 \end{bmatrix} \begin{bmatrix} \dot{x}_1 \\ \dot{x}_2 \\ \dot{x}_3 \end{bmatrix} = \begin{bmatrix} -1 & 3 & 0 \\ 0 & -2 & 0 \\ 0 & 0 & 1 \end{bmatrix} \begin{bmatrix} x_1 \\ x_2 \\ x_3 \end{bmatrix} + \begin{bmatrix} 1.5 \\ 4 \\ 1 \end{bmatrix} u_1, \\ y_1 = \begin{bmatrix} 0.5 \; 3.2 \; 1 \end{bmatrix} \begin{bmatrix} x_1 \\ x_2 \\ x_3 \end{bmatrix} + u_1, \end{cases} \tag{3.11}$$

$$G_2 := \begin{cases} \begin{bmatrix} 1\,0\,0 \\ 0\,1\,0 \\ 0\,0\,0 \end{bmatrix} \begin{bmatrix} \dot{\hat{x}}_1 \\ \dot{\hat{x}}_2 \\ \dot{\hat{x}}_3 \end{bmatrix} = \begin{bmatrix} -3 & 0 & 0 \\ 2 & -2 & 0 \\ 0 & 0 & 1 \end{bmatrix} \begin{bmatrix} \hat{x}_1 \\ \hat{x}_2 \\ \hat{x}_3 \end{bmatrix} + \begin{bmatrix} 3.3 \\ 5 \\ 1 \end{bmatrix} u_2, \\ y_2 = \begin{bmatrix} 1.7 \; 2 \; 1 \end{bmatrix} \begin{bmatrix} \hat{x}_1 \\ \hat{x}_2 \\ \hat{x}_3 \end{bmatrix} + u_2, \end{cases} \tag{3.12}$$

Obviously, for (3.11) and (3.12), we have $I - D\hat{D} = 0$. In our opinion, we should not consider the interconnection of (3.11)-(3.12) to be well-posed in the sense that transfer functions between any two points of the closed-loop

[1] We thank an anonymous expert for providing the example and improving the definition.

may not be well-defined. This is because that we have introduced redundant variables, namely x_3 and $\hat{x}_3$, in (3.11) and (3.12). ◊◊◊

Consider a descriptor system described by the following set of differential-algebraic equations

$$G : \begin{cases} \boldsymbol{E}\dot{x} = \boldsymbol{A}\,\boldsymbol{x} + \boldsymbol{B}\,\boldsymbol{u} \\ \boldsymbol{y} = \boldsymbol{C}\,\boldsymbol{x} + \boldsymbol{D}\,\boldsymbol{u} \end{cases} \quad \boldsymbol{Ex}(0) \in \Im m\boldsymbol{E} \quad \text{given}, \tag{3.13}$$

or in more compact packed matrix notation

$$\boldsymbol{G} \triangleq \left\{ \boldsymbol{E}, \left[\begin{array}{c|c} \boldsymbol{A} & \boldsymbol{B} \\ \hline \boldsymbol{C} & \boldsymbol{D} \end{array} \right] \right\},$$

where $\boldsymbol{x} \in \mathbb{R}^n$ denotes the state, $\boldsymbol{u} \in \mathbb{R}^m$ the input, and $\boldsymbol{y} \in \mathbb{R}^p$ the output; $\boldsymbol{A}$, $\boldsymbol{B}$, $\boldsymbol{C}$, and $\boldsymbol{D}$ are constant matrices with compatible dimensions and $\boldsymbol{E}$ is a square matrix of rank $r < n$. The pencil $sE - A$ is assumed to be regular. Again, the measured output $\boldsymbol{y}$ is fed back to $\boldsymbol{u}$ by $\boldsymbol{u} = \boldsymbol{K}\,\boldsymbol{y}$. Suppose the controller $\boldsymbol{K}$ is, in general, also a descriptor system described by

$$K \triangleq \begin{cases} \hat{\boldsymbol{E}}_0\dot{\hat{\boldsymbol{x}}}_0 = \hat{\boldsymbol{A}}_0\hat{\boldsymbol{x}}_0 + \hat{\boldsymbol{B}}_0\boldsymbol{y}, \\ \boldsymbol{u} = \hat{\boldsymbol{C}}_0\hat{\boldsymbol{x}}_0 + \hat{\boldsymbol{D}}_0\boldsymbol{y} \end{cases} \tag{3.14}$$

The objective of the control considered in this section is to internally stabilize $\boldsymbol{G}$. Recall that internal stability means that the resulting closed-loop system is regular, impulse-free, and that the states of G and K go to zero from all initial values. We seek all controllers K that internally stabilize $\boldsymbol{G}$.

For simplicity, we first make a change of variable: $\boldsymbol{y}_1 \triangleq \boldsymbol{y} - \boldsymbol{D}\boldsymbol{u}$. The problem of finding all internally stabilizing controllers $\boldsymbol{K}$ of $\boldsymbol{G}$ is then transferred to an equivalent problem of finding all internally stabilizing controllers $\hat{\boldsymbol{K}}$ of $\hat{G}$ (See Fig. 3.5), where

$$\hat{\boldsymbol{G}} \triangleq \begin{cases} \boldsymbol{E}\dot{\boldsymbol{x}} = \boldsymbol{A}\boldsymbol{x} + \boldsymbol{B}\boldsymbol{u}, \quad \boldsymbol{Ex}(0) \in \Im m\boldsymbol{E} \quad \text{given}, \\ \boldsymbol{y}_1 = \boldsymbol{C}\boldsymbol{x}. \end{cases}$$

That is, $\boldsymbol{K}$ internally stabilizes $\boldsymbol{G}$ if and only if $\hat{\boldsymbol{K}}$ internally stabilizes $\hat{\boldsymbol{G}}$. Note that the $\boldsymbol{D}$-matrix of $\hat{\boldsymbol{G}}$ is zero. This will greatly simplify the complexity of computation. Once we find internally stabilizing controllers $\hat{\boldsymbol{K}}$ for $\hat{\boldsymbol{G}}$, it follows from Fig. 3.5 that

$$\boldsymbol{K} = (I + \hat{\boldsymbol{K}}\boldsymbol{D})^{-1}\hat{\boldsymbol{K}} = \hat{\boldsymbol{K}}(\boldsymbol{I} + \boldsymbol{D}\hat{\boldsymbol{K}})^{-1}. \tag{3.15}$$

Thus, if $\hat{\boldsymbol{K}}$ has a state-space realization of the following form

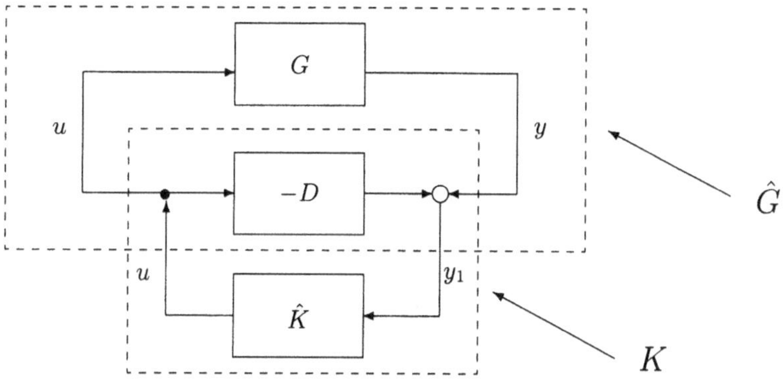

Fig. 3.5. Block diagram for Plant G with $D \neq 0$

$$\hat{\boldsymbol{K}} \triangleq \begin{cases} \hat{\boldsymbol{E}}_k \dot{\hat{\boldsymbol{x}}}_k = \hat{\boldsymbol{A}}_k \hat{\boldsymbol{x}}_k + \hat{\boldsymbol{B}}_k \boldsymbol{y}_1, \\ \boldsymbol{u} = \hat{\boldsymbol{C}}_k \hat{\boldsymbol{x}}_k + \hat{\boldsymbol{D}}_k \boldsymbol{y}_1. \end{cases}$$

Then, simple algebra shows that a state-space realization of $\boldsymbol{K}$ is given by

$$\boldsymbol{K} \triangleq \begin{cases} \hat{\boldsymbol{E}}_0 \dot{\hat{\boldsymbol{x}}}_0 = \hat{\boldsymbol{A}}_0 \hat{\boldsymbol{x}}_0 + \hat{\boldsymbol{B}}_0 \boldsymbol{y}, \\ \boldsymbol{u} = \hat{\boldsymbol{C}}_0 \hat{\boldsymbol{x}}_0 + \hat{\boldsymbol{D}}_0 \boldsymbol{y}, \end{cases}$$

$$= \begin{cases} \hat{\boldsymbol{E}}_k \dot{\hat{\boldsymbol{x}}}_0 = \left(\hat{\boldsymbol{A}}_k - \hat{\boldsymbol{B}}_k \hat{\boldsymbol{S}}_1^{-1} \boldsymbol{D} \hat{\boldsymbol{C}}_k \right) \hat{\boldsymbol{x}}_0 + \hat{\boldsymbol{B}}_k \hat{\boldsymbol{S}}_1^{-1} \boldsymbol{y}, \\ \boldsymbol{u} = \hat{\boldsymbol{S}}_2^{-1} \hat{\boldsymbol{C}}_k \hat{\boldsymbol{x}}_0 + \hat{\boldsymbol{D}}_k \hat{\boldsymbol{S}}_1^{-1} \boldsymbol{y}, \end{cases}$$

where $\hat{\boldsymbol{S}}_1 \triangleq \boldsymbol{I} + \boldsymbol{D}\hat{\boldsymbol{D}}_k$ and $\hat{\boldsymbol{S}}_2 \triangleq \boldsymbol{I} + \hat{\boldsymbol{D}}_k \boldsymbol{D}$ are invertible provided by the well-posedness. Thus, without loss of generality, in what follows we shall restrict ourselves to the consideration of systems with zero "$\boldsymbol{D}$" matrix.

The following theorem is the main result of this section, which may be considered as the generalization of the celebrated Youla parameterization given by Youla *et al.*[127] and Doyle[23].

Theorem 3.6. *Suppose that $\{\boldsymbol{E}, \boldsymbol{A}, \boldsymbol{B}\}$ is finite dynamic stabilizable and impulse controllable, and $\{\boldsymbol{E}, \boldsymbol{A}, \boldsymbol{C}\}$ is finite dynamic detectable and impulse observable. Let $\boldsymbol{F}$ and L be such that $\{\boldsymbol{E}, \boldsymbol{A} + \boldsymbol{B}\,\boldsymbol{F}\}$ and $\{\boldsymbol{E}, \boldsymbol{A} + L\boldsymbol{C}\}$ are admissible. Then all controllers that internally stabilize $\hat{\boldsymbol{G}}$ can be parameterized as $\hat{\boldsymbol{K}} = \mathcal{LFT}(\boldsymbol{J}, \boldsymbol{Q})$, where*

$$\boldsymbol{J} = \left\{ \boldsymbol{E}, \left[\begin{array}{c|cc} \boldsymbol{A} + \boldsymbol{B}\,\boldsymbol{F} + \boldsymbol{L}\,\boldsymbol{C} & -\boldsymbol{L} & \boldsymbol{B} \\ \hline \boldsymbol{F} & 0 & \boldsymbol{I} \\ -\boldsymbol{C} & \boldsymbol{I} & 0 \end{array} \right] \right\},$$

with any $\boldsymbol{Q} \in \mathcal{RH}^\infty$. □

Proof. We first show that $\hat{\boldsymbol{K}} = \mathcal{LFT}(\boldsymbol{J},\, \boldsymbol{Q})$ are stabilizing controllers. Take an admissible realization of $\boldsymbol{Q}$:

$$\boldsymbol{Q} = \left\{ \boldsymbol{E}_Q,\ \left[\begin{array}{c|c} \boldsymbol{A}_Q & \boldsymbol{B}_Q \\ \hline \boldsymbol{C}_Q & \boldsymbol{D}_Q \end{array}\right] \right\}.$$

Since $\boldsymbol{Q} \in \mathcal{RH}^\infty$, $\{\boldsymbol{E}_Q,\, \boldsymbol{A}_Q\}$ is admissible. Denote the states of $\boldsymbol{J}$ and $\boldsymbol{Q}$ by $\boldsymbol{\xi}$ and $\boldsymbol{\eta}$, respectively. Set $\boldsymbol{e} = \boldsymbol{x} - \boldsymbol{\xi}$. Then the overall closed-loop system can be written as:

$$\begin{bmatrix} \boldsymbol{E} & 0 & 0 \\ 0 & \boldsymbol{E}_Q & 0 \\ 0 & 0 & \boldsymbol{E} \end{bmatrix} \begin{bmatrix} \dot{\boldsymbol{x}} \\ \dot{\boldsymbol{\eta}} \\ \dot{\boldsymbol{e}} \end{bmatrix} = \begin{bmatrix} \boldsymbol{A}+\boldsymbol{B}\boldsymbol{F} & \boldsymbol{B}\boldsymbol{C}_Q & -\boldsymbol{B}\boldsymbol{F}+\boldsymbol{B}\boldsymbol{D}_Q\boldsymbol{C} \\ 0 & \boldsymbol{A}_Q & \boldsymbol{B}_Q\boldsymbol{C} \\ 0 & 0 & \boldsymbol{A}+\boldsymbol{L}\boldsymbol{C} \end{bmatrix} \begin{bmatrix} \boldsymbol{x} \\ \boldsymbol{\eta} \\ \boldsymbol{e} \end{bmatrix}.$$

Since $\{\boldsymbol{E},\, \boldsymbol{A}+\boldsymbol{B}\boldsymbol{F}\}$, $\{\boldsymbol{E},\, \boldsymbol{A}+\boldsymbol{L}\boldsymbol{C}\}$ and $\{\boldsymbol{E}_Q,\, \boldsymbol{A}_Q\}$ are all admissible, it is concluded that the closed-loop system is internally stable. Next, we show that for any internally stabilizing controller $\boldsymbol{K}_0$, a $\boldsymbol{Q}_0 \in \mathcal{RH}^\infty$ exists such that $\boldsymbol{K}_0 = \mathcal{LFT}(\boldsymbol{J},\, \boldsymbol{Q}_0)$. For this purpose, set

$$\boldsymbol{J}_0 = \left\{ \boldsymbol{E},\ \left[\begin{array}{c|cc} \boldsymbol{A} & -\boldsymbol{L} & \boldsymbol{B} \\ \hline -\boldsymbol{F} & 0 & \boldsymbol{I} \\ \boldsymbol{C} & \boldsymbol{I} & 0 \end{array}\right] \right\},$$

and set $\boldsymbol{Q}_0 \triangleq \mathcal{LFT}(\boldsymbol{J}_0,\, \boldsymbol{K}_0)$. Note that $\boldsymbol{J}_0$ and $\hat{\boldsymbol{G}}$ have the same internal structure, so that $\boldsymbol{Q}_0 \in \mathcal{RH}^\infty$. Calculate $\mathcal{LFT}(\boldsymbol{J}, \boldsymbol{Q}_0)$ to obtain

$$\mathcal{LFT}(\boldsymbol{J},\, \boldsymbol{Q}_0) = \mathcal{LFT}(\boldsymbol{J},\, \mathcal{LFT}(\boldsymbol{J}_0,\, \boldsymbol{K}_0)) = \mathcal{LFT}(\boldsymbol{J}_1,\, \boldsymbol{K}_0),$$

where

$$\boldsymbol{J}_1 \triangleq \left\{ \begin{bmatrix} \boldsymbol{E} & 0 \\ 0 & \boldsymbol{E} \end{bmatrix},\ \left[\begin{array}{cc|cc} \boldsymbol{A}+\boldsymbol{B}\boldsymbol{F}+\boldsymbol{L}\boldsymbol{C} & -\boldsymbol{B}\boldsymbol{F} & -\boldsymbol{L} & \boldsymbol{B} \\ \boldsymbol{L}\boldsymbol{C} & \boldsymbol{A} & -\boldsymbol{L} & \boldsymbol{B} \\ \hline \boldsymbol{F} & -\boldsymbol{F} & 0 & \boldsymbol{I} \\ -\boldsymbol{C} & -\boldsymbol{C} & \boldsymbol{I} & 0 \end{array}\right] \right\}.$$

Conjugate the states of $\boldsymbol{J}_1$ by $\begin{bmatrix} \boldsymbol{I} & 0 \\ -\boldsymbol{I} & \boldsymbol{I} \end{bmatrix}$ on the left and $\begin{bmatrix} \boldsymbol{I} & 0 \\ -\boldsymbol{I} & \boldsymbol{I} \end{bmatrix}^{-1} = \begin{bmatrix} \boldsymbol{I} & 0 \\ \boldsymbol{I} & \boldsymbol{I} \end{bmatrix}$ on the right to get

$$\begin{aligned} \boldsymbol{J}_1 &= \left\{ \begin{bmatrix} \boldsymbol{E} & 0 \\ 0 & \boldsymbol{E} \end{bmatrix},\ \left[\begin{array}{cc|cc} \boldsymbol{A}+\boldsymbol{L}\boldsymbol{C} & \boldsymbol{B}\boldsymbol{F} & -\boldsymbol{L} & \boldsymbol{B} \\ 0 & \boldsymbol{A}+\boldsymbol{B}\boldsymbol{F} & 0 & 0 \\ \hline 0 & \boldsymbol{F} & 0 & \boldsymbol{I} \\ 0 & -\boldsymbol{C} & \boldsymbol{I} & 0 \end{array}\right] \right\} \\ &= \begin{bmatrix} 0 & \boldsymbol{I} \\ \boldsymbol{I} & 0 \end{bmatrix}. \end{aligned}$$

Thus $\mathcal{LFT}(\boldsymbol{J},\, \boldsymbol{Q}_0) = \mathcal{LFT}(\boldsymbol{J}_1,\, \boldsymbol{K}_0) = \boldsymbol{K}_0$. This completes the proof.

Q.E.D.

3.3 An Application of the Youla Parameterization

To serve as a simple application of the Youla parameterization, we consider the positive real control problems for descriptor systems in this section. The most general motivation for studying this problem stems from robust and nonlinear control: If the system uncertainties (linear and/or nonlinear) can be characterized by positive real property, then the classical results in stability theory can be used to guarantee robust stability provided an appropriate closed-loop system is strictly positive real[95][137][138]. Other materials come from circuit theory [2], adaptive control[5] and stability analysis[95]. It is beyond the scope of this monograph to review the vast literature associated with positive realness theory. For a more extensive bibliography and review of the literature, see, e.g., Anderson *et al.*[2], Safonov *et al.*[98], Sun *et al.*[109], and the references cited therein.

3.3.1 Positive Real Control Problem

Let's begin with the definition of positive real systems.

Definition 3.7. *Consider the descriptor system*

$$\begin{aligned} \boldsymbol{E}\dot{\boldsymbol{x}} &= \boldsymbol{A}\boldsymbol{x} + \boldsymbol{B}\boldsymbol{u}, \quad \boldsymbol{E}\boldsymbol{x}(0) \in \Im m\boldsymbol{E} \quad \text{given} \\ \boldsymbol{y} &= \boldsymbol{C}\boldsymbol{x} + \boldsymbol{D}\boldsymbol{u}, \end{aligned} \tag{3.16}$$

Denote by $\boldsymbol{G}(s)$ its transfer function matrix. Then
(1) *The system (3.16) is said to be positive real (PR) if $\boldsymbol{G}(s)$ is analytic in $\Re e(s) > 0$ and satisfies $\boldsymbol{G}(s) + \boldsymbol{G}^T(s^*) \geq 0, \ \forall \Re e(s) > 0$,*

(2) *The system (3.16) is strictly positive real (SPR) if $\boldsymbol{G}(s)$ is analytic in $\Re e(s) \geq 0$ and satisfies $\boldsymbol{G}(\mathrm{j}\omega) + \boldsymbol{G}^T(-\mathrm{j}\omega) > 0, \ \forall \omega \in [0, \infty)$,*

(3) *The system (3.16) is extended strictly positive real (ESPR) if it is SPR and $\boldsymbol{G}(\mathrm{j}\infty) + \boldsymbol{G}^T(-\mathrm{j}\infty) > 0$.* □

The following lemma is a generalization of the well know Kalman-Yacubovich-Popov positive real lemma to the descriptor system case. It plays a crucial role in our later proofs. This lemma connects the ESPR property of descriptor systems with solutions of GARE and GARI. The proof of the lemma is pretty much similar to the proof of the Bounded Real Lemma given in Chapter 2 and is thus omitted.

Lemma 3.8. Generalized Positive Real Lemma
Consider the system (3.16). Suppose that $\boldsymbol{D}+\boldsymbol{D}^T \triangleq \boldsymbol{\Delta} > 0$ and $p = m$. Then the following statements are equivalent.
(i) *The pair $\{\boldsymbol{E}, \boldsymbol{A}\}$ is admissible and $\boldsymbol{G}(s)$ is ESPR,*

(ii) *The pair* $\{E, A\}$ *is admissible and the Hamiltonian pencil* $\hat{P}(s) \triangleq s\hat{E} - \hat{H}$ *with*

$$\hat{E} \triangleq \begin{bmatrix} E & 0 \\ 0 & E^T \end{bmatrix}, \quad and \quad \hat{H} \triangleq \begin{bmatrix} A - B\Delta^{-1}C & B\Delta^{-1}B^T \\ -C^T\Delta^{-1}C & -A^T + C^T\Delta^{-1}B^T \end{bmatrix},$$

has no pure imaginary zeros and satisfies

$$\Im m\, \hat{E} + \hat{H}(\aleph(\hat{E})) = \mathbb{R}^{2n}, \tag{3.17}$$

(iii) *The GARE*

$$\begin{cases} A^T X + X^T A + (X^T B - C^T)\Delta^{-1}(X^T B - C^T)^T = 0, \\ E^T X = X^T E \end{cases}, \tag{3.18}$$

has an admissible solution X_E *with* $E^T X_E = X_E^T E \geq 0$,

(iv) *The GARI*

$$\begin{cases} A^T P + P^T A + (P^T B - C^T)\Delta^{-1}(P^T B - C^T)^T < 0, \\ E^T P = P^T E \end{cases}, \tag{3.19}$$

has a nonsingular solution P *with* $E^T P = P^T E \geq 0$. □

Example. Positive Real Systems
Consider an SISO descritpor system $G(s)$, which is described by the following set of differential-algebraic equations

$$G(s) := \begin{cases} E\dot{x}(t) = Ax(t) + Bu(t), \\ y(t) = Cx(t) \end{cases}$$
$$\triangleq \begin{cases} \begin{bmatrix} 1 & 0 & 0 \\ 0 & 1 & 0 \\ 0 & 0 & 0 \end{bmatrix} \dot{x}(t) = \begin{bmatrix} -2 & 0 & 0 \\ 0 & -1 & 0 \\ 0 & 0 & 1 \end{bmatrix} x(t) + \begin{bmatrix} -10 \\ 40 \\ 1 \end{bmatrix} u(t), \\ y(t) = \begin{bmatrix} 1 & 0 & 0 \end{bmatrix} x(t) + u(t). \end{cases}$$

It is easily verified that the matrix X_P defined by

$$X_P \triangleq \begin{bmatrix} \frac{1}{50} & \frac{1}{50} & 0 \\ \frac{1}{50} & \frac{9}{500} & 0 \\ 0 & 0 & 0 \end{bmatrix}$$

is an admissible solution satisfying GARE (3.18) with $E^T X_P = X_P^T E \geq 0$. This shows that $G(s)$ is ESPR by Lemma 3.8. Moreover, it can be shown that the function $\Phi(s)$ defined by

$$\Phi(s) \triangleq \Delta^{\frac{1}{2}} \left(1 - (B^T X_P - C)(sE - A)^{-1} B\right) = \sqrt{2}\frac{(s+3)(s+4)}{(s+1)(s+2)}$$

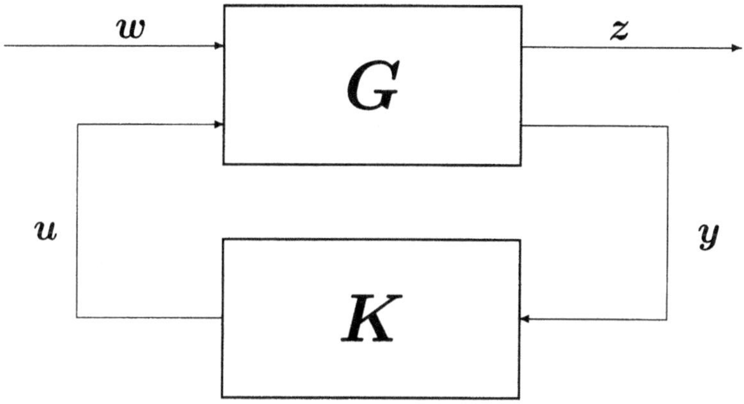

Fig. 3.6. Standard block diagram for positive real control problems

is a spectral factor satisfying

$$G(s) + G^*(s) = \Phi^*(s)\Phi(s).$$

◊◊◊

Consider the standard block diagram shown in Fig. 3.6. In this section, the to-be-controlled plant $\boldsymbol{G}$ is a descriptor system described by the following set of differential-algebraic equations:

$$\boldsymbol{G} = \boldsymbol{G}_{OF} \triangleq \begin{cases} \boldsymbol{E}\dot{\boldsymbol{x}} = \boldsymbol{A}\boldsymbol{x} + \boldsymbol{B}_1\boldsymbol{w} + \boldsymbol{B}_2\boldsymbol{u}, \\ \boldsymbol{z} = \boldsymbol{C}_1\boldsymbol{x} + \boldsymbol{D}_{11}\boldsymbol{w} + \boldsymbol{D}_{12}\boldsymbol{u}, \qquad \boldsymbol{E}\boldsymbol{x}(0) \in \Im m\boldsymbol{E} \quad \text{given}, \\ \boldsymbol{y} = \boldsymbol{C}_2\boldsymbol{x} + \boldsymbol{D}_{21}\boldsymbol{w}, \end{cases} \tag{3.20}$$

(the subscript "OF" stands for output feedback) where $x \in \mathbb{R}^n$ is the state, and $w \in \mathbb{R}^p$ represents a set of exogenous inputs which includes disturbances to be rejected and/or reference commands to be tracked. $z \in \mathbb{R}^p$ is the output to be controlled and $y \in \mathbb{R}^l$ is the measured output. $u \in \mathbb{R}^m$ is the control input. $\boldsymbol{A}, \boldsymbol{B}_1, \boldsymbol{B}_2, \boldsymbol{C}_1, \boldsymbol{C}_2, \boldsymbol{D}_{11}, \boldsymbol{D}_{12}$, and $\boldsymbol{D}_{21}$ are constant matrices with compatible dimensions. $\boldsymbol{E} \in \mathbb{R}^{n\times n}$ and rank$\boldsymbol{E} = r < n$.

The controller $\boldsymbol{K}$ is, in general, also a descriptor system described by

$$\boldsymbol{K} = \boldsymbol{K}_{OF} \triangleq \begin{cases} \hat{\boldsymbol{E}}\dot{\hat{\boldsymbol{x}}}_0 = \hat{\boldsymbol{A}}_0\hat{\boldsymbol{x}}_0 + \hat{\boldsymbol{B}}_0\boldsymbol{y}, \\ \boldsymbol{u} = \hat{\boldsymbol{C}}_0\hat{\boldsymbol{x}}_0, \end{cases} \tag{3.21}$$

where $\hat{\boldsymbol{x}}_0$ is the state of the controller. The objective of the control is to internally stablizes $\boldsymbol{G}$ such that the closed-loop transfer function $\boldsymbol{T}_{zw}$ is extended strictly positive real.

This problem, referred to as the ESPR output feedback control problem, has been recently addressed and extensively studied in Sun *et al.*[109] for linear time-invariant (LTI) plant and controller in state-space model, in which several necessary and sufficient conditions in terms of solutions to algebraic Riccati equations or inequalities(ARE or ARI) were proposed for the solvability of the ESPR control problem. State-Space formulas for the controller design were also given in Sun *et al.*[109]. Most recently, Yung[128] have characterized all solutions to this problem. He also adopted state-space formulation.

Throughout this section, we make the following standing assumptions.

(A1) $(\boldsymbol{D}_{11}+\boldsymbol{D}_{11}^T) \triangleq \boldsymbol{R}_0 > 0$,
(A2) $\{\boldsymbol{E}, \boldsymbol{A}, \boldsymbol{B}_2\}$ is finite dynamic stabilizable and impulse controllable,
(A3) The matrix $\boldsymbol{D}_{12}$ has full column rank,
(A4) The matrix pencil $\boldsymbol{P}_1(s) \triangleq s\boldsymbol{E}_1 + \boldsymbol{H}_1$, where

$$\boldsymbol{E}_1 = \begin{bmatrix} -\boldsymbol{E} & 0 \\ 0 & 0 \end{bmatrix}, \qquad \boldsymbol{H}_1 = \begin{bmatrix} \boldsymbol{A} & \boldsymbol{B}_2 \\ \boldsymbol{C}_1 & \boldsymbol{D}_{12} \end{bmatrix}$$

is of full column rank on the jω-axis and is column reduced (i.e., $\boldsymbol{H}_1^{-1}(\Im m\boldsymbol{E}_1) \cap \aleph(\boldsymbol{E}_1) = \{0\}$).
(A5) $\{\boldsymbol{E}, \boldsymbol{A}, \boldsymbol{C}_2\}$ is finite dynamic detectable and impulse observable
(A6) The matrix $\boldsymbol{D}_{21}$ has full row rank,
(A7) The matrix pencil $\boldsymbol{P}_2(s) \triangleq s\boldsymbol{E}_2 + \boldsymbol{H}_2$, where

$$\boldsymbol{E}_2 = \begin{bmatrix} -\boldsymbol{E} & 0 \\ 0 & 0 \end{bmatrix}, \qquad \boldsymbol{H}_2 = \begin{bmatrix} \boldsymbol{A} & \boldsymbol{B}_1 \\ \boldsymbol{C}_2 & \boldsymbol{D}_{21} \end{bmatrix}$$

is of full row rank on the jω-axis and is row reduced (i.e., $\Im m\boldsymbol{E}_2 + \boldsymbol{H}_2(\aleph(\boldsymbol{E}_2)) = \mathbb{R}^{n+p}$).

Note that $\boldsymbol{E}_1$ is, in general, not equal to $\boldsymbol{E}_2$ (They may have different sizes).

Before proceeding to the main problem of this section, we first investigate the state feedback case which is stated in the following theorem.

Theorem 3.9. *Consider the plant $\boldsymbol{G}_{OF}$ with $\boldsymbol{y} = \boldsymbol{x}$ and suppose that Assumptions (A1) to (A4) are satisfied. Then there exists a state feedback controller $\boldsymbol{u} = \boldsymbol{F}\boldsymbol{x}$ such that the resulting closed-loop system is internally stable and ESPR if and only if the GARE*

$$\begin{aligned} \mathrm{R}ic_1(\boldsymbol{X}) \triangleq\ & [\boldsymbol{A} - \boldsymbol{B}_1\boldsymbol{R}_0^{-1}\boldsymbol{C}_1 - (\boldsymbol{B}_2 - \boldsymbol{B}_1\boldsymbol{R}_0^{-1}\boldsymbol{D}_{12})\boldsymbol{\Delta}_1^{-1}\boldsymbol{D}_{12}^T\boldsymbol{R}_0^{-1}\boldsymbol{C}_1]^T\boldsymbol{X} \\ & + \boldsymbol{X}^T[\boldsymbol{A} - \boldsymbol{B}_1\boldsymbol{R}_0^{-1}\boldsymbol{C}_1 - (\boldsymbol{B}_2 - \boldsymbol{B}_1\boldsymbol{R}_0^{-1}\boldsymbol{D}_{12})\boldsymbol{\Delta}_1^{-1}\boldsymbol{D}_{12}^T\boldsymbol{R}_0^{-1}\boldsymbol{C}_1] \\ & + \boldsymbol{X}^T[\boldsymbol{B}_1\boldsymbol{R}_0^{-1}\boldsymbol{B}_1^T \\ & \quad -(\boldsymbol{B}_2 - \boldsymbol{B}_1\boldsymbol{R}_0^{-1}\boldsymbol{D}_{12})\boldsymbol{\Delta}_1^{-1}(\boldsymbol{B}_2 - \boldsymbol{B}_1\boldsymbol{R}_0^{-1}\boldsymbol{D}_{12})^T]\boldsymbol{X} \\ & + \boldsymbol{C}_1^T\boldsymbol{R}_0^{-1}(\boldsymbol{R}_0 - \boldsymbol{D}_{12}\boldsymbol{\Delta}_1^{-1}\boldsymbol{D}_{12}^T)\boldsymbol{R}_0^{-1}\boldsymbol{C}_1 \\ & = 0, \qquad \boldsymbol{E}^T\boldsymbol{X} = \boldsymbol{X}^T\boldsymbol{E} \end{aligned}$$

has an admissible solution $\boldsymbol{X}_E$ *with* $\boldsymbol{E}^T\boldsymbol{X}_E = \boldsymbol{X}_E^T\boldsymbol{E} \geq 0$, *where* $\boldsymbol{\Delta}_1 \triangleq \boldsymbol{D}_{12}^T\boldsymbol{R}_0^{-1}\boldsymbol{D}_{12}$. *Moreover, when this condition holds, all ESPR state feedback controllers can be parameterized as* $\boldsymbol{K}_{SF} = \boldsymbol{\mathcal{LFT}}(\boldsymbol{M}_{SF}, \boldsymbol{Q})$, *where*

$$\boldsymbol{M}_{SF} = \left\{ \boldsymbol{E}, \left[\begin{array}{c|cc} \boldsymbol{A} + \boldsymbol{B}_1\boldsymbol{F}_1 + \boldsymbol{B}_2\boldsymbol{F}_2 & 0 & \boldsymbol{B}_2 \\ \hline 0 & \boldsymbol{F}_2 & \boldsymbol{I} \\ -\boldsymbol{I} & \boldsymbol{I} & 0 \end{array} \right] \right\},$$

$\boldsymbol{Q} \in \mathcal{RH}^\infty$, *and* $\boldsymbol{D}_{11} + \boldsymbol{D}_{12}\boldsymbol{Q}\boldsymbol{P}$ *is ESPR with*

$$\boldsymbol{P} = \left\{ E, \left[\begin{array}{c|c} \boldsymbol{A} + \boldsymbol{B}_1\boldsymbol{F}_1 + \boldsymbol{B}_2\boldsymbol{F}_2 & \boldsymbol{B}_1 \\ \hline \boldsymbol{I} & 0 \end{array} \right] \right\},$$

in which

$$\boldsymbol{F}_2 \triangleq \boldsymbol{\Delta}_1^{-1}\boldsymbol{D}_{12}^T\boldsymbol{R}_0^{-1}(\boldsymbol{B}_1^T\boldsymbol{X}_E - \boldsymbol{C}_1) - \boldsymbol{\Delta}_1^{-1}\boldsymbol{B}_2^T\boldsymbol{X}_E \tag{3.22}$$

$$\boldsymbol{F}_1 \triangleq \boldsymbol{R}_0^{-1}(\boldsymbol{B}_1^T\boldsymbol{X}_E - \boldsymbol{C}_1 - \boldsymbol{D}_{12}\boldsymbol{F}_2) \tag{3.23}$$

□

Remark. The proof of the previous theorem is fairly straightforward and is omitted. See also Yung[128].

The following theorem provides a necessary and sufficient condition for the solvability of ESPR output feedback control problem.

Theorem 3.10. *Consider the standard system connection (Fig. 3.6) with* $\boldsymbol{G} = \boldsymbol{G}_{OF}$. *Suppose that Assumptions (A1) to (A7) hold. Then the following statements are equivalent.*

(I) *There exists a controller of the form (3.21) such that the resulting closed-loop system is internally stable and ESPR.*

(II) *1. the GARE* $\mathrm{R}ic_1(\boldsymbol{X}) = 0$ *has an admissible solution* $\boldsymbol{X}_E$ *with* $\boldsymbol{E}^T\boldsymbol{X}_E = \boldsymbol{X}_E^T\boldsymbol{E} \geq 0$.

2. the GARE

$$\begin{aligned} \mathrm{R}ic_3(\boldsymbol{Z}) \triangleq\ & \boldsymbol{A}_Z\boldsymbol{Z} + \boldsymbol{Z}^T\boldsymbol{A}_Z^T - \boldsymbol{Z}^T(\boldsymbol{C}_{2Z}^T\boldsymbol{\Delta}_2^{-1}\boldsymbol{C}_{2Z} \\ & -\boldsymbol{F}_2^T\boldsymbol{\Delta}_1\boldsymbol{F}_2)\boldsymbol{Z} + \boldsymbol{B}_{1Z}\boldsymbol{R}_0^{-1}\boldsymbol{B}_{1Z}^T = 0, \\ & \boldsymbol{E}\boldsymbol{Z} = \boldsymbol{Z}^T\boldsymbol{E}^T \end{aligned}$$

has an admissible solution $\boldsymbol{Z}_E$ *with* $\boldsymbol{E}\boldsymbol{Z}_E = \boldsymbol{Z}_E^T\boldsymbol{E}^T \geq 0$, *where*

$$\begin{aligned} \boldsymbol{\Delta}_2 &\triangleq \boldsymbol{D}_{21}\boldsymbol{R}_0^{-1}\boldsymbol{D}_{21}^T \\ \boldsymbol{A}_Z &\triangleq \boldsymbol{A} + \boldsymbol{B}_1\boldsymbol{R}_0^{-1}(\boldsymbol{B}_1^T\boldsymbol{X}_E - \boldsymbol{C}_1) - \boldsymbol{B}_1\boldsymbol{R}_0^{-1}\boldsymbol{D}_{21}^T\boldsymbol{\Delta}_2^{-1}\boldsymbol{C}_{2Z} \\ \boldsymbol{B}_{1Z} &\triangleq \boldsymbol{B}_1(\boldsymbol{I} - \boldsymbol{R}_0^{-1}\boldsymbol{D}_{21}^T\boldsymbol{\Delta}_2^{-1}\boldsymbol{D}_{21}) \\ \boldsymbol{C}_{2Z} &\triangleq \boldsymbol{C}_2 + \boldsymbol{D}_{21}\boldsymbol{R}_0^{-1}(\boldsymbol{B}_1^T\boldsymbol{X}_E - \boldsymbol{C}_1) \end{aligned}$$

Moreover, when these conditions are satisfied, one such controller is given as in the form (3.21) with

$$\begin{aligned}\hat{A}_0 &= A + B_2 \hat{C}_0 - \hat{B}_0 C_2 + (B_1 - \hat{B}_0 D_{21}) R_0^{-1} (B_1^T X_E - C_1 - D_{12} \hat{C}_0)\\ \hat{B}_0 &= (Z_E^T C_{2Z}^T + B_1 R_0^{-1} D_{21}^T) \Delta_2^{-1}\\ \hat{C}_0 &= F_2\end{aligned}$$

□

Before giving the proof of theorem, we note that the hierarchically coupled pair of GAREs given above can be further manipulated to yield two mutually decoupled GAREs, with a separated additional spectral radius condition. This is summarized in the following statement.

Lemma 3.11. *Suppose that GARE* $\mathbf{R}ic_1(X) = 0$ *has an admissible solution* X_E *with* $E^T X_E - X_E^T E \geq 0$. *Suppose also that the following conditions are satisfied.*

1. *the GARE*

$$\begin{aligned}\mathbf{R}ic_2(Y) \triangleq{}& [A - B_1 R_0^{-1} C_1 - B_1 R_0^{-1} D_{21}^T \Delta_2^{-1} (C_2 - D_{21} R_0^{-1} C_1)] Y\\ &+ Y^T [A - B_1 R_0^{-1} C_1 -\\ &B_1 R_0^{-1} D_{21}^T \Delta_2^{-1} (C_2 - D_{21} R_0^{-1} C_1)]^T\\ &+ Y^T [C_1^T R_0^{-1} C_1 -\\ &(C_2 - D_{21} R_0^{-1} C_1)^T \Delta_2^{-1} (C_2 - D_{21} R_0^{-1} C_1)] Y\\ &+ B_1 R_0^{-1} (R_0 - D_{21}^T \Delta_2^{-1} D_{21}) R_0^{-1} B_1^T = 0\\ EY ={}& Y^T E^T,\end{aligned}$$

   has an admissible solution Y_E *with* $EY_E = Y_E^T E^T \geq 0$
2. *the spectral radius* $\rho(Y_E X_E) < 1$

Then the condition (II) of Theorem 3.10 holds. Moreover, when these conditions are satisfied, the matrices X_E, Y_E *and* Z_E *have the following relationship:*

$$Z_E = (I - Y_E X_E)^{-1} Y_E = Y_E (I - X_E Y_E)^{-1}.$$

□

Proof of Sufficiency of Theorem 3.10.
Observe that the closed-loop system (3.20)-(3.21) can be written as

$$\begin{aligned}\mathcal{E}_C \begin{bmatrix} \dot{x} \\ \dot{e} \end{bmatrix} &= \mathcal{A}_C \begin{bmatrix} x \\ e \end{bmatrix} + \mathcal{B}_C w\\ z &= \mathcal{C}_C \begin{bmatrix} x \\ e \end{bmatrix} + \mathcal{D}_C w\end{aligned}$$

where

$$\mathcal{E}_C \triangleq \begin{bmatrix} \boldsymbol{E} & 0 \\ 0 & \boldsymbol{E} \end{bmatrix}$$

$$\mathcal{A}_C \triangleq \begin{bmatrix} \boldsymbol{A} + \boldsymbol{B}_2\hat{\boldsymbol{C}} & -\boldsymbol{B}_2\hat{\boldsymbol{C}} \\ \boldsymbol{A} - \hat{\boldsymbol{A}} + \boldsymbol{B}_2\hat{\boldsymbol{C}} - \hat{\boldsymbol{B}}\boldsymbol{C}_2 & \hat{\boldsymbol{A}} - \boldsymbol{B}_2\hat{\boldsymbol{C}} \end{bmatrix}$$

$$\mathcal{B}_C \triangleq \begin{bmatrix} \boldsymbol{B}_1 \\ \boldsymbol{B}_1 - \hat{\boldsymbol{B}}\boldsymbol{D}_{21} \end{bmatrix}$$

$$\mathcal{C}_C \triangleq \begin{bmatrix} \boldsymbol{C}_1 + \boldsymbol{D}_{12}\hat{\boldsymbol{C}} & -\boldsymbol{D}_{12}\hat{\boldsymbol{C}} \end{bmatrix}$$

$$\mathcal{D}_C \triangleq \boldsymbol{D}_{11}$$

and $\boldsymbol{e} \triangleq \boldsymbol{x} - \hat{\boldsymbol{x}}_0$. The GARE $\mathbf{R}ic_3(\boldsymbol{Z}) = 0$ can be rewritten as:

$$\begin{aligned} \mathbf{R}ic_3(\boldsymbol{Z}) &= (\bar{\boldsymbol{A}}_Z - \bar{\boldsymbol{B}}_Z\boldsymbol{R}_0^{-1}\bar{\boldsymbol{C}}_Z)\boldsymbol{Z} + \boldsymbol{Z}^T(\bar{\boldsymbol{A}}_Z - \bar{\boldsymbol{B}}_Z\boldsymbol{R}_0^{-1}\bar{\boldsymbol{C}}_Z)^T \\ &\quad + \bar{\boldsymbol{B}}_Z\boldsymbol{R}_0^{-1}\bar{\boldsymbol{B}}_Z^T + \boldsymbol{Z}^T\bar{\boldsymbol{C}}_Z^T\boldsymbol{R}_0^{-1}\bar{\boldsymbol{C}}_Z\boldsymbol{Z} = 0, \\ \boldsymbol{E}\boldsymbol{Z} &= \boldsymbol{Z}^T\boldsymbol{E}^T \end{aligned} \tag{3.24}$$

where

$$\bar{\boldsymbol{A}}_Z \triangleq \boldsymbol{A}_Z - \boldsymbol{Z}_E^T\boldsymbol{C}_{2Z}^T\boldsymbol{\Delta}_2^{-1}\boldsymbol{C}_{2Z} + \bar{\boldsymbol{B}}_Z\boldsymbol{R}_0^{-1}\bar{\boldsymbol{C}}_Z$$

$$\bar{\boldsymbol{B}}_Z \triangleq \boldsymbol{B}_{1Z} - \boldsymbol{Z}_E^T\boldsymbol{C}_{2z}^T\boldsymbol{\Delta}_2^{-1}\boldsymbol{D}_{21}$$

$$\bar{\boldsymbol{C}}_Z \triangleq \boldsymbol{R}_0^{\frac{1}{2}}\boldsymbol{\Delta}_1^{\frac{1}{2}}\boldsymbol{F}_2$$

Note that $\boldsymbol{Z}_E$ is also an admissible solution to GARE (3.24). By Generalized Positive Real Lemma, we can conclude that $\{\boldsymbol{E}^T, \bar{\boldsymbol{A}}_Z^T\}$ is admissible and $\bar{\boldsymbol{B}}_Z^T(s\boldsymbol{E}^T - \bar{\boldsymbol{A}}_Z^T)^{-1}\bar{\boldsymbol{C}}_Z^T + \boldsymbol{D}_{11}^T$ is ESPR. This, in turn, implies that $\{\boldsymbol{E}, \bar{\boldsymbol{A}}_Z\}$ is admissible and $\bar{\boldsymbol{C}}_Z(s\boldsymbol{E} - \bar{\boldsymbol{A}}_Z)^{-1}\bar{\boldsymbol{B}}_Z + \boldsymbol{D}_{11}$ is ESPR. Again by Generalized Positive Real Lemma, the GARE

$$\begin{aligned} \mathbf{R}ic_4(\boldsymbol{W}) &= (\bar{\boldsymbol{A}}_Z - \bar{\boldsymbol{B}}_Z\boldsymbol{R}_0^{-1}\bar{\boldsymbol{C}}_Z)^T\boldsymbol{W} + \boldsymbol{W}^T(\bar{\boldsymbol{A}}_Z - \bar{\boldsymbol{B}}_Z\boldsymbol{R}_0^{-1}\bar{\boldsymbol{C}}_Z) \\ &\quad + \bar{\boldsymbol{C}}_Z^T\boldsymbol{R}_0^{-1}\bar{\boldsymbol{C}}_Z + \boldsymbol{W}^T\bar{\boldsymbol{B}}_Z\boldsymbol{R}_0^{-1}\bar{\boldsymbol{B}}_Z^T\boldsymbol{W} = 0 \\ \boldsymbol{E}^T\boldsymbol{W} &= \boldsymbol{W}^T\boldsymbol{E} \end{aligned}$$

has an admissible solution $\boldsymbol{W}_E$ with $\boldsymbol{E}^T\boldsymbol{W}_E = \boldsymbol{W}_E^T\boldsymbol{E} \geq 0$.

Now set

$$\mathcal{P}_C \triangleq \begin{bmatrix} \boldsymbol{X}_E & 0 \\ 0 & \boldsymbol{W}_E \end{bmatrix}.$$

Clearly, $\mathcal{E}_C^T\mathcal{P}_C = \mathcal{P}_C^T\mathcal{E}_C$. Lengthy but otherwise routine calculation shows that $\mathcal{P}_C$ is an admissible solution to the GARE

$$\begin{aligned} &\mathcal{A}_C^T\mathcal{P} + \mathcal{P}^T\mathcal{A}_C + (\mathcal{P}^T\mathcal{B}_C - \mathcal{C}_C^T)R_0^{-1}(\mathcal{P}^T\mathcal{B}_C - \mathcal{C}_C^T)^T = 0, \\ &\mathcal{E}_C^T\mathcal{P} = \mathcal{P}^T\mathcal{E}_C, \end{aligned}$$

with $\mathcal{E}_C^T \mathcal{P}_C = \mathcal{P}_C^T \mathcal{E}_C \geq 0$. It follows, again from Generalized Positive Real Lemma, that (3.21) is a stabilizing controller such that T_{zw} is ESPR. This completes the proof of sufficiency.

Proof of Necessity of Theorem 3.10.
To prove the necessity, we need a preliminary result which is stated in the following theorem.

Theorem 3.12. *Consider the standard system connection (Fig. 3.6) with*

$$G = G_{FI} \triangleq \begin{cases} \boldsymbol{E}\dot{\boldsymbol{x}} = \boldsymbol{A}\boldsymbol{x} + \boldsymbol{B}_1\boldsymbol{w} + \boldsymbol{B}_2\boldsymbol{u}, \\ \boldsymbol{z} = \boldsymbol{C}_1\boldsymbol{x} + \boldsymbol{D}_{11}\boldsymbol{w} + \boldsymbol{D}_{12}\boldsymbol{u}, \qquad \boldsymbol{E}\boldsymbol{x}(0^-) = \boldsymbol{E}\boldsymbol{x}_0 \quad \text{given}, \\ \boldsymbol{y} = \begin{bmatrix} \boldsymbol{x} \\ \boldsymbol{w} \end{bmatrix}, \end{cases} \tag{3.25}$$

("FI" stands for Full Information) Suppose that Assumptions (A1) to (A4) hold. Then the ESPR full information control problem for descriptor systems is solvable if and only if the GARE $\mathbf{R}ic_1(\boldsymbol{X}) = 0$ *has an admissible solution* $\boldsymbol{X}_E$ *with* $\boldsymbol{E}^T\boldsymbol{X}_E = \boldsymbol{X}_E^T\boldsymbol{E} \geq 0$*. Moreover, if this condition is satisfied, one such controller is given by* $\boldsymbol{u} = \begin{bmatrix} \boldsymbol{F}_2 & 0 \end{bmatrix} \boldsymbol{y}$. □

We can now prove the necessary part of Theorem 3.10. Suppose that $\boldsymbol{K}_{OF}$ is one solution. Then the controller

$$\boldsymbol{K}_{FI} = \boldsymbol{K}_{OF} \begin{bmatrix} \boldsymbol{C}_2 & \boldsymbol{D}_{21} \end{bmatrix}$$

solves the FI problem. From Theorem 3.12, it follows that GARE $\mathbf{R}ic_1(\boldsymbol{X}) = 0$ has an admissible solution. Consequently, we see that $\boldsymbol{F}_1, \boldsymbol{F}_2$ are well defined. We shall make a change of variables described by

$$\begin{aligned} \boldsymbol{r} &\triangleq \boldsymbol{w} - \boldsymbol{F}_1\boldsymbol{x} \\ \boldsymbol{q} &\triangleq \boldsymbol{D}_{11}\boldsymbol{r} + \boldsymbol{D}_{12}(\boldsymbol{u} - \boldsymbol{F}_2\boldsymbol{x}) \end{aligned} \tag{3.26}$$

Then

$$\boldsymbol{q} = -(\boldsymbol{D}_{11}\,\boldsymbol{F}_1 + \boldsymbol{D}_{12}\,\boldsymbol{F}_2)x + \boldsymbol{D}_{11}\,\boldsymbol{w} + \boldsymbol{D}_{12}\,\boldsymbol{u}\,. \tag{3.27}$$

Pre-multiplying (3.27) by $\boldsymbol{D}_{12}^T\boldsymbol{R}_0^{-1}$ and solving it for $\boldsymbol{u}$, we obtain

$$\boldsymbol{u} = (\boldsymbol{F}_2 + \boldsymbol{\Delta}_1^{-1}\,\boldsymbol{D}_{12}^T\,\boldsymbol{R}_0^{-1}\,\boldsymbol{D}_{11}\,\boldsymbol{F}_1)\boldsymbol{x} - \boldsymbol{\Delta}_1^{-1}\,\boldsymbol{D}_{12}^T\,\boldsymbol{R}_0^{-1}\,\boldsymbol{D}_{11}\,\boldsymbol{w} + \boldsymbol{\Delta}_1^{-1}\boldsymbol{D}_{12}^T\boldsymbol{R}_0^{-1}\boldsymbol{q}\,.$$

Then the closed-loop system in Fig. 3.6 can be decomposed as two interconnected subsystems. This is illustrated in Fig. 3.7, where $\hat{P}$ is given by

$$\hat{\boldsymbol{P}} = \left\{ \boldsymbol{E}, \left[\begin{array}{c|cc} \boldsymbol{A} + \boldsymbol{B}_2\boldsymbol{F}_2 & \boldsymbol{B}_1 & \boldsymbol{B}_2\boldsymbol{\Delta}_1^{-1}\boldsymbol{D}_{12}^T\boldsymbol{R}_0^{-1} \\ +\boldsymbol{B}_2\boldsymbol{\Delta}_1^{-1}\boldsymbol{D}_{12}^T\boldsymbol{R}_0^{-1}\boldsymbol{D}_{11}\boldsymbol{F}_1 & -\boldsymbol{B}_2\boldsymbol{\Delta}_1^{-1}\boldsymbol{D}_{12}^T\boldsymbol{R}_0^{-1}\boldsymbol{D}_{11} & \\ \hline \boldsymbol{C}_1 + \boldsymbol{D}_{11}\boldsymbol{F}_1 + \boldsymbol{D}_{12}\boldsymbol{F}_2 & 0 & \boldsymbol{I} \\ -\boldsymbol{F}_1 & \boldsymbol{I} & 0 \end{array} \right] \right\} \tag{3.28}$$

and

$$\hat{G}_{OF} = \left\{ E, \left[\begin{array}{c|cc} A + B_1 F_1 & B_1 & B_2 \\ \hline -D_{12} F_2 & D_{11} & D_{12} \\ C_2 + D_{21} F_1 & D_{21} & 0 \end{array} \right] \right\}.$$

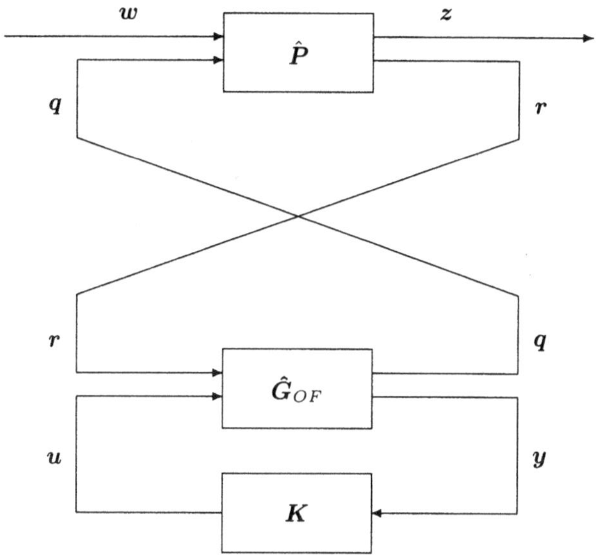

Fig. 3.7. Decomposition of the closed-loop system in output feedback case.

The following lemma is useful. Its proof involves only repeated uses of Lemma 3.8, which is pretty straightforward, and is thus omitted.

Lemma 3.13. *Suppose that the GARE* $\mathbf{Ric}_1(X) = 0$ *has an admissible solution* X_E *with* $E^T X_E = X_E^T E \geq 0$. *Then* K *internally stabilizes* G_{OF} *and* T_{zw} *is ESPR if and only if* K *internally stabilizes* $\hat{G}_{OF}$ *and* T_{qr} *is ESPR.* □

The rest of necessity proof follows by observing first that K internally stabilizes $\hat{G}_{OF}$ and T_{qr} is ESPR if and only if K^T internally stabilizes $\hat{G}_{OF}^T$ and T_{qr}^T is ESPR. Now introduce

$$G_b \triangleq \hat{G}_{OF}{}^T = \left\{ E^T, \left[\begin{array}{c|cc} (A + B_1 F_1)^T & -F_2^T D_{12}^T & (C_2 + D_{21} F_1)^T \\ \hline B_1^T & D_{11}^T & D_{21}^T \\ B_2^T & D_{12}^T & 0 \end{array} \right] \right\}$$

$$\triangleq \left\{ \tilde{\boldsymbol{E}}, \left[\begin{array}{c|cc} \tilde{\boldsymbol{A}} & \tilde{\boldsymbol{B}}_1 & \tilde{\boldsymbol{B}}_2 \\ \hline \tilde{\boldsymbol{C}}_1 & \tilde{\boldsymbol{D}}_{11} & \tilde{\boldsymbol{D}}_{12} \\ \tilde{\boldsymbol{C}}_2 & \tilde{\boldsymbol{D}}_{21} & 0 \end{array} \right] \right\}.$$

As a consequence of Lemma 3.13 and the above observation, the ESPR OF problem for $\boldsymbol{G}_b$ is also solvable. Again, suppose that $\boldsymbol{K}_b$ solves the ESPR OF problem for $\boldsymbol{G}_b$ then the controller

$$\boldsymbol{K}_b \left[\tilde{\boldsymbol{C}}_2 \; \tilde{\boldsymbol{D}}_{21} \right]$$

solves the FI problem inherited from $\boldsymbol{G}_b$. It follows, again form Theorem 3.12, that GARE

$$\begin{aligned} \tilde{\mathbf{R}}ic_1(\boldsymbol{X}) \triangleq\; & [\tilde{\boldsymbol{A}} - \tilde{\boldsymbol{B}}_1 \tilde{\boldsymbol{R}}_0^{-1} \tilde{\boldsymbol{C}}_1 - (\tilde{\boldsymbol{B}}_2 - \tilde{\boldsymbol{B}}_1 \tilde{\boldsymbol{R}}_0^{-1} \tilde{\boldsymbol{D}}_{12}) \tilde{\boldsymbol{\Delta}}_1^{-1} \tilde{\boldsymbol{D}}_{12}^T \tilde{\boldsymbol{R}}_0^{-1} \tilde{\boldsymbol{C}}_1]^T \boldsymbol{X} \\ & + \boldsymbol{X}^T [\tilde{\boldsymbol{A}} - \tilde{\boldsymbol{B}}_1 \tilde{\boldsymbol{R}}_0^{-1} \tilde{\boldsymbol{C}}_1 - (\tilde{\boldsymbol{B}}_2 - \tilde{\boldsymbol{B}}_1 \tilde{\boldsymbol{R}}_0^{-1} \tilde{\boldsymbol{D}}_{12}) \tilde{\boldsymbol{\Delta}}_1^{-1} \tilde{\boldsymbol{D}}_{12}^T \boldsymbol{R}_0^{-1} \tilde{\boldsymbol{C}}_1] \\ & + \boldsymbol{X} [\tilde{\boldsymbol{B}}_1 \tilde{\boldsymbol{R}}_0^{-1} \tilde{\boldsymbol{B}}_1^T - \\ & \quad (\tilde{\boldsymbol{B}}_2 - \tilde{\boldsymbol{B}}_1 \tilde{\boldsymbol{R}}_0^{-1} \tilde{\boldsymbol{D}}_{12}) \tilde{\boldsymbol{\Delta}}_1^{-1} (\tilde{\boldsymbol{B}}_2 - \tilde{\boldsymbol{B}}_1 \tilde{\boldsymbol{R}}_0^{-1} \tilde{\boldsymbol{D}}_{12})^T] \boldsymbol{X} \\ & + \tilde{\boldsymbol{C}}_1^T \tilde{\boldsymbol{R}}_0^{-1} (\tilde{\boldsymbol{R}}_0 - \tilde{\boldsymbol{D}}_{12} \tilde{\boldsymbol{\Delta}}_1^{-1} \tilde{\boldsymbol{D}}_{12}^T) \tilde{\boldsymbol{R}}_0^{-1} \tilde{\boldsymbol{C}}_1 \\ & \tilde{\boldsymbol{E}}^T \boldsymbol{X} = \boldsymbol{X}^T \tilde{\boldsymbol{E}} \end{aligned}$$

has an admissible solution $\tilde{\boldsymbol{X}}$ with $\tilde{\boldsymbol{E}}^T \tilde{\boldsymbol{X}} = \tilde{\boldsymbol{X}}^T \tilde{\boldsymbol{E}} \geq 0$ where $\tilde{\boldsymbol{R}}_0 = \tilde{\boldsymbol{D}}_{11}^T + \tilde{\boldsymbol{D}}_{11} = \boldsymbol{R}_0$ and $\tilde{\boldsymbol{\Delta}}_1 = \tilde{\boldsymbol{D}}_{12}^T \tilde{\boldsymbol{R}}_0^{-1} \tilde{\boldsymbol{D}}_{12} = \boldsymbol{\Delta}_2$. It is easy to verify that $\tilde{\boldsymbol{R}}_1(\boldsymbol{X})$ is exactly $\mathbf{R}ic_3(\boldsymbol{X})$. This shows that $\tilde{\boldsymbol{X}} \triangleq \boldsymbol{Z}_E$ indeed exists. This completes the necessity proof. **Q.E.D.**

3.3.2 Characterization of All ESPR Output Feedback Controllers

The following theorem, which is the main result of this section, characterizes all output feedback controllers that achieve closed-loop internal stability and ESPR. The proof involves only simple ideas beginning with a change of variables and the Youla parameterization.

Theorem 3.14. *Suppose that Assumptions (A1) to (A7) are satisfied. Suppose also that Condition (II) of Theorem 3.10 holds. Then the set of all output feedback controllers that achieve closed-loop internal stability and ESPR can be parameterized as* $\boldsymbol{K} = \boldsymbol{\mathcal{LFT}}(\boldsymbol{M}_{OF}, \boldsymbol{Q})$, *where*

$$\boldsymbol{M}_{OF} = \left\{ \boldsymbol{E}, \left[\begin{array}{c|cc} \hat{\boldsymbol{A}}_0 & \hat{\boldsymbol{B}}_0 & \hat{\boldsymbol{B}}_1 \\ \hline \hat{\boldsymbol{C}}_0 & 0 & \boldsymbol{I} \\ -(\boldsymbol{C}_2 + \boldsymbol{D}_{21} \boldsymbol{F}_1) & \boldsymbol{I} & 0 \end{array} \right] \right\},$$

$\hat{\boldsymbol{A}}_0, \hat{\boldsymbol{B}}_0, \hat{\boldsymbol{C}}_0$ *are defined as in Theorem 3.10,*

$$\hat{B}_1 \triangleq (I + X_E Z_E)^T (B_2 - B_1 R_0^{-1} D_{12}) + (C_1 Z_E + D_{21}^T \hat{B}_0^T)^T R_0^{-1} D_{12} \tag{3.29}$$

and $Q \in \mathcal{RH}^\infty$ *with* $D_{11} + D_{12} Q D_{21}$ *ESPR.* □

Proof. From Section 5.1.2, the problem of parameterizing all internally stabilizing controller for G_{OF} such that T_{zw} is ESPR is equivalent to the problem of parameterizing all internally stabilizing controllers K_b for G_b such that T_{qr}^T is ESPR. In what follows, we shall focus on the problem of parameterization of all internally stabilizing controllers K_b for G_b such that T_{qr}^T is ESPR.

Denote by $\tilde{x}$, $\{\tilde{w}, \tilde{u}\}$ and $\{\tilde{z}, \tilde{y}\}$ the state, inputs and outputs of G_b, respectively, as shown in Fig. 3.8. Next, make a change of variables as follows:

$$\begin{aligned} \tilde{r} &\triangleq \tilde{w} - \tilde{F}_1 \tilde{x} \\ \tilde{q} &\triangleq \tilde{D}_{11} \tilde{r} + \tilde{D}_{12}(\tilde{u} - \tilde{F}_2 \tilde{x}) \end{aligned}$$

where

$$\begin{aligned} \tilde{F}_1 &\triangleq \tilde{R}_0^{-1}(\tilde{B}_1^T \tilde{X} - \tilde{C}_1 - \tilde{D}_{12}\tilde{F}_2) \\ \tilde{F}_2 &\triangleq \tilde{\Delta}_1^{-1} \tilde{D}_{12}^T \tilde{R}_0^{-1} (\tilde{B}_1^T \tilde{X} - \tilde{C}_1) - \tilde{\Delta}_1^{-1} \tilde{B}_2^T \tilde{X} \, . \end{aligned}$$

Then it follows, by direct verification using the same arguments as before, that the closed-loop system in Fig. 3.8 can be decomposed as two interconnected subsystems. This is shown pictorically in Fig. 3.9, where

$$P_b =$$

$$\left\{ \tilde{E}, \left[\begin{array}{c|cc} \tilde{A} + \tilde{B}_2 \tilde{F}_2 & \tilde{B}_1 & \tilde{B}_2 \tilde{\Delta}_1^{-1} \tilde{D}_{12}^T \tilde{R}_0^{-1} \\ + \tilde{B}_2 \tilde{\Delta}_1^{-1} \tilde{D}_{12}^T \tilde{R}_0^{-1} \tilde{D}_{11} \tilde{F}_1 & - \tilde{B}_2 \tilde{\Delta}_1^{-1} \tilde{D}_{12}^T \tilde{R}_0^{-1} \tilde{D}_{11} & \\ \hline \tilde{C}_1 + \tilde{D}_{11} \tilde{F}_1 + \tilde{D}_{12} \tilde{F}_2 & 0 & I \\ \tilde{F}_1 & I & 0 \end{array} \right] \right\}$$

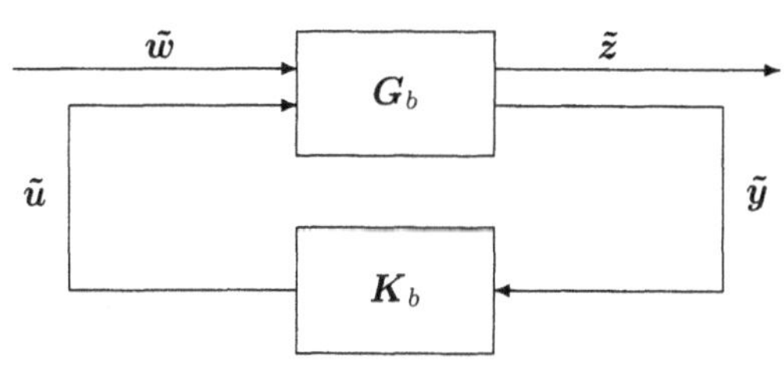

Fig. 3.8. Block diagram for $\mathcal{LFT}(G_b, K_b)$

and

$$\hat{G}_b = \left\{ \tilde{E}, \left[\begin{array}{c|cc} \tilde{A} + \tilde{B}_1\tilde{F}_1 & \tilde{B}_1 & \tilde{B}_2 \\ \hline -\tilde{D}_{12}\tilde{F}_2 & \tilde{D}_{11} & \tilde{D}_{12} \\ \tilde{C}_2 + \tilde{D}_{21}\tilde{F}_1 & \tilde{D}_{21} & 0 \end{array} \right] \right\}.$$

Again, applying Lemma 3.13 shows that K_b internally stabilizes G_b and $T_{qr}^T = T_{\tilde{z}\tilde{w}}$ is ESPR if and only if K_b internally stabilizes $\hat{G}_b$ and $T_{\tilde{q}\tilde{r}}$ is ESPR.

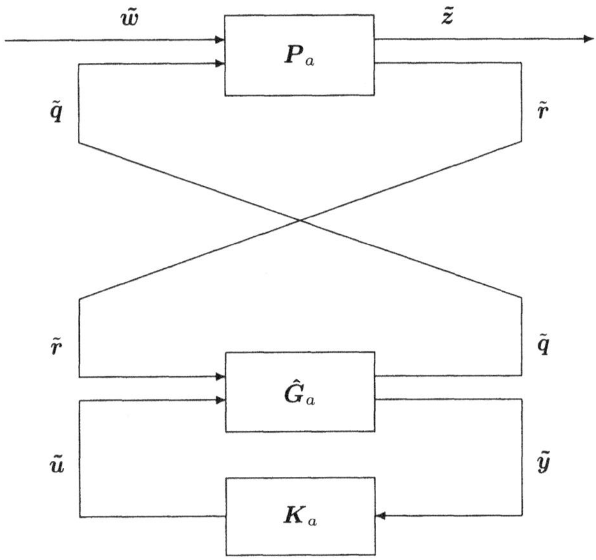

Fig. 3.9. Decomposition of $\mathcal{LFT}(G_b, K_b)$.

Now let $L \triangleq F_2^T = \hat{C}_0^T$, then simple algebra shows that $\{\tilde{E}, \tilde{A} + \tilde{B}_1\tilde{F}_1 + L(\tilde{C}_2 + \tilde{D}_{21}\tilde{F}_1)\} = \{E^T, (A + B_1F_1 + B_2F_2)^T\}$ which is admissible. Also note that if we let $F \triangleq -\hat{B}_0^T$, then it is straightforward to verify that $\{\tilde{E}, \tilde{A} + \tilde{B}_1\tilde{F}_1 + \tilde{B}_2F\} = \{E^T, A_Z^T - C_{2Z}^T\Delta_2^{-1}C_{2Z}Z_E + F_2^T\Delta_1F_2Z_E\}$, which is also admissible since Z_E is an admissible solution to $\mathbf{R}ic_3(Z) = 0$. Then, from Lemma 3.6, all internally stabilizing controllers for $\hat{G}_b$ can be parameterized as $K_b = \mathcal{LFT}(M_b, Q_b)$ with $Q_b \in \mathcal{RH}^\infty$, where

$$M_b = \left\{ \tilde{E}, \left[\begin{array}{c|cc} \tilde{A} + \tilde{B}_1\tilde{F}_1 - \tilde{B}_2\hat{B}_0^T + \hat{C}_0^T(\tilde{C}_2 + \tilde{D}_{21}\tilde{F}_1) & -\hat{C}_0^T & \tilde{B}_2 \\ \hline -\hat{B}_0^T & 0 & I \\ -(\tilde{C}_2 + \tilde{D}_{21}\tilde{F}_1) & I & 0 \end{array} \right] \right\}.$$

It is easy to see that M_b can be rewritten as

$$\boldsymbol{M}_b = \left\{ \boldsymbol{E}^T, \left[\begin{array}{c|cc} \hat{\boldsymbol{A}}_0^T & -\hat{\boldsymbol{C}}_0^T & (\boldsymbol{C}_2 + \boldsymbol{D}_{21}\boldsymbol{F}_1)^T \\ \hline -\hat{\boldsymbol{B}}_0^T & 0 & \boldsymbol{I} \\ -\hat{\boldsymbol{B}}_1^T & \boldsymbol{I} & 0 \end{array} \right] \right\},$$

where $\hat{\boldsymbol{B}}_1$ is as defined by (3.29).

Moreover, a little bit of algebra shows that

$$\boldsymbol{T}_{\tilde{q}\tilde{r}} = \mathcal{LFT}(\hat{\boldsymbol{G}}_b, \mathcal{LFT}(\boldsymbol{M}_b, \boldsymbol{Q}_b)) = \boldsymbol{D}_{11}^T + \boldsymbol{D}_{21}^T \boldsymbol{Q}_b \boldsymbol{D}_{12}^T .$$

So $\boldsymbol{T}_{\tilde{q}\tilde{r}}$ is ESPR if and only if $\boldsymbol{D}_{11}^T + \boldsymbol{D}_{21}^T \boldsymbol{Q}_b \boldsymbol{D}_{12}^T$ is ESPR.

Thus all internally stabilizing controllers for $\hat{\boldsymbol{G}}_{OF}^T$ such that $\boldsymbol{T}_{qr}^T$ is ESPR can be parameterized as $\boldsymbol{K}_b = \mathcal{LFT}(\boldsymbol{M}_b, \boldsymbol{Q}_b)$, where $\boldsymbol{Q}_b \in \mathcal{RH}^\infty$ with $\boldsymbol{D}_{11}^T + \boldsymbol{D}_{21}^T \boldsymbol{Q}_b \boldsymbol{D}_{12}^T$ ESPR. Consequently, all internally stabilizing controllers for $\hat{\boldsymbol{G}}_{OF}$ such that $\boldsymbol{T}_{qr}$ is ESPR can be parameterized as $\boldsymbol{K} = \boldsymbol{K}_b^T = \mathcal{LFT}(\boldsymbol{M}_b^T, \boldsymbol{Q})$ where $\boldsymbol{Q} \triangleq \boldsymbol{Q}_b^T \in \mathcal{RH}^\infty$ with $\boldsymbol{D}_{11} + \boldsymbol{D}_{12}\boldsymbol{Q}\boldsymbol{D}_{21}$ ESPR. It is easy to see that $\boldsymbol{M}_b^T = \boldsymbol{M}_{OF}$. This completes the proof. **Q.E.D.**

The parameterization given above is an implicit one. We leave this form here for, in our opinion, it shows more intrinsic feature of the ESPR control problem. However, the parameterization can be explicitly defined via simple algebraic manipulations.

Theorem 3.15. *Suppose that Assumptions (A1) to (A7) are satisfied. Suppose also that Condition (II) of Theorem 3.10 holds. Then the set of all output feedback controllers that achieve closed-loop internal stability and ESPR can be parameterized as* $\boldsymbol{K} = \mathcal{LFT}(\hat{\boldsymbol{M}}_{OF}, \hat{\boldsymbol{Q}})$, *where*

$$\boldsymbol{M}_{OF} = \left\{ \boldsymbol{E}, \left[\begin{array}{c|cc} \hat{\boldsymbol{A}}_0 - \hat{\boldsymbol{B}}_1 \boldsymbol{D}_{12}^{-L} \boldsymbol{D}_{11} (\boldsymbol{D}_{21}^{-R} + \boldsymbol{F}_1) & \hat{\boldsymbol{B}}_0 + \hat{\boldsymbol{B}}_1 \boldsymbol{D}_{12}^{-L} \boldsymbol{D}_{11} \boldsymbol{D}_{21}^{-R} & \hat{\boldsymbol{B}}_1 \boldsymbol{D}_{12}^{-L} \\ \hline \hat{\boldsymbol{C}}_0 - \boldsymbol{D}_{12}^{-L} \boldsymbol{D}_{11} (\boldsymbol{D}_{21}^{-R} + \boldsymbol{F}_1) & \boldsymbol{D}_{12}^{-L} \boldsymbol{D}_{11} \boldsymbol{D}_{21}^{-R} & \boldsymbol{D}_{12}^{-L} \\ -(\boldsymbol{D}_{21}^{-R} \boldsymbol{C}_2 + \boldsymbol{F}_1) & \boldsymbol{D}_{21}^{-R} & 0 \end{array} \right] \right\},$$

with $\hat{\boldsymbol{Q}}$ *ESPR, where* $\boldsymbol{D}_{12}^{-L}$ *and* $\boldsymbol{D}_{21}^{-R}$ *denote any left and right inverses of* $\boldsymbol{D}_{12}$ *and* $\boldsymbol{D}_{21}$, *respectively.* □

Proof. The proof is fairly straightforward, and is thus omitted.

Notes and References

1. Youla parameterization, first introduced by Youla *et al.*[127], has become a powerful tool in the synthesis of linear feedback systems, for it provides a systematic way to select an (optimal) internally stabilizing controller which also meets some performance requirements. By using a frequency domain approach, Takaba *et al.*[111] extended the work of Youla

et al.[127] to the descriptor system case. In both cases, the coprime factorization technique was used. Recently, Liu *et al.*[72] have considered the latter case by using a stable factorization method, in which the improper transfer matrices were factorized over proper stable rational functions. Basically, their computation was also a frequency domain approach. In the paper of Kawamoto and Katayama[57], a parameterization of all controllers solving the standard $\mathcal{H}_\infty$ control problem for descriptor systems has been provided. Nevertheless, only the result was presented in that paper, and no further details or proofs were discussed.

Recently, a state-space approach to the Youla parameterization for linear and nonlinear systems in state-space model has been reported in Lu[76] (see also Moore *et al.*[85], and Doyle[23] for some related work). The Youla parameterization approach to the characterization of all ESPR output feedback controllers was first proposed by Yung[128]. His treatment was mainly based on a time-domain viewpoint.

2. In this chapter, it is shown that, for a given descriptor system, its stabilizing controller is, in general, also described by a set of differential-algebraic equations. Specifically, the existence of a descriptor-form controller, which has a identical $\boldsymbol{E}$-matrix with the plant, is a necessary condition (and is sufficient as well) for the stabilizing control problem to be solvable. Hence, if we want to construct a controller in conventional state-variable form (i.e., $\boldsymbol{E} = \boldsymbol{I}$), we need to show that a descriptor-form controller is indeed obtainable. Later on, in Chapter 6, we will give a way to find controllers in conventional state-variable form for a given descriptor system. In the book of Zhou[140], it has been demonstrated that, for a conventional state-space system, there may exist descriptor-form controllers, but the resulting controller is, in fact, impulse-free. Therefore it can be replaced by a controller in conventional state-space form without violating the closed-loop performance.

4

The $\mathcal{H}_\infty$ Control Problems

Consider the standard feedback configuration shown in Fig. 4.1. Let $\mathcal{G}_P$ be a nonlinear system described by the following DAE

$$\mathcal{G}_P \triangleq \begin{cases} \boldsymbol{E}\dot{\boldsymbol{x}} = \boldsymbol{F}(\boldsymbol{x}, \boldsymbol{w}, \boldsymbol{u}), \\ \boldsymbol{z} = \boldsymbol{Z}(\boldsymbol{x}, \boldsymbol{w}, \boldsymbol{u}), \\ \boldsymbol{y} = \boldsymbol{Y}(\boldsymbol{x}, \boldsymbol{w}, \boldsymbol{u}), \end{cases} \tag{4.1}$$

where $\boldsymbol{x} \in \mathcal{X}$, $\boldsymbol{w} \in \mathbb{W} \subset \mathbb{R}^l$, $\boldsymbol{u} \in \mathbb{U} \subset \mathbb{R}^m$, $\boldsymbol{z} \in \mathbb{Z} \subset \mathbb{R}^s$, and $\boldsymbol{y} \in \mathbb{Y} \subset \mathbb{R}^p$. Here $\boldsymbol{u}$ stands for the vector of control inputs, $\boldsymbol{w}$ is the exogenous input (disturbances to-be-rejected or signals to-be-tracked), $\boldsymbol{y}$ is the measured output, and finally $\boldsymbol{z}$ denotes the to-be-controlled outputs (tracking errors, cost variables). It is assumed throughout that $\boldsymbol{F}(0, 0, 0) = 0$, $\boldsymbol{Z}(0, 0, 0) = 0$ and $\boldsymbol{Y}(0, 0, 0) = 0$. The standard $\mathcal{H}_\infty$ control problem consists of finding, if possible, a controller Γ such that the resulting closed-loop system is locally asymptotically stable (or internally stable) and has $\mathcal{L}_2$ gain (from $\boldsymbol{w}$ to $\boldsymbol{z}$) less than or equal to a prescribed positive real number γ. Here the closed-loop internal stability can be defined in the following way. Consider a nonlinear descriptor system $\mathcal{G}_P$ described by equation of the form

$$\begin{aligned} \boldsymbol{E}\dot{\boldsymbol{x}}(t) &= \boldsymbol{F}(\boldsymbol{x}(t), \boldsymbol{u}(t)), \\ \boldsymbol{y}(t) &= \boldsymbol{H}(\boldsymbol{x}(t)), \end{aligned} \tag{4.2}$$

with $\boldsymbol{x} \in \mathbb{R}^n$, $\boldsymbol{u} \in \mathbb{R}^m$ and $\boldsymbol{y} \in \mathbb{R}^p$. Let $\mathbb{B}_\rho$ denote the open ball in $\mathbb{R}^n$ of radius ρ centered at $\boldsymbol{x} = 0$ and $\mathcal{U}_\delta$ the set of all piecewise continuous functions $\boldsymbol{u} : \mathbb{R} \to \mathbb{R}^m$ satisfying $\|\boldsymbol{u}(t)\| < \delta$ for all $t \in \mathbb{R}$. Then $\boldsymbol{x} = 0$ is said to be a locally asymptotically stable (or internally stable) equilibrium of $\mathcal{G}_P$ if for each $\epsilon > 0$ there exists $\delta(\epsilon)$ such that for every input $\boldsymbol{u}(\bullet) \in \mathcal{U}_{\delta(\epsilon)}$ the state response $\boldsymbol{x}(\bullet)$ of $\mathcal{G}_P$ from the initial state $\boldsymbol{x}(0) = 0$ exists and satisfies $\boldsymbol{x}(t) \in \mathbb{B}_\epsilon$ for all $t \geq 0$.

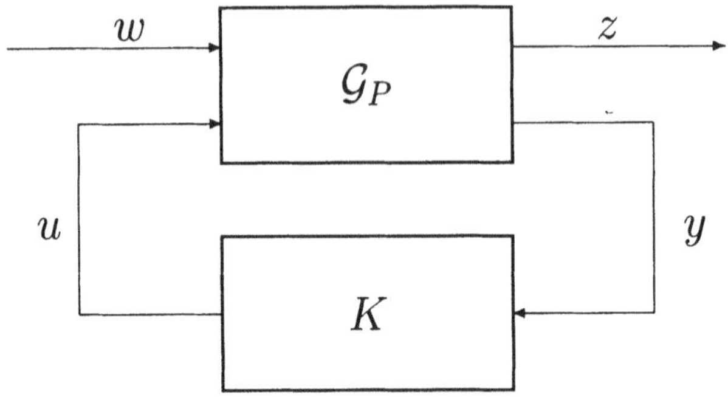

Fig. 4.1. Block diagram for standard $\mathcal{H}_\infty$ control problem

4.1 State Feedback $\mathcal{H}_\infty$ Control

In the state feedback $\mathcal{H}_\infty$ control problem we assume that $\boldsymbol{y} = \boldsymbol{x}$ in (4.1), i.e., that the whole state is available for measurement. We suppose the following.

(NA1) The matrix $\boldsymbol{D}_{12}$ has rank m and the matrix $\boldsymbol{D}_{11}^T\boldsymbol{D}_{11}-\gamma^2\boldsymbol{I}$ is negative definite, where $\boldsymbol{D}_{12} = (\frac{\partial \boldsymbol{Z}}{\partial \boldsymbol{u}})_{(\boldsymbol{x},\boldsymbol{w},\boldsymbol{u})=(0,0,0)}$ and $\boldsymbol{D}_{11} = (\frac{\partial \boldsymbol{Z}}{\partial \boldsymbol{w}})_{(\boldsymbol{x},\boldsymbol{w},\boldsymbol{u})=(0,0,0)}$.

(NA2) Any bounded trajectory $\boldsymbol{x}(t)$ of the system $\boldsymbol{E}\dot{\boldsymbol{x}}(t) = \boldsymbol{F}(\boldsymbol{x}(t),\, 0,\, \boldsymbol{u}(t))$ satisfying $\boldsymbol{Z}(\boldsymbol{x}(t),\, 0,\, \boldsymbol{u}(t)) = 0$ for all $t \geq 0$ is such that $\lim_{t\to\infty} \boldsymbol{x}(t) = 0$.

(NA3) The matrix pencil

$$\begin{bmatrix} \boldsymbol{A} - \mathrm{j}\omega\boldsymbol{E} & \boldsymbol{B}_2 \\ \boldsymbol{C}_1 & \boldsymbol{D}_{12} \end{bmatrix}$$

has full column rank for all $\omega \in \mathbb{R}\cup\{\infty\}$, where $\boldsymbol{A} = (\frac{\partial \boldsymbol{F}}{\partial \boldsymbol{x}})_{(\boldsymbol{x},\boldsymbol{w},\boldsymbol{u})=(0,0,0)}$, $\boldsymbol{B}_2 = (\frac{\partial \boldsymbol{F}}{\partial \boldsymbol{u}})_{(\boldsymbol{x},\boldsymbol{w},\boldsymbol{u})=(0,0,0)}$ and $\boldsymbol{C}_1 = (\frac{\partial \boldsymbol{Z}}{\partial \boldsymbol{x}})_{(\boldsymbol{x},\boldsymbol{w},\boldsymbol{u})=(0,0,0)}$.

Lemma 4.1. *Consider the DAE (4.1). Assume that Assumptions NA1-NA3 are satisfied. Suppose the following hypothesis also holds.*

H1: *There exists a smooth function $\mathsf{V}(\boldsymbol{x}(t))$ vanishing at $\boldsymbol{E}\boldsymbol{x} = 0$ and positive elsewhere, locally defined on a neighborhood of the equilibrium point 0 in $\mathcal{X}$, such that the function*

$$Y_1(\boldsymbol{x}(t)) \;=\; \mathfrak{H}(\boldsymbol{x},\, \boldsymbol{V}(\boldsymbol{x}(t),\, \boldsymbol{\alpha}_1(\boldsymbol{x}(t)),\, \boldsymbol{\alpha}_2(\boldsymbol{x}(t)))$$

is negative semidefinite near $\boldsymbol{x} = 0$, where the function $\mathfrak{H} : \mathcal{X} \times \mathbb{R}^n \times \mathbb{R}^l \times \mathbb{R}^m \longrightarrow \mathbb{R}$ is defined on a neighborhood of $(\boldsymbol{x},\, \boldsymbol{p},\, \boldsymbol{w},\, \boldsymbol{u}) = (0,0,0,0)$ as

$$\mathfrak{H}(\boldsymbol{x},\boldsymbol{p},\boldsymbol{w},\boldsymbol{u}) \;=\; \boldsymbol{p}^T\boldsymbol{F}(\boldsymbol{x},\, \boldsymbol{w},\, \boldsymbol{u}) + \|\boldsymbol{Z}(\boldsymbol{x},\, \boldsymbol{w},\, \boldsymbol{u})\|^2 - \gamma^2\|\boldsymbol{w}\|^2, \qquad (4.3)$$

$\frac{\partial \mathsf{V}}{\partial \boldsymbol{x}} = \boldsymbol{V}^T(\boldsymbol{x}(t))\boldsymbol{E}$ *defined as before, $\boldsymbol{\alpha}_1(\boldsymbol{x}) = \boldsymbol{w}^*(\boldsymbol{x},\, \boldsymbol{V}(\boldsymbol{x}))$ and $\boldsymbol{\alpha}_2(\boldsymbol{x}) = \boldsymbol{u}^*(\boldsymbol{x},\, \boldsymbol{V}(\boldsymbol{x}))$, and $\boldsymbol{w}^*(\boldsymbol{x},\, \boldsymbol{p})$ and $\boldsymbol{u}^*(\boldsymbol{x},\, \boldsymbol{p})$ are defined on a neighborhood of $(\boldsymbol{x},\, \boldsymbol{p}) = (0,0)$ satisfying*

$$\frac{\partial \mathfrak{H}}{\partial \boldsymbol{w}}(\boldsymbol{x},\, \boldsymbol{p},\, \boldsymbol{w}^*(\boldsymbol{x},\, \boldsymbol{p}),\, \boldsymbol{u}^*(\boldsymbol{x},\, \boldsymbol{p})) = 0,$$
$$\frac{\partial \mathfrak{H}}{\partial \boldsymbol{u}}(\boldsymbol{x},\, \boldsymbol{p},\, \boldsymbol{w}^*(\boldsymbol{x},\, \boldsymbol{p}),\, \boldsymbol{u}^*(\boldsymbol{x},\, \boldsymbol{p})) = 0$$

with $\boldsymbol{w}^*(0,0) = 0$ *and* $\boldsymbol{u}^*(0,0) = 0$.

Then the feedback law $\boldsymbol{u} = \boldsymbol{\alpha}_2(\boldsymbol{x})$ *solves the* $\mathcal{H}_\infty$ *state feedback control problem.* □

Proof. Firstly, observe that $Y_1 \leq 0$ implies that

$$\frac{\mathrm{dV}}{\mathrm{d}t} + \|\boldsymbol{Z}(\boldsymbol{x},\, \boldsymbol{w},\, \boldsymbol{\alpha}_2(\boldsymbol{x}))\|^2 - \gamma^2\|\boldsymbol{w}\|^2 \leq 0, \tag{4.4}$$

which, in turn, implies that the closed-loop system has $\mathcal{L}_2$-gain$\leq \gamma$. Next, in view of the Theorem 2.25, we need to show that the closed-loop system has index one. Taking the value of $\dfrac{\partial^2 Y_1}{\partial \boldsymbol{x}^2}$ on $(\boldsymbol{x}, \boldsymbol{p}, \boldsymbol{w}, \boldsymbol{u}) = (0,0,0,0)$ and rearranging terms yields the following linearization

$$(\boldsymbol{A}+\boldsymbol{B}_2\,\boldsymbol{F}_2)^T\boldsymbol{V}_{\boldsymbol{x}}(0)+\boldsymbol{V}_{\boldsymbol{x}}^T(0)(\boldsymbol{A}+\boldsymbol{B}_2\,\boldsymbol{F}_2)+(\boldsymbol{C}_1+\boldsymbol{D}_{12}\,\boldsymbol{F}_2)^T(\boldsymbol{C}_1+\boldsymbol{D}_{12}\,\boldsymbol{F}_2) \leq 0 \tag{4.5}$$

locally around $(\boldsymbol{x}, \boldsymbol{p}, \boldsymbol{w}, \boldsymbol{u}) = (0,0,0,0)$, where $\boldsymbol{F}_2 = (\frac{\partial}{\partial \boldsymbol{x}}\boldsymbol{\alpha}_2(\boldsymbol{x}))_{\boldsymbol{x}=0}$. In Equation (4.5), we have set $\boldsymbol{w}(t) \equiv 0$. By assumption (NA3), it is easy to see that $\{\boldsymbol{E},\, \boldsymbol{A} + \boldsymbol{B}_2\,\boldsymbol{F}_2,\, \boldsymbol{C}_1 + \boldsymbol{D}_{12}\,\boldsymbol{F}_2\}$ is impulse observable; this in turn implies, from equation (4.5), that $\{\boldsymbol{E},\, \boldsymbol{A} + \boldsymbol{B}_2\,\boldsymbol{F}_2\}$ is impulse-free by the generalized Lyapunov theorem (Lemma 2.17). We can, then, conclude that DAE (4.1) is of index one by using the state feedback law $\boldsymbol{u} = \boldsymbol{\alpha}_2(\boldsymbol{x})$. From the above discussion as well as assumptions (NA1) and (NA2), we can conclude, by Theorem 2.25, that $\boldsymbol{u} = \boldsymbol{\alpha}_2(\boldsymbol{x})$ is indeed a state feedback law rendering the closed-loop system asymptotically stable with index one and $\mathcal{L}_2$-gain $\leq \gamma$.

Q.E.D.

4.2 Output Feedback $\mathcal{H}_\infty$ Control

In this section, we consider the case in which the state $\boldsymbol{x}$ of the DAE (4.1) is not available for direct measurement. Motivated by the works of Isidori and Kang[49] and Yung *et al.*[135], we consider a dynamic controller of the form

$$\begin{cases} \hat{\boldsymbol{E}}\dot{\boldsymbol{\xi}} = \boldsymbol{F}(\boldsymbol{\xi}, \boldsymbol{\alpha}_1(\boldsymbol{\xi}), \boldsymbol{\alpha}_2(\boldsymbol{\xi})) + \boldsymbol{G}(\boldsymbol{\xi})(\boldsymbol{y} - \boldsymbol{Y}(\boldsymbol{\xi}, \boldsymbol{\alpha}_1(\boldsymbol{\xi}), \boldsymbol{\alpha}_2(\boldsymbol{\xi})), \\ \quad \boldsymbol{u} = \boldsymbol{\alpha}_2(\boldsymbol{\xi}) \end{cases}$$
$$\triangleq \begin{cases} \boldsymbol{E}_k = \boldsymbol{F}_k(\boldsymbol{\xi}) + \boldsymbol{G}_k(\boldsymbol{\xi}), \\ \quad \boldsymbol{u} = \boldsymbol{H}_k(\boldsymbol{\xi}), \end{cases} \tag{4.6}$$

where $\boldsymbol{\xi} = [\boldsymbol{\xi}_1, \ldots, \boldsymbol{\xi}_n]$ are local coordinates for the state-space manifold $\mathcal{X}_c$ of the controller Γ. The matrix $\boldsymbol{G}(\boldsymbol{\xi})$, called the *output injection gain*, is to be determined. Substitute the controller (4.6) in (4.1) to obtain the corresponding closed-loop system as

$$\boldsymbol{E}^e\dot{\boldsymbol{x}}^e = \boldsymbol{F}^e(\boldsymbol{x}^e,\,\boldsymbol{w}), \qquad \boldsymbol{z} = \boldsymbol{Z}^e(\boldsymbol{x}^e,\,\boldsymbol{w}) = \boldsymbol{Z}(\boldsymbol{x},\,\boldsymbol{w},\boldsymbol{\alpha}_2(\boldsymbol{\xi})), \tag{4.7}$$

where $\boldsymbol{E}^e = \begin{bmatrix} \boldsymbol{E} & 0 \\ 0 & \hat{\boldsymbol{E}} \end{bmatrix}$, $\boldsymbol{x}^e = \begin{bmatrix} \boldsymbol{x} \\ \boldsymbol{\xi} \end{bmatrix}$, and

$$\boldsymbol{F}^e(\boldsymbol{x}^e,\boldsymbol{w}) = \begin{bmatrix} \boldsymbol{F}(\boldsymbol{x},\boldsymbol{w},\boldsymbol{\alpha}_2(\boldsymbol{\xi})) \\ \boldsymbol{F}(\boldsymbol{\xi},\boldsymbol{\alpha}_1(\boldsymbol{\xi}),\boldsymbol{\alpha}_2(\boldsymbol{\xi})) + \boldsymbol{G}(\boldsymbol{\xi})(\boldsymbol{y} - \boldsymbol{Y}(\boldsymbol{\xi},\boldsymbol{\alpha}_1(\boldsymbol{\xi}),\boldsymbol{\alpha}_2(\boldsymbol{\xi})) \end{bmatrix}.$$

Again, we try to render the closed-loop system locally dissipative with respect to the supply rate $\gamma^2\|\boldsymbol{w}\|^2 - \|\boldsymbol{z}\|^2$. Clearly, it suffices to show that there exists a smooth nonnegative function $\mathsf{U}(\boldsymbol{x}^e)$ with $\dfrac{\partial \mathsf{U}}{\partial \boldsymbol{x}^e} = \boldsymbol{U}^T\boldsymbol{E}^e$ and $\boldsymbol{E}^{eT}\boldsymbol{U}_{\boldsymbol{x}^e} - \boldsymbol{U}_{\boldsymbol{x}^e}^T\boldsymbol{E}^e$ satisfying

$$\boldsymbol{U}^T\boldsymbol{F}^e(\boldsymbol{x}^e,\,\boldsymbol{w}) + \|\boldsymbol{Z}(\boldsymbol{x}^e,\,\boldsymbol{w})\|^2 - \gamma^2\|\boldsymbol{w}\|^2 \le 0, \qquad \text{for all } \boldsymbol{w}, \tag{4.8}$$

such that the closed-loop system is locally asymptotically stable. As a matter of fact, we have the following result.

Lemma 4.2. *Consider (4.7). Suppose that Assumptions NA1-NA3 hold. Suppose the DAE*

$$\hat{\boldsymbol{E}}\dot{\boldsymbol{\xi}} = \boldsymbol{F}(\boldsymbol{\xi},\boldsymbol{\alpha}_1(\boldsymbol{\xi}),0) - \boldsymbol{G}(\boldsymbol{\xi})\boldsymbol{Y}(\boldsymbol{\xi},\boldsymbol{\alpha}_1(\boldsymbol{\xi}),0) \tag{4.9}$$

has index one and a locally asymptotically stable equilibrium at $\boldsymbol{\xi} = 0$. Furthermore, suppose that there exists a smooth function $\mathsf{U}(\boldsymbol{x}^e)$, vanishing at $\boldsymbol{E}^e\boldsymbol{x}^e = 0$ and positive elsewhere, which satisfies the inequality (4.8) for all $\boldsymbol{w}$. Then the closed-loop system (4.7) has an $\mathcal{L}_2$-gain no greater than γ and a locally asymptotically stable equilibrium at $\boldsymbol{x}^e = 0$. □

Proof. In view of Theorem 2.25, we need only to show that the closed-loop system has index one. Observe that the function $\mathsf{U}(\boldsymbol{x}^e)$ defined above satisfies

$$\frac{\mathrm{d}\mathsf{U}(\boldsymbol{x}^e)}{\mathrm{d}t} + \|\boldsymbol{Z}(\boldsymbol{x}^e,\,\boldsymbol{w})\|^2 \le 0. \tag{4.10}$$

This shows that the closed-loop system is stable near $\boldsymbol{x}^e = 0$. Now, consider any trajectory $\boldsymbol{x}^e(t)$ of $\boldsymbol{E}\dot{\boldsymbol{x}}^e = \boldsymbol{F}^e(\boldsymbol{x}^e,\,0)$ satisfying $\dot{\mathsf{U}}(\boldsymbol{x}^e(t)) = 0$ for all $t \ge 0$. Then, from Inequality (4.10), it is necessary that $\boldsymbol{Z}(\boldsymbol{x}(t),\,0,\,\boldsymbol{\alpha}_2(\boldsymbol{\xi}(t))) = 0$ for all $t \ge 0$. Hence, we can conclude that the DAE $\boldsymbol{E}\dot{\boldsymbol{x}} = \boldsymbol{F}(\boldsymbol{x},\,0,\,\boldsymbol{\alpha}_2(\boldsymbol{\xi}))$ has index one by NA3, and $\lim_{t\to\infty}\boldsymbol{x}(t) = 0$ by NA2. Furthermore, from NA1, we see that there exists a unique smooth function $\boldsymbol{u} = \boldsymbol{u}(\boldsymbol{x})$ defined around $\boldsymbol{x} = 0$ such that $\boldsymbol{Z}(\boldsymbol{x},\,0,\,\boldsymbol{u}(\boldsymbol{x})) = 0$ and $\boldsymbol{u}(0) = 0$. Hence, $\lim_{t\to\infty}\boldsymbol{x}(t) = 0$ and $\boldsymbol{Z}(\boldsymbol{x}(t),\,0,\,\boldsymbol{\alpha}_2(\boldsymbol{\xi}(t))) = 0$ imply that $\lim_{t\to\infty}\boldsymbol{\alpha}_2(\boldsymbol{\xi}(t)) = 0$. This shows that the set $\mathcal{M} = \{(\boldsymbol{x},\,\boldsymbol{\xi}) :\ \boldsymbol{x} = 0,\,\boldsymbol{\alpha}_2(\boldsymbol{\xi}) = 0\}$ contains the ω-limit set of the trajectory satisfying Equation (4.10). It follows that any initial condition in this set yields a trajectory with $\boldsymbol{x} = 0$, and, at the same time, that $\boldsymbol{\xi}(t)$ is necessarily a trajectory satisfying Equation (4.9) which has index one by hypothesis. This amounts to showing that the closed-loop system has index one. **Q.E.D.**

The inequality (4.8), in fact, contains undetermined function $\boldsymbol{G}(\bullet)$, and involves $2n$ independent variables. The next theorem shows how the conditions in Lemma 4.2 can be met, while reducing the number of independent variables of (4.8). A further assumption is needed.

(NA4) The matrix $\boldsymbol{D}_{21} = (\frac{\partial \boldsymbol{Y}}{\partial \boldsymbol{w}})_{(\boldsymbol{x},\boldsymbol{w},\boldsymbol{u})=(0,0,0)}$ has rank p.

The next treatment is essentially following the one given in [49]. Consider the Hamiltonian function $\mathfrak{K}(\boldsymbol{x}, \boldsymbol{p}, \boldsymbol{w}, \boldsymbol{y}) : \mathbb{R}^n \times \mathbb{R}^n \times \mathbb{R}^r \times \mathbb{R}^p \longrightarrow \mathbb{R}$ defined as

$$\mathfrak{K}(\boldsymbol{x}, \boldsymbol{p}, \boldsymbol{w}, \boldsymbol{y}) = \boldsymbol{p}^T \boldsymbol{F}(\boldsymbol{x}, \boldsymbol{w}, 0) - \boldsymbol{y}^T \boldsymbol{Y}(\boldsymbol{x}, \boldsymbol{w}, 0) + \|\boldsymbol{Z}(\boldsymbol{x}, \boldsymbol{w}, 0)\|^2 - \gamma^2 \|\boldsymbol{w}\|^2.$$

It is easy to see that

$$\left(\frac{\partial^2 \mathfrak{K}(\boldsymbol{x}, \boldsymbol{p}, \boldsymbol{w}, \boldsymbol{y})}{\partial \boldsymbol{w}^2}\right)_{(\boldsymbol{x},\boldsymbol{p},\boldsymbol{w},\boldsymbol{y})=(0,0,0,0)} = 2(\boldsymbol{D}_{11}^T \boldsymbol{D}_{11} - \gamma^2 \boldsymbol{I}). \tag{4.11}$$

Then there exists a smooth function $\hat{\boldsymbol{w}}(\boldsymbol{x}, \boldsymbol{p}, \boldsymbol{y})$, defined in a neighborhood of $(0, 0, 0)$ such that

$$\left(\frac{\partial \mathfrak{K}(\boldsymbol{x}, \boldsymbol{p}, \boldsymbol{w}, \boldsymbol{y})}{\partial \boldsymbol{w}}\right)_{\boldsymbol{w}=\hat{\boldsymbol{w}}(\boldsymbol{x},\boldsymbol{p},\boldsymbol{y})} = 0, \quad \hat{\boldsymbol{w}}(0, 0, 0) = 0.$$

Next, it can also be easily derived that

$$\left(\frac{\partial^2 \mathfrak{K}(\boldsymbol{x}, \boldsymbol{p}, \hat{\boldsymbol{w}}(\boldsymbol{x}, \boldsymbol{p}, \boldsymbol{y}), \boldsymbol{y})}{\partial \boldsymbol{y}^2}\right)_{(\boldsymbol{x},\boldsymbol{p},\boldsymbol{y})=(0,0,0)} = \frac{1}{2\gamma^2} \boldsymbol{D}_{21} \boldsymbol{D}_{21}^T. \tag{4.12}$$

Thus, there exists a smooth function $\boldsymbol{y}_*(\boldsymbol{x}, \boldsymbol{p})$, defined in a neighborhood of $(0, 0)$ such that

$$\left(\frac{\partial \mathfrak{K}(\boldsymbol{x}, \boldsymbol{p}, \hat{\boldsymbol{w}}(\boldsymbol{x}, \boldsymbol{p}, \boldsymbol{y}), \boldsymbol{y})}{\partial \boldsymbol{y}}\right)_{\boldsymbol{y}=\boldsymbol{y}_*(\boldsymbol{x},\boldsymbol{p})} = 0, \quad \boldsymbol{y}_*(0, 0) = 0.$$

Set $\boldsymbol{w}_{**}(\boldsymbol{x}, \boldsymbol{p}) = \hat{\boldsymbol{w}}(\boldsymbol{x}, \boldsymbol{p}, \boldsymbol{y}_*(\boldsymbol{x}, \boldsymbol{p}))$. Because the definiteness of Equations (4.11) and (4.12), we have

$$\mathfrak{K}(\boldsymbol{x}, \boldsymbol{p}, \boldsymbol{w}, \boldsymbol{y}) \leq \mathfrak{K}(\boldsymbol{x}, \boldsymbol{p}, \hat{\boldsymbol{w}}(\boldsymbol{x}, \boldsymbol{p}, \boldsymbol{y}), \boldsymbol{y})$$

for all $(\boldsymbol{x}, \boldsymbol{p}, \boldsymbol{w}, \boldsymbol{y})$ in a neighborhood of $(0,0,0,0)$ and

$$\mathfrak{K}(\boldsymbol{x}, \boldsymbol{p}, \hat{\boldsymbol{w}}(\boldsymbol{x}, \boldsymbol{p}, \boldsymbol{y}), \boldsymbol{y}) \geq \mathfrak{K}(\boldsymbol{x}, \boldsymbol{p}, \boldsymbol{w}_{**}(\boldsymbol{x}, \boldsymbol{p}), \boldsymbol{y}_*(\boldsymbol{x}, \boldsymbol{p}))$$

for all $(\boldsymbol{x}, \boldsymbol{p}, \boldsymbol{y})$ in a neighborhood of $(0,0,0)$. The following theorem is readily obtained.

Theorem 4.3. *Consider (4.7). Suppose Assumptions NA1-NA4 are satisfied. Suppose hypothesis H1 of Lemma 4.1 holds. Suppose the following hypothesis also holds.*

H2: *There exists a smooth function $\mathsf{Q}(\boldsymbol{x})$ vanishing at $\boldsymbol{E}\boldsymbol{x} = 0$ and positive elsewhere, locally defined on a neighborhood of $\boldsymbol{x} = 0$, such that the function*

$$Y_2(\boldsymbol{x}) = \mathfrak{K}(\boldsymbol{x}, \boldsymbol{Q}(\boldsymbol{x}), \hat{\boldsymbol{w}}(\boldsymbol{x}, \boldsymbol{Q}(\boldsymbol{x}), \hat{\boldsymbol{y}}(\boldsymbol{x}, \boldsymbol{Q}(\boldsymbol{x}))), \hat{\boldsymbol{y}}(\boldsymbol{x}, \boldsymbol{Q}(\boldsymbol{x}))) - Y_1(\boldsymbol{x})$$

is negative definite near $\boldsymbol{x} = 0$, and its Hessian matrix is nonsingular at $\boldsymbol{x} = 0$. Here $\boldsymbol{Q}$ is defined by $\dfrac{\partial \mathsf{Q}}{\partial \boldsymbol{x}} = \boldsymbol{Q}^T\boldsymbol{E}$ with $\boldsymbol{E}^T\boldsymbol{Q}_{\boldsymbol{x}} = \boldsymbol{Q}_{\boldsymbol{x}}^T\boldsymbol{E}$, the function $\mathfrak{K} : \mathcal{X} \times \mathbb{R}^n \times \mathbb{R}^l \times \mathbb{R}^p \to \mathbb{R}$ is defined on a neighborhood of the origin as

$$\mathfrak{K}(\boldsymbol{x}, \boldsymbol{p}, \boldsymbol{w}, \boldsymbol{y}) = \boldsymbol{p}^T\boldsymbol{F}(\boldsymbol{x}, \boldsymbol{w}, 0) - \boldsymbol{y}^T\boldsymbol{Y}(\boldsymbol{x}, \boldsymbol{w}, 0) + \|\boldsymbol{Z}(\boldsymbol{x}, \boldsymbol{w}, 0)\|^2 - \gamma^2\|\boldsymbol{w}\|^2.$$

and the function $\hat{\boldsymbol{w}}(\boldsymbol{x}, \boldsymbol{p}, \boldsymbol{y})$, respectively $\hat{\boldsymbol{y}}(\boldsymbol{x}, \boldsymbol{p})$, defined on a neighborhood of $(0,0,0)$, respectively $(0,0)$, satisfies

$$\left(\frac{\partial \mathfrak{K}(\boldsymbol{x}, \boldsymbol{p}, \boldsymbol{w}, \boldsymbol{y})}{\partial \boldsymbol{w}}\right)_{\boldsymbol{w}=\hat{\boldsymbol{w}}(\boldsymbol{x},\boldsymbol{p},\boldsymbol{y})} = 0, \quad \hat{\boldsymbol{w}}(0,0,0) = 0,$$

respectively,

$$\left(\frac{\partial \mathfrak{K}(\boldsymbol{x}, \boldsymbol{p}, \hat{\boldsymbol{w}}(\boldsymbol{x}, \boldsymbol{p}, \boldsymbol{y}), \boldsymbol{y})}{\partial \boldsymbol{y}}\right)_{\boldsymbol{y}=\hat{\boldsymbol{y}}(\boldsymbol{x},\boldsymbol{p})} = 0, \quad \hat{\boldsymbol{y}}(0,0) = 0.$$

Then, if $\mathsf{Q}(\boldsymbol{x}) - \mathsf{V}(\boldsymbol{x}) > 0$ for all $\boldsymbol{x}$ such that $\boldsymbol{E}\boldsymbol{x} \neq 0$, and if the equation

$$(\boldsymbol{Q}(\boldsymbol{x}) - \boldsymbol{V}(\boldsymbol{x}))^T\boldsymbol{G}(\boldsymbol{x}) = \hat{\boldsymbol{y}}(\boldsymbol{x}, \boldsymbol{Q}(\boldsymbol{x})) \tag{4.13}$$

has a smooth solution $\boldsymbol{G}(\boldsymbol{x})$ near $\boldsymbol{x} = 0$, the nonlinear $\mathcal{H}^\infty$ output feedback control problem is solved by the output feedback controller (4.6) with $\hat{\boldsymbol{E}} = \boldsymbol{E}$. □

Proof. This follows easily by directly verifying that the function $\mathsf{U}(\boldsymbol{x}^e) \equiv \mathsf{Q}(\boldsymbol{x} - \boldsymbol{\xi}) + \mathsf{V}(\boldsymbol{x})$ satisfies the hypotheses of Lemma 4.2. **Q.E.D.**

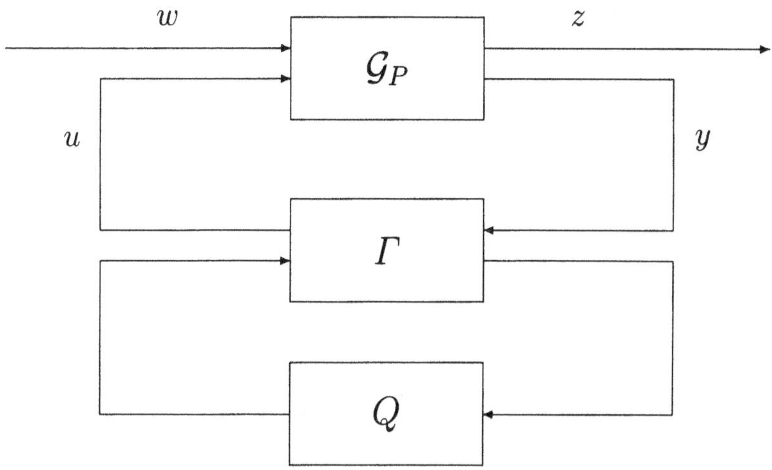

Fig. 4.2. Standard closed-loop configuration for controller parameterization

4.3 Parameterization of Output Feedback Controllers

For affine nonlinear systems, state-space formulas of a family of output feedback controllers were given in Lu and Doyle [75] and Yung *et al.* [130]. Recently, Yung *et al.*[135] have derived a set of parameterized solutions for general nonlinear systems. They have considered both output feedback and state feedback cases. Indeed, we can extend the technique developed in [135] to give a family of nonlinear $\mathcal{H}_\infty$ controllers for differential-algebraic systems.

We begin by providing another solution for output feedback problem. Define

$$r_{11}(\boldsymbol{x}) = \frac{1}{2}\left(\frac{\partial^2 \mathfrak{H}(\boldsymbol{x}, \boldsymbol{V}^T(\boldsymbol{x}), \boldsymbol{w}, \boldsymbol{u})}{\partial \boldsymbol{w}^2}\right)_{\boldsymbol{w}=\boldsymbol{\alpha}_1(\boldsymbol{x}), \boldsymbol{u}=\boldsymbol{\alpha}_2(\boldsymbol{x})},$$

$$r_{12}(\boldsymbol{x}) = \frac{1}{2}\left(\frac{\partial^2 \mathfrak{H}(\boldsymbol{x}, \boldsymbol{V}^T(\boldsymbol{x}), \boldsymbol{w}, \boldsymbol{u})}{\partial \boldsymbol{u}\partial \boldsymbol{w}}\right)_{\boldsymbol{w}=\boldsymbol{\alpha}_1(\boldsymbol{x}), \boldsymbol{u}=\boldsymbol{\alpha}_2(\boldsymbol{x})},$$

$$r_{21}(\boldsymbol{x}) = \frac{1}{2}\left(\frac{\partial^2 \mathfrak{H}(\boldsymbol{x}, \boldsymbol{V}^T(\boldsymbol{x}), \boldsymbol{w}, \boldsymbol{u})}{\partial \boldsymbol{w}\partial \boldsymbol{u}}\right)_{\boldsymbol{w}=\boldsymbol{\alpha}_1(\boldsymbol{x}), \boldsymbol{u}=\boldsymbol{\alpha}_2(\boldsymbol{x})},$$

$$r_{22}(\boldsymbol{x}) = \frac{1}{2}\left(\frac{\partial^2 \mathfrak{H}(\boldsymbol{x}, \boldsymbol{V}^T(\boldsymbol{x}), \boldsymbol{w}, \boldsymbol{u})}{\partial \boldsymbol{u}^2}\right)_{\boldsymbol{w}=\boldsymbol{\alpha}_1(\boldsymbol{x}), \boldsymbol{u}=\boldsymbol{\alpha}_2(\boldsymbol{x})},$$

and set

$$\boldsymbol{R}_\epsilon(\boldsymbol{x}) = \begin{bmatrix} (1-\epsilon_1)r_{11}(\boldsymbol{x}) & r_{12}(\boldsymbol{x}) \\ r_{21}(\boldsymbol{x}) & (1+\epsilon_2)r_{22}(\boldsymbol{x}) \end{bmatrix}$$

where ϵ_1 and ϵ_2 are any real numbers satisfying $0 < \epsilon_1 < 1$ and $\epsilon_2 > 0$, respectively. The following result will be used in the subsequent development.

Theorem 4.4. *Consider (4.7). Suppose Assumptions NA1-NA4 are satisfied. Suppose Hypothesis H1 of Lemma 4.1 holds. Suppose the following hypothesis also holds.*

H2a: *There exists a smooth real-valued function* $\mathsf{Q}(\boldsymbol{x})$, *locally defined on a neighborhood of* $\boldsymbol{x} = 0$, *which is vanishing at the points where* $\boldsymbol{E}\boldsymbol{x} = 0$ *and positive elsewhere such that the function*

$$Y_{2\epsilon}(\boldsymbol{x}) = K(\boldsymbol{x}, \boldsymbol{Q}(\boldsymbol{x}), \hat{\boldsymbol{w}}(\boldsymbol{x}, \boldsymbol{Q}(\boldsymbol{x}), \hat{\boldsymbol{y}}(\boldsymbol{x}, \boldsymbol{Q}(\boldsymbol{x}))), \hat{\boldsymbol{y}}(\boldsymbol{x}, \boldsymbol{Q}(\boldsymbol{x})))$$

is negative definite near $\boldsymbol{x} = 0$, *and its Hessian matrix is nonsingular at* $\boldsymbol{x} = 0$. *Here* $\boldsymbol{Q} : \mathbb{R}^n \to \mathbb{R}^n$ *is a smooth function defined by* $\dfrac{\partial \mathsf{Q}}{\partial \boldsymbol{x}} = \boldsymbol{Q}^T\boldsymbol{E}$ *with* $\boldsymbol{E}^T\boldsymbol{Q}_{\boldsymbol{x}} = \boldsymbol{Q}_{\boldsymbol{x}}^T\boldsymbol{E}$, *the function* $K : \mathbb{R}^n \times \mathbb{R}^n \times \mathbb{R}^l \times \mathbb{R}^p \to \mathbb{R}$ *is defined on a neighborhood of the origin as*

$$K(\boldsymbol{x}, \boldsymbol{p}, \boldsymbol{w}, \boldsymbol{y}) = \boldsymbol{p}^T\boldsymbol{F}(\boldsymbol{x}, \boldsymbol{w} + \boldsymbol{\alpha}_1(\boldsymbol{x}), 0) - \boldsymbol{y}^T\boldsymbol{Y}(\boldsymbol{x}, \boldsymbol{w} + \boldsymbol{\alpha}_1(\boldsymbol{x}), 0) + \begin{bmatrix} \boldsymbol{w} \\ -\boldsymbol{\alpha}_2(\boldsymbol{x}) \end{bmatrix}^T \boldsymbol{R}_\epsilon(\boldsymbol{x}) \begin{bmatrix} \boldsymbol{w} \\ -\boldsymbol{\alpha}_2(\boldsymbol{x}) \end{bmatrix}.$$

and the function $\hat{\boldsymbol{w}}(\boldsymbol{x}, \boldsymbol{p}, \boldsymbol{y})$, *respectively* $\hat{\boldsymbol{y}}(\boldsymbol{x}, \boldsymbol{p})$, *defined on a neighborhood of* $(0, 0, 0)$, *respectively* $(0, 0)$, *satisfies*

$$\left(\frac{\partial K(\boldsymbol{x}, \boldsymbol{p}, \boldsymbol{w}, \boldsymbol{y})}{\partial \boldsymbol{w}} \right)_{\boldsymbol{w} = \hat{\boldsymbol{w}}(\boldsymbol{x}, \boldsymbol{p}, \boldsymbol{y})} = 0, \quad \hat{\boldsymbol{w}}(0, 0, 0) = 0,$$

respectively,

$$\left(\frac{\partial K(\boldsymbol{x}, \boldsymbol{p}, \hat{\boldsymbol{w}}(\boldsymbol{x}, \boldsymbol{p}, \boldsymbol{y}), \boldsymbol{y})}{\partial \boldsymbol{y}} \right)_{\boldsymbol{y} = \hat{\boldsymbol{y}}(\boldsymbol{x}, \boldsymbol{p})} = 0, \quad \hat{\boldsymbol{y}}(0, 0) = 0.$$

Then, if the equation

$$\boldsymbol{Q}(\boldsymbol{x})\boldsymbol{G}(\boldsymbol{x}) = \hat{\boldsymbol{y}}^T(\boldsymbol{x}, \boldsymbol{Q}(\boldsymbol{x})) \tag{4.14}$$

has a smooth solution $\boldsymbol{G}(\boldsymbol{x})$ *near* $\boldsymbol{x} = 0$, *the nonlinear* $\mathcal{H}_\infty$ *output feedback control problem is solved by the output feedback controller (4.6) with* $\hat{\boldsymbol{E}} = \boldsymbol{E}$. □

Proof. The proof of the theorem is pretty much similar to the proof of Theorem 4.3, we omit the proof here for brevity. **Q.E.D.**

Motivated by the works of [130][135], we consider the family of controllers described by the following DAEs (See Fig. 4.2)

$$
\begin{aligned}
\hat{\boldsymbol{E}}\dot{\boldsymbol{\xi}} &= \boldsymbol{F}(\boldsymbol{\xi},\boldsymbol{\alpha}_1(\boldsymbol{\xi}),\boldsymbol{\alpha}_2(\boldsymbol{\xi})+\boldsymbol{c}(\boldsymbol{\eta}))+\boldsymbol{G}(\boldsymbol{\xi})(\boldsymbol{y}-\boldsymbol{Y}(\boldsymbol{\xi},\boldsymbol{\alpha}_1(\boldsymbol{\xi}),\boldsymbol{\alpha}_2(\boldsymbol{\xi})+\boldsymbol{c}(\boldsymbol{\eta})))\\
&\quad+\hat{g}_1(\boldsymbol{\xi})\boldsymbol{c}(\boldsymbol{\eta})+\hat{g}_2(\boldsymbol{\xi})d(\boldsymbol{\eta}),\\
\boldsymbol{E}_Q\dot{\boldsymbol{\eta}} &= \boldsymbol{a}(\boldsymbol{\eta},\boldsymbol{y}-Y(\boldsymbol{\xi},\boldsymbol{\alpha}_1(\boldsymbol{\xi}),\boldsymbol{\alpha}_2(\boldsymbol{\xi})+\boldsymbol{c}(\boldsymbol{\eta}))),\\
\boldsymbol{u} &= \boldsymbol{\alpha}_2(\boldsymbol{\xi})+\boldsymbol{c}(\boldsymbol{\eta}),
\end{aligned}
\tag{4.15}
$$

where $\boldsymbol{\xi}$ and $\boldsymbol{\eta}$ are defined on some neighborhoods of the origins in $\mathcal{X}_c$ and $\mathbb{R}^q$, respectively. $\boldsymbol{G}(\bullet)$ satisfies (4.14). $\boldsymbol{a}(\bullet,\bullet)$ and $\boldsymbol{c}(\bullet)$ are smooth functions with $\boldsymbol{a}(0,0)=0$ and $\boldsymbol{c}(0)=0$. $\hat{g}_1(\bullet)$, $\hat{g}_2(\bullet)$ and $d(\bullet)$ are C^k functions $(k\geq 1)$. E_Q is a constant matrix, and, in general, is singular. The functions $\boldsymbol{a}(\bullet,\bullet)$, $\boldsymbol{c}(\bullet)$, $\hat{g}_1(\bullet)$, $\hat{g}_2(\bullet)$, $d(\bullet)$, and the matrix E_Q are to-be-determined variables such that the closed-loop system (4.1) -(4.15) is dissipative with respect to the supply rate $\gamma^2\|\boldsymbol{w}\|^2-\|\boldsymbol{z}\|^2$, and is locally asymptotically stable with index one.

Observe first that the DAEs describing the closed-loop system (4.1)-(4.15) can be put in the form

$$
\begin{aligned}
\boldsymbol{E}_a\dot{\boldsymbol{x}}_a &= \boldsymbol{F}_a(\boldsymbol{x}_a,\,\boldsymbol{w}),\\
\boldsymbol{z} &= \boldsymbol{Z}(\boldsymbol{x},\,\boldsymbol{w},\,\boldsymbol{\alpha}_2(\boldsymbol{\xi})+\boldsymbol{c}(\boldsymbol{\eta})),
\end{aligned}
$$

where $\boldsymbol{x}_a = col[\boldsymbol{x},\boldsymbol{\xi},\boldsymbol{\eta}] \triangleq \begin{bmatrix}\boldsymbol{x}^T & \boldsymbol{\xi}^T & \boldsymbol{\eta}^T\end{bmatrix}^T$, $\boldsymbol{E}_a \triangleq \begin{bmatrix}\boldsymbol{E} & 0 & 0\\ 0 & \hat{\boldsymbol{E}} & 0\\ 0 & 0 & \boldsymbol{E}_Q\end{bmatrix}$, and

$$
\boldsymbol{F}_a \triangleq \begin{bmatrix}
\boldsymbol{F}(\boldsymbol{x},\,\boldsymbol{w},\,\boldsymbol{\alpha}_2(\boldsymbol{\xi})+\boldsymbol{c}(\boldsymbol{\eta}))\\
\tilde{\boldsymbol{F}}(\boldsymbol{\xi},\boldsymbol{\eta})+\boldsymbol{G}(\boldsymbol{\xi})\boldsymbol{Y}(\boldsymbol{x},\boldsymbol{w},\boldsymbol{\alpha}_2(\boldsymbol{\xi})+\boldsymbol{c}(\boldsymbol{\eta}))+\hat{g}_1(\boldsymbol{\xi})\boldsymbol{c}(\boldsymbol{\eta})+\hat{g}_2(\boldsymbol{\xi})d(\boldsymbol{\eta})\\
\boldsymbol{a}(\boldsymbol{\eta},\boldsymbol{Y}(\boldsymbol{x},\boldsymbol{w},\boldsymbol{\alpha}_2(\boldsymbol{\xi})+\boldsymbol{c}(\boldsymbol{\eta}))-\boldsymbol{Y}(\boldsymbol{\xi},\boldsymbol{\alpha}_1(\boldsymbol{\xi}),\boldsymbol{\alpha}_2(\boldsymbol{\xi})+\boldsymbol{c}(\boldsymbol{\eta})))
\end{bmatrix}.
$$

In the above equation,

$$
\tilde{\boldsymbol{F}}(\boldsymbol{\xi},\boldsymbol{\eta}) \triangleq \boldsymbol{F}(\boldsymbol{\xi},\boldsymbol{\alpha}_1(\boldsymbol{x}),\boldsymbol{\alpha}_2(\boldsymbol{\xi})+\boldsymbol{c}(\boldsymbol{\eta}))-\boldsymbol{G}(\boldsymbol{\xi})\boldsymbol{Y}(\boldsymbol{\xi},\boldsymbol{\alpha}_1(\boldsymbol{\xi}),\boldsymbol{\alpha}_2(\boldsymbol{\xi})+\boldsymbol{c}(\boldsymbol{\eta})).
$$

Consider a Hamiltonian function $\mathfrak{M}:\mathbb{R}^{2n+q}\times\mathbb{R}^{2n+q}\times\mathbb{R}^r\rightarrow\mathbb{R}$ defined as follows.

$$
\begin{aligned}
\mathfrak{M}(\boldsymbol{x}_a,\boldsymbol{p}_a,\boldsymbol{w}) &= \boldsymbol{p}_a^T\boldsymbol{F}_a(\boldsymbol{x}_a,\boldsymbol{w})\\
&\quad+\begin{bmatrix}\boldsymbol{w}-\boldsymbol{\alpha}_2(\boldsymbol{x})\\ \boldsymbol{\alpha}_2(\boldsymbol{\xi})+\boldsymbol{c}(\boldsymbol{\eta})-\boldsymbol{\alpha}_2(\boldsymbol{x})\end{bmatrix}^T\boldsymbol{R}(\boldsymbol{x})\begin{bmatrix}\boldsymbol{w}-\boldsymbol{\alpha}_2(\boldsymbol{x})\\ \boldsymbol{\alpha}_2(\boldsymbol{\xi})+\boldsymbol{c}(\boldsymbol{\eta})-\boldsymbol{\alpha}_2(\boldsymbol{x})\end{bmatrix}.
\end{aligned}
\tag{4.16}
$$

It is easy to check that

$$
\left(\frac{\partial^2\mathfrak{M}(\boldsymbol{x}_a,\boldsymbol{p}_a,\boldsymbol{w})}{\partial\boldsymbol{w}^2}\right)_{(\boldsymbol{x}_a,\boldsymbol{p}_a,\boldsymbol{w})=(0,0,0)} = 2(1-\epsilon_1)(\boldsymbol{D}_{11}^T\boldsymbol{D}_{11}-\gamma^2\boldsymbol{I}),
$$

which is negative definite by NA1. Then, by the Implicit Function Theorem, there exists a unique smooth function $\tilde{\boldsymbol{w}}(\boldsymbol{x}_a, \boldsymbol{p}_a)$, defined on a neighborhood of the origin, satisfying

$$\left(\frac{\partial \mathfrak{M}(\boldsymbol{x}_a, \boldsymbol{p}_a, \boldsymbol{w})}{\partial \boldsymbol{w}}\right)_{\boldsymbol{w}=\tilde{\boldsymbol{w}}(\boldsymbol{x}_a, \boldsymbol{p}_a)} = 0, \quad \tilde{\boldsymbol{w}}(0,0) = 0.$$

Lemma 4.5. *Consider (4.1) and (4.15). Suppose Assumptions NA1-NA4 are satisfied. Suppose Hypotheses H1 of Lemma 4.1 and H2a of Theorem 4.4 hold. Furthermore, suppose that the following hypothesis also holds.*

H3: *there exists a smooth function* $\mathtt{M}(\boldsymbol{x}_a)$, *locally defined on a neighborhood of the origin in* $\mathbb{R}^{2n+q}$, *vanishing at* $\boldsymbol{x}_a = col[\boldsymbol{E}\boldsymbol{x}, \boldsymbol{E}\boldsymbol{x}, 0]$ *and positive elsewhere which satisfies* $\dfrac{\partial \mathtt{M}(\boldsymbol{x}_a)}{\partial \boldsymbol{x}_a} = \boldsymbol{M}^T(\boldsymbol{x}_a)\boldsymbol{E}_a$ *with* $\boldsymbol{M}^T_{\boldsymbol{x}_a}(\boldsymbol{x}_a)\boldsymbol{E}_a = \boldsymbol{E}_a^T \boldsymbol{M}_{\boldsymbol{x}_a}(x_a)$, *and is such that the function*

$$Y_3(\boldsymbol{x}_a) = \mathfrak{M}(\boldsymbol{x}_a, \boldsymbol{M}(\boldsymbol{x}_a), \tilde{\boldsymbol{w}}(\boldsymbol{x}_a, \boldsymbol{M}(\boldsymbol{x}_a)))$$

vanishes at $\boldsymbol{x}_a = col[\boldsymbol{x}, \boldsymbol{x}, 0]$ *and is negative elsewhere*

Then the family of controllers (4.15) with $\hat{\boldsymbol{E}} = \boldsymbol{E}$ *solves the* $\mathcal{H}_\infty$ *output feedback control problem.* □

Proof. Set $\mathtt{U}(\boldsymbol{x}_a) = \mathtt{V}(\boldsymbol{x}) + \mathtt{M}(\boldsymbol{x}_a)$. It follows that $\dfrac{\partial \mathtt{U}}{\partial \boldsymbol{x}_a} = \boldsymbol{U}^T(\boldsymbol{x}_a)\boldsymbol{E}_a$ with $\boldsymbol{E}_a^T \boldsymbol{U}_{\boldsymbol{x}_a}(\boldsymbol{x}_a) = \boldsymbol{U}^T_{\boldsymbol{x}_a}(\boldsymbol{x}_a)\boldsymbol{E}_a$. With the equations $Y_1(\boldsymbol{x})$ and $Y_3(\boldsymbol{x})$ in hand, we have the following Hamiltonian equation by using Taylor series expansion,

$$\begin{aligned}
&\frac{\mathrm{d}\mathtt{U}}{\mathrm{d}t} + \|\boldsymbol{Z}(\boldsymbol{x}, \boldsymbol{w}, \boldsymbol{\alpha}_2(\boldsymbol{\xi}) + \boldsymbol{c}(\boldsymbol{\eta}))\|^2 - \gamma^2\|\boldsymbol{w}\|^2 \\
&= Y_1(\boldsymbol{x}) + Y_3(\boldsymbol{x}_a) &(4.17)\\
&+ \begin{bmatrix} \boldsymbol{w} - \boldsymbol{\alpha}_2(\boldsymbol{x}) \\ \boldsymbol{\alpha}_2(\boldsymbol{\xi}) + \boldsymbol{c}(\boldsymbol{\eta}) - \boldsymbol{\alpha}_2(\boldsymbol{x}) \end{bmatrix}^T \begin{bmatrix} \epsilon_1 r_{11}(\boldsymbol{x}) & 0 \\ 0 & -\epsilon_2 r_{22}(\boldsymbol{x}) \end{bmatrix} \begin{bmatrix} \boldsymbol{w} - \boldsymbol{\alpha}_2(\boldsymbol{x}) \\ \boldsymbol{\alpha}_2(\boldsymbol{\xi}) + \boldsymbol{c}(\boldsymbol{\eta}) - \boldsymbol{\alpha}_2(\boldsymbol{x}) \end{bmatrix} \\
&+ \|\boldsymbol{w} - \tilde{\boldsymbol{w}}(\boldsymbol{x}_a, \boldsymbol{M}^T(\boldsymbol{x}_a))\|^2_{\tilde{\boldsymbol{R}}(\boldsymbol{x}_a)} + o(\left\| \begin{matrix} \boldsymbol{w} - \boldsymbol{\alpha}_1(\boldsymbol{x}) \\ \boldsymbol{\alpha}_2(\boldsymbol{\xi}) + \boldsymbol{c}(\boldsymbol{\eta}) - \boldsymbol{\alpha}_2(\boldsymbol{x}) \end{matrix} \right\|^3) &(4.18)\\
&+ o(\|\boldsymbol{w} - \tilde{\boldsymbol{w}}(\boldsymbol{x}_a, \boldsymbol{M}^T(\boldsymbol{x}_a))\|^3), &(4.19)
\end{aligned}$$

where

$$\tilde{\boldsymbol{R}}(\boldsymbol{x}_a) \triangleq \frac{1}{2}\left(\frac{\partial^2 \mathfrak{M}(\boldsymbol{x}_a, \boldsymbol{M}^T(\mathrm{x}_a), w)}{\partial \boldsymbol{w}^2}\right)_{\boldsymbol{w}=\tilde{\boldsymbol{w}}(\boldsymbol{x}_a, \boldsymbol{M}^T(\boldsymbol{x}_a))}$$

and the notation $\|v\|^2_{\tilde{R}}$ stands for $v^T\tilde{R}v$. It is easy to verify that $\tilde{R}(0) = (1-\epsilon_1)(\boldsymbol{D}_{11}^T\boldsymbol{D}_{11} - \gamma^2\boldsymbol{I})$. Since $Y_1(\boldsymbol{x})$ and $Y_3(\boldsymbol{x}_a)$ are nonpositive, Equation (4.19) implies that

$$\frac{\mathrm{dU}}{\mathrm{d}t} + \|\boldsymbol{Z}(\boldsymbol{x},\, \boldsymbol{w},\, \boldsymbol{\alpha}_2(\boldsymbol{\xi}) + \boldsymbol{c}(\boldsymbol{\eta}))\|^2 - \gamma^2\|\boldsymbol{w}\|^2 \leq 0, \tag{4.20}$$

which, in turn, implies that the closed-loop system has an $\mathcal{L}_2$-gain less than or equal to γ. Set $\boldsymbol{w} = 0$, rearrange terms, and use Equation (4.19) to get

$$\begin{aligned}\frac{\mathrm{dU}}{\mathrm{d}t} &= -\|\boldsymbol{Z}(\boldsymbol{x}, 0, \boldsymbol{\alpha}_2(\boldsymbol{\xi}) + \boldsymbol{c}(\boldsymbol{\eta}))\|^2 + Y_{2\epsilon}(\boldsymbol{x}) + Y_3(\boldsymbol{x}_a) \\ &+ \begin{bmatrix} -\boldsymbol{\alpha}_2(\boldsymbol{x}) \\ \boldsymbol{\alpha}_2(\boldsymbol{\xi}) + \boldsymbol{c}(\boldsymbol{\eta}) - \boldsymbol{\alpha}_2(\boldsymbol{x}) \end{bmatrix}^T \begin{bmatrix} \epsilon_1 r_{11}(\boldsymbol{x}) & 0 \\ 0 & -\epsilon_2 r_{22}(\boldsymbol{x}) \end{bmatrix} \cdot \\ &\begin{bmatrix} -\boldsymbol{\alpha}_2(\boldsymbol{x}) \\ \boldsymbol{\alpha}_2(\boldsymbol{\xi}) + \boldsymbol{c}(\boldsymbol{\eta}) - \boldsymbol{\alpha}_2(\boldsymbol{x}) \end{bmatrix} \\ &+ \|\tilde{\boldsymbol{w}}(\boldsymbol{x}_a, \boldsymbol{M}^T(\boldsymbol{x}_a))\|^2_{\tilde{R}(\boldsymbol{x}_a)} + o(\left\| \begin{matrix} -\boldsymbol{\alpha}_1(\boldsymbol{x}) \\ \boldsymbol{\alpha}_2(\boldsymbol{\xi}) + \boldsymbol{c}(\boldsymbol{\eta}) - \boldsymbol{\alpha}_2(\boldsymbol{x}) \end{matrix} \right\|^3) \\ &+ o(\|\tilde{\boldsymbol{w}}(\boldsymbol{x}_a, \boldsymbol{M}^T(\boldsymbol{x}_a))\|^3),\end{aligned}$$

which is negative semidefinite near $\boldsymbol{x}_a = 0$ by hypothesis. This shows that the closed-loop system is stable locally around the equilibrium point. We claim that the DAE (4.15) has index one. To see this, observe that any trajectory satisfying $\frac{\mathrm{dU}}{\mathrm{d}t}(\boldsymbol{x}(t),\, \boldsymbol{\xi}(t),\, \boldsymbol{\eta}(t)) = 0$ for all $t \geq 0$ is necessarily a trajectory of

$$\boldsymbol{E}\dot{\boldsymbol{x}}(t) = \boldsymbol{F}(\boldsymbol{x},\, 0,\, \boldsymbol{\alpha}_2(\boldsymbol{\xi}) + \boldsymbol{c}(\boldsymbol{\eta})) \tag{4.21}$$

such that $\boldsymbol{x}(t)$ is bounded and $\boldsymbol{Z}(\boldsymbol{x},\, 0,\, \boldsymbol{\alpha}_2(\boldsymbol{\xi}) + \boldsymbol{c}(\boldsymbol{\eta})) = 0$ for all $t \geq 0$. This shows that the previous DAE has index one. Moreover, Hypotheses H1 and H3 along with Assumption (NA1) imply that the trajectory satisfying $\frac{\mathrm{dU}}{\mathrm{d}t}(\boldsymbol{x}(t), \boldsymbol{\xi}(t), \boldsymbol{\eta}(t)) = 0$ for all $t \geq 0$ is necessarily a trajectory such that $\boldsymbol{x}(t) = \boldsymbol{\xi}(t)$ and $\boldsymbol{\eta}(t) = 0$ for all $t \geq 0$. Setting $\boldsymbol{x}(t) = \boldsymbol{\xi}(t) \equiv 0$ and $\boldsymbol{w}(t) = 0$ in (4.16), we have

$$(\boldsymbol{M}_\eta)^T \boldsymbol{a}(\boldsymbol{\eta}, \boldsymbol{Y}(0,\, 0,\, \boldsymbol{c}(\boldsymbol{\eta})) - \boldsymbol{Y}(0,\, 0,\, \boldsymbol{c}(\boldsymbol{\eta}))) < 0,$$

for all $\boldsymbol{\eta} \neq 0$, where $\boldsymbol{M}^T(\boldsymbol{x}_a) = [(\boldsymbol{M}_{\boldsymbol{x}})^T\ (\boldsymbol{M}_\xi)^T\ (\boldsymbol{M}_\eta)^T]$. This shows that the DAE

$$\boldsymbol{E}_Q\dot{\boldsymbol{\eta}} = \boldsymbol{a}(\boldsymbol{\eta}, \boldsymbol{Y}(0,\, 0,\, \boldsymbol{c}(\boldsymbol{\eta})) - \boldsymbol{Y}(0,\, 0,\, \boldsymbol{c}(\boldsymbol{\eta}))) \tag{4.22}$$

has index one and is asymptotically stable. Hence, by Hypothesis H2 and the fact that DAE (4.21) and (4.22) have index one, we can conclude that the closed-loop system (4.1)-(4.15) has index one. Asymptotical stability then easily follows by Theorem 2.25. **Q.E.D.**

The previous lemma gives a general form of the output feedback controllers. However, it does not explicitly specify how we can choose the free system parameters $\boldsymbol{E}_Q$,, $\boldsymbol{a}(\bullet,\, \bullet)$ and $\boldsymbol{c}(\bullet)$ to meet the hypothesis in Lemma 4.5. In the sequel, we give a way to meet the condition in Lemma 4.5, and in the mean time, to reduce the number of independent variables.

Consider the following DAE.

$$\boldsymbol{E}_Q\dot{\boldsymbol{\eta}} = \boldsymbol{a}(\boldsymbol{\eta}, \bullet). \tag{4.23}$$

If there exists a smooth function $\mathrm{L}(\boldsymbol{\eta})$, locally defined on a neighborhood of $\boldsymbol{\eta} = 0$, which vanishes at $\boldsymbol{E}_Q\boldsymbol{\eta} = 0$, is positive elsewhere, and satisfies that $\dfrac{\partial \mathrm{L}(\boldsymbol{\eta})}{\partial \boldsymbol{\eta}} = \boldsymbol{L}(\boldsymbol{\eta})^T\boldsymbol{E}_Q$ with $\boldsymbol{E}_Q^T\boldsymbol{L}_\eta(\boldsymbol{\eta}) = \boldsymbol{L}_\eta(\boldsymbol{\eta})^T\boldsymbol{E}_Q$, such that

$$\boldsymbol{L}^T(\boldsymbol{\eta})\boldsymbol{a}(\boldsymbol{\eta}, \bullet) < 0,$$

then we can conclude from Theorem 2.15 that DAE (4.23) is locally asymptotically stable and has index one. Henceforth, if somc further hypotheses are imposed in the above inequality, the condition in Lemma 4.5 can be met. This is summarized in the following theorem.

Theorem 4.6. *Consider (4.1) and (4.15). Suppose Assumptions NA1-NA4 are satisfied. Suppose Hypotheses H1 of Lemma 4.1 and H2a of Theorem 4.4 hold. Suppose also that the following hypothesis holds.*

H4: *There exists a smooth function $\boldsymbol{L}(\boldsymbol{\eta})$ defined as above such that the function*

$$\mathrm{Y}_4(\boldsymbol{\eta}, w) = \boldsymbol{L}^T(\boldsymbol{\eta})\boldsymbol{a}(\boldsymbol{\eta}, \boldsymbol{Y}(0, w, 0)) + \begin{bmatrix} w \\ \boldsymbol{c}(\boldsymbol{\eta}) \end{bmatrix}^T \boldsymbol{R}(0) \begin{bmatrix} w \\ \boldsymbol{c}(\boldsymbol{\eta}) \end{bmatrix}$$

at $\boldsymbol{w} = \boldsymbol{w}^+(\boldsymbol{\eta})$, viewed as a function of $\boldsymbol{\eta}$, is negative definite near $\boldsymbol{\eta} = 0$, and its Hessian matrix is nonsingular at $\boldsymbol{\eta} = 0$. The function $\boldsymbol{w}^+(\boldsymbol{\eta})$ is defined on a neighborhood of $\boldsymbol{\eta} = 0$, which satisfies $(\dfrac{\partial \mathrm{Y}_4(\boldsymbol{\eta}, \boldsymbol{w})}{\partial \boldsymbol{w}})_{\boldsymbol{w}=\boldsymbol{w}^+(\boldsymbol{\eta})} = 0$ with $\boldsymbol{w}^+(0) = 0$ (This function exists, for $\boldsymbol{R}(0)$ is nonsingular).

Then, if $\hat{\boldsymbol{g}}_1(\bullet)$ and $\hat{\boldsymbol{g}}_2(\bullet)$ satisfy

$$\boldsymbol{Q}(\boldsymbol{x})\hat{\boldsymbol{g}}_1(\boldsymbol{x}) = 2\boldsymbol{\beta}^T(\boldsymbol{x}, 0, 0)\boldsymbol{r}_{12}(\boldsymbol{x}) - 2(1 + \epsilon_2)\boldsymbol{\alpha}_2^T(\boldsymbol{x})\boldsymbol{r}_{22}(\boldsymbol{x})$$

and

$$\boldsymbol{Q}(\boldsymbol{x})\hat{\boldsymbol{g}}_2(\boldsymbol{x}) = \boldsymbol{a}(0, \boldsymbol{Y}^T(\boldsymbol{x}, \boldsymbol{\alpha}_1(\boldsymbol{x}) + \boldsymbol{\beta}(\boldsymbol{x}, 0, 0), 0)$$

respectively, where $\boldsymbol{\beta}(\boldsymbol{x}, \boldsymbol{\xi}, \boldsymbol{\eta}) = \tilde{\boldsymbol{w}}(\boldsymbol{x}_a, \begin{bmatrix} \boldsymbol{Q}(\boldsymbol{x} - \boldsymbol{\xi}) & -\boldsymbol{Q}(\boldsymbol{x} - \boldsymbol{\xi})\ \boldsymbol{L}(\boldsymbol{\eta}) \end{bmatrix})$, the family of controllers (4.15) with $\boldsymbol{d}(\boldsymbol{\eta}) \triangleq \boldsymbol{L}(\boldsymbol{\eta})$ solves the $\mathcal{H}_\infty$ output feedback control problem. □

Proof. It is straightforward to verify that $\mathtt{M}(\boldsymbol{x}_a) \triangleq \mathtt{Q}(\boldsymbol{x} - \boldsymbol{\xi}) + \boldsymbol{L}(\boldsymbol{\eta})$ satisfies the hypothesis of Lemma 4.5. **Q.E.D.**

4.4 Linear Case

Let us now get back to the linear case. Consider the standard feedback configuration shown in Fig. 4.1. Let the plant $\mathcal{G}_P$ be described by

$$\begin{aligned} \boldsymbol{E}\dot{\boldsymbol{x}} &= \boldsymbol{A}\boldsymbol{x} + \boldsymbol{B}_1\boldsymbol{w} + \boldsymbol{B}_2\boldsymbol{u} \\ \boldsymbol{z} &= \boldsymbol{C}_1\boldsymbol{x} + \boldsymbol{D}_{12}\boldsymbol{u} \\ \boldsymbol{y} &= \boldsymbol{C}_2\boldsymbol{x} + \boldsymbol{D}_{21}\boldsymbol{w} \end{aligned} \tag{4.24}$$

where $\boldsymbol{x} \in \mathbb{R}^n$ is the state, and $\boldsymbol{w} \in \mathbb{R}^m$ represents a set of exogenous inputs which includes disturbances to be rejected and/or reference commands to be tracked. $\boldsymbol{z} \in \mathbb{R}^p$ is the output to be controlled and $\boldsymbol{y} \in \mathbb{R}^q$ is the measured output. $\boldsymbol{u} \in \mathbb{R}^l$ is the control input. $\boldsymbol{A}, \boldsymbol{B}_1, \boldsymbol{B}_2, \boldsymbol{C}_1, \boldsymbol{C}_2, \boldsymbol{D}_{12}$, and $\boldsymbol{D}_{21}$ are constant matrices with compatible dimensions. $\boldsymbol{E} \in \mathbb{R}^{n\times n}$ and $\mathrm{rank}\boldsymbol{E} = r < n$.

The standard $\mathcal{H}_\infty$ control problem consists of finding a controller $\boldsymbol{K}$ of the form

$$\begin{aligned} \hat{\boldsymbol{E}}\dot{\boldsymbol{\xi}} &= \hat{\boldsymbol{A}}\boldsymbol{\xi} + \hat{\boldsymbol{B}}\boldsymbol{y} \\ \boldsymbol{u} &= \hat{\boldsymbol{C}}\boldsymbol{\xi} \end{aligned} \tag{4.25}$$

where $\hat{\boldsymbol{E}}$, $\hat{\boldsymbol{A}} \in \mathbb{R}^{n\times n}$, $\hat{\boldsymbol{B}} \in \mathbb{R}^{n\times q}$ and $\hat{\boldsymbol{C}} \in \mathbb{R}^{l\times n}$, such that the resulting closed-loop system is internally stable and $\boldsymbol{T}_{\boldsymbol{zw}}$, the closed-loop transfer matrix from $\boldsymbol{w}$ to $\boldsymbol{z}$, has an $\mathcal{H}_\infty$ norm strictly less than a prescribed positive number γ. Note that we do not assume *a priori* structure of the matrix $\hat{\boldsymbol{E}}$; it may be singular or nonsingular, equal to $\boldsymbol{E}$ or not.

The system (4.24) is assumed to satisfy the following assumptions, see also[57][111].

(A1) $\{\boldsymbol{E}, \boldsymbol{A}\}$ is regular.
(A2) $\{\boldsymbol{E}, \boldsymbol{A}, \boldsymbol{B}_2\}$ is finite dynamics stabilizable and impulse controllable.
(A3) $\{\boldsymbol{E}, \boldsymbol{A}, \boldsymbol{C}_2\}$ is finite dynamics detectable and impulse observable.
(A4) $\begin{bmatrix} \boldsymbol{A} - \mathrm{j}\omega\boldsymbol{E} & \boldsymbol{B}_1 \\ \boldsymbol{C}_2 & \boldsymbol{D}_{21} \end{bmatrix}$ has full row rank for all $\omega \in \mathbb{R}$ and is row reduced.
(A5) $\begin{bmatrix} \boldsymbol{A} - \mathrm{j}\omega\boldsymbol{E} & \boldsymbol{B}_2 \\ \boldsymbol{C}_1 & \boldsymbol{D}_{12} \end{bmatrix}$ has full column rank for all $\omega \in \mathbb{R}$ and is column reduced.
(A6) $\boldsymbol{R}_1 \triangleq \boldsymbol{D}_{12}^T\boldsymbol{D}_{12} > 0$.
(A7) $\boldsymbol{R}_2 \triangleq \boldsymbol{D}_{21}\boldsymbol{D}_{21}^T > 0$.

The main result of this section is summarized in the following statements.

Theorem 4.7. *Consider (4.24). Suppose that Assumptions (A1)-(A7) hold. Then there exists a controller of the form (4.25) that internally stabilizes*

(4.24) and render $\|\boldsymbol{T}_{\boldsymbol{zw}}\|_\infty < \gamma$ if and only if the following conditions are satisfied.

(i) *There exists an admissible solution $\boldsymbol{X}_\infty$ to the GARE*

$$\begin{aligned}\mathbf{R}ic_1(\boldsymbol{X}) = (&\boldsymbol{A} - \boldsymbol{B}_2\boldsymbol{R}_1^{-1}\boldsymbol{D}_{12}^T\boldsymbol{C}_1)^T\boldsymbol{X} + \boldsymbol{X}^T(\boldsymbol{A} - \boldsymbol{B}_2\boldsymbol{R}_1^{-1}\boldsymbol{D}_{12}^T\boldsymbol{C}_1)\\ &+\boldsymbol{C}_1^T(\boldsymbol{I} - \boldsymbol{D}_{12}\boldsymbol{R}_1^{-1}\boldsymbol{D}_{12}^T)\boldsymbol{C}_1\\ &+\boldsymbol{X}^T(\frac{1}{\gamma^2}\boldsymbol{B}_1\boldsymbol{B}_1^T - \boldsymbol{B}_2\boldsymbol{R}_1^{-1}\boldsymbol{B}_2^T)\boldsymbol{X} = 0,\\ &\boldsymbol{E}^T\boldsymbol{X} = \boldsymbol{X}^T\boldsymbol{E},\end{aligned}$$

with $\boldsymbol{E}^T\boldsymbol{X}_\infty = \boldsymbol{X}_\infty^T\boldsymbol{E} \geq 0$.

(ii) *There exists an admissible solution $\boldsymbol{Z}_\infty$ to the GARE*

$$\begin{aligned}\mathbf{R}ic_3(\boldsymbol{Z}) = (&\tilde{\boldsymbol{A}} - \boldsymbol{B}_1\boldsymbol{D}_{21}^T\boldsymbol{R}_2^{-1}\tilde{\boldsymbol{C}}_2)\boldsymbol{Z} + \boldsymbol{Z}^T(\tilde{\boldsymbol{A}} - \boldsymbol{B}_1\boldsymbol{D}_{21}^T\boldsymbol{R}_2^{-1}\tilde{\boldsymbol{C}}_2)^T\\ &-\boldsymbol{Z}^T(\tilde{\boldsymbol{C}}_2^T\boldsymbol{R}_2^{-1}\tilde{\boldsymbol{C}}_2 - \frac{1}{\gamma^2}\boldsymbol{F}_\infty^T\boldsymbol{R}_1\boldsymbol{F}_\infty)\boldsymbol{Z} + \tilde{\boldsymbol{B}}_1\tilde{\boldsymbol{B}}_1^T = 0,\\ &\boldsymbol{E}\boldsymbol{Z} = \boldsymbol{Z}^T\boldsymbol{E}^T,\end{aligned} \tag{4.26}$$

with $\boldsymbol{E}\boldsymbol{Z}_\infty = \boldsymbol{Z}_\infty^T\boldsymbol{E}^T \geq 0$, where

$$\begin{aligned}\tilde{\boldsymbol{A}} &= \boldsymbol{A} + \frac{1}{\gamma^2}\boldsymbol{B}_1\boldsymbol{B}_1^T\boldsymbol{X}_\infty,\\ \tilde{\boldsymbol{B}}_1 &= \boldsymbol{B}_1(\boldsymbol{I} - \boldsymbol{D}_{21}^T\boldsymbol{R}_2^{-1}\boldsymbol{D}_{21}),\\ \tilde{\boldsymbol{C}}_2 &= \boldsymbol{C}_2 + \frac{1}{\gamma^2}\boldsymbol{D}_{21}\boldsymbol{B}_1^T\boldsymbol{X}_\infty,\end{aligned}$$

and

$$\boldsymbol{F}_\infty = -\boldsymbol{R}_1^{-1}(\boldsymbol{B}_2^T\boldsymbol{X}_\infty + \boldsymbol{D}_{12}^T\boldsymbol{C}_1).$$

When these conditions hold, one such controller is given by

$$\begin{aligned}&\hat{\boldsymbol{E}} = \boldsymbol{E}\\ &\hat{\boldsymbol{A}} = \boldsymbol{A}_k \triangleq \boldsymbol{A} + \boldsymbol{B}_2\hat{\boldsymbol{C}} - \hat{\boldsymbol{B}}\boldsymbol{C}_2 + \tfrac{1}{\gamma^2}(\boldsymbol{B}_1 - \hat{\boldsymbol{B}}\boldsymbol{D}_{21})\boldsymbol{B}_1^T\boldsymbol{X}_\infty,\\ &\hat{\boldsymbol{B}} = \boldsymbol{B}_k \triangleq (\boldsymbol{Z}_\infty^T\boldsymbol{C}_2^T + (\boldsymbol{I} + \tfrac{1}{\gamma^2}\boldsymbol{X}_\infty\boldsymbol{Z}_\infty)^T\boldsymbol{B}_1\boldsymbol{D}_{21}^T)\boldsymbol{R}_2^{-1},\\ &\hat{\boldsymbol{C}} = \boldsymbol{C}_k \triangleq \boldsymbol{F}_\infty = -\boldsymbol{R}_1^{-1}(\boldsymbol{B}_2^T\boldsymbol{X}_\infty + \boldsymbol{D}_{12}^T\boldsymbol{C}_1).\end{aligned} \tag{4.27}$$

□

Proof.
Sufficiency. We will show that the controller given in (4.27) both stabilizes the system and makes $\|\boldsymbol{T}_{\boldsymbol{zw}}\|_\infty < \gamma$. To this end, observe that the resulting closed-loop system can be expressed as

$$\underbrace{\begin{bmatrix} \boldsymbol{E} & 0 \\ 0 & \boldsymbol{E} \end{bmatrix}}_{\boldsymbol{E}_c} \begin{bmatrix} \dot{\boldsymbol{x}} \\ \dot{\boldsymbol{e}} \end{bmatrix} = \underbrace{\begin{bmatrix} \boldsymbol{A} + \boldsymbol{B}_2\boldsymbol{C}_k & -\boldsymbol{B}_2\boldsymbol{C}_k \\ \boldsymbol{A} - \boldsymbol{A}_k + \boldsymbol{B}_2\boldsymbol{C}_k - \boldsymbol{B}_k\boldsymbol{C}_2 & \boldsymbol{A}_k - \boldsymbol{B}_2\boldsymbol{C}_k \end{bmatrix}}_{\boldsymbol{A}_c} \begin{bmatrix} \boldsymbol{x} \\ \boldsymbol{e} \end{bmatrix}$$

$$+ \underbrace{\begin{bmatrix} \boldsymbol{B}_1 \\ \boldsymbol{B}_1 - \boldsymbol{B}_k\boldsymbol{D}_{21} \end{bmatrix}}_{\boldsymbol{B}_c} \boldsymbol{w},$$

$$\boldsymbol{z} = \underbrace{\begin{bmatrix} \boldsymbol{C}_1 + \boldsymbol{D}_{12}\boldsymbol{C}_k & -\boldsymbol{D}_{12}\boldsymbol{C}_k \end{bmatrix}}_{\boldsymbol{C}_c} \begin{bmatrix} \boldsymbol{x} \\ \boldsymbol{e} \end{bmatrix},$$

where $\boldsymbol{e} \stackrel{\triangle}{=} \boldsymbol{x} - \boldsymbol{\xi}$. Note that the GARE $R_3(\boldsymbol{Z})$ with $\boldsymbol{Z} = \boldsymbol{Z}_\infty$ can be rewritten as $\boldsymbol{A}_0\boldsymbol{Z}_\infty + \boldsymbol{Z}_\infty^T\boldsymbol{A}_0^T + \frac{1}{\gamma^2}\boldsymbol{Z}_\infty^T\boldsymbol{C}_0^T\boldsymbol{C}_0\boldsymbol{Z}_\infty + \boldsymbol{B}_0\boldsymbol{B}_0^T = 0$ with $\boldsymbol{E}\boldsymbol{Z}_\infty = \boldsymbol{Z}_\infty^T\boldsymbol{E}^T$, where

$$\begin{aligned} \boldsymbol{A}_0 &= \tilde{\boldsymbol{A}} - \boldsymbol{B}_1\boldsymbol{D}_{21}^T\boldsymbol{R}_2^{-1}\tilde{\boldsymbol{C}}_2 - \boldsymbol{Z}_\infty^T\tilde{\boldsymbol{C}}_2^T\boldsymbol{R}_2^{-1}\tilde{\boldsymbol{C}}_2, \\ \boldsymbol{B}_0 &= \tilde{\boldsymbol{B}}_1 - \boldsymbol{Z}_\infty^T\tilde{\boldsymbol{C}}_2^T\boldsymbol{R}_2^{-1}\boldsymbol{D}_{21}, \\ \boldsymbol{C}_0 &= -\boldsymbol{R}_1^{-\frac{1}{2}}(\boldsymbol{B}_2^T\boldsymbol{X}_\infty + \boldsymbol{D}_{21}^T\boldsymbol{C}_1) = \boldsymbol{R}_1^{\frac{1}{2}}\boldsymbol{F}_\infty. \end{aligned}$$

Moreover, it is easy to verify that $\{\boldsymbol{E}^T, \boldsymbol{A}_0^T + \dfrac{1}{\gamma^2}\boldsymbol{C}_0^T\boldsymbol{C}_0\boldsymbol{Z}_\infty\}$ is admissible because $\boldsymbol{Z}_\infty$ is an admissible solution to the GARE $\mathbf{R}ic_3(\boldsymbol{Z}) = 0$. Thus, it is concluded that $\boldsymbol{Z}_\infty$ is also an admissible solution to the GARE

$$\begin{aligned} &\boldsymbol{A}_0\boldsymbol{Z} + \boldsymbol{Z}^T\boldsymbol{A}_0^T + \frac{1}{\gamma^2}\boldsymbol{Z}^T\boldsymbol{C}_0^T\boldsymbol{C}_0\boldsymbol{Z} + \boldsymbol{B}_0\boldsymbol{B}_0^T = 0, \\ &\boldsymbol{E}\boldsymbol{Z} = \boldsymbol{Z}^T\boldsymbol{E}^T. \end{aligned} \tag{4.28}$$

Then it follows from Lemma 2.26, that $\{\boldsymbol{E}^T, \boldsymbol{A}_0^T\}$ is stable, impulse-free and $\|\boldsymbol{B}_0^T(s\boldsymbol{E}^T - \boldsymbol{A}_0^T)^{-1}\boldsymbol{C}_0^T\|_\infty < \gamma$. This, in turn, implies that $\{\boldsymbol{E}, \boldsymbol{A}_0\}$ is stable, impulse-free and $\|\boldsymbol{C}_0(s\boldsymbol{E} - \boldsymbol{A}_0)^{-1}\boldsymbol{B}_0\|_\infty < \gamma$. Again by Lemma 2.26, the GARE

$$\begin{aligned} \mathbf{R}ic_4(\boldsymbol{W}) &= \boldsymbol{A}_0^T\boldsymbol{W} + \boldsymbol{W}^T\boldsymbol{A}_0 + \frac{1}{\gamma^2}\boldsymbol{W}^T\boldsymbol{B}_0\boldsymbol{B}_0^T\boldsymbol{W} + \boldsymbol{C}_0^T\boldsymbol{C}_0 = 0, \\ &\boldsymbol{W}^T\boldsymbol{E} = \boldsymbol{E}^T\boldsymbol{W} \end{aligned} \tag{4.29}$$

has an admissible solution $\boldsymbol{W}_\infty$ with $\boldsymbol{W}_\infty^T\boldsymbol{E} = \boldsymbol{E}^T\boldsymbol{W}_\infty \geq 0$. Now Set

$$\boldsymbol{P}_c \stackrel{\triangle}{=} \begin{bmatrix} \boldsymbol{X}_\infty & 0 \\ 0 & \boldsymbol{W}_\infty \end{bmatrix}.$$

Lengthy but otherwise routine calculation shows that $\boldsymbol{P}_c$ is an admissible solution to the GARE

$$\begin{aligned} &\boldsymbol{A}_c^T\boldsymbol{P} + \boldsymbol{P}^T\boldsymbol{A}_c + \boldsymbol{C}_c^T\boldsymbol{C}_c + \tfrac{1}{\gamma^2}\boldsymbol{P}^T\boldsymbol{B}_c\boldsymbol{B}_c^T\boldsymbol{P} = 0, \\ &\boldsymbol{E}_c^T\boldsymbol{P} = \boldsymbol{P}^T\boldsymbol{E}_c \end{aligned}$$

with $\boldsymbol{E}_c^T\boldsymbol{P}_c = \boldsymbol{P}_c^T\boldsymbol{E}_c \geq 0$.It follows, again from Lemma 2.26, that (4.27) is an admissible controller such that $\|\boldsymbol{T}_{zw}\|_\infty < \gamma$. This completes the proof of sufficiency.

Necessity. To prove the necessity part, we need the following lemma. See [82] for proof.

Lemma 4.8. *Consider the standard system diagram shown in Fig. 4.1. Suppose that there exists a controller of the form (4.25) that internally stabilizes (4.24) and render $\|\boldsymbol{T}_{zw}\|_\infty < \gamma$. Then the following conditons hold.*

(i) *There exists a state feedback matrix $\boldsymbol{K}$ and a matrix $\boldsymbol{P}$ such that*

$$\begin{cases} (\boldsymbol{A}+\boldsymbol{B}_2\boldsymbol{K})^T\boldsymbol{P}+\boldsymbol{P}^T(\boldsymbol{A}+\boldsymbol{B}_2\boldsymbol{K})+\boldsymbol{P}^T\boldsymbol{B}_1\boldsymbol{B}_1^T\boldsymbol{P} \\ \qquad +(\boldsymbol{C}_1\boldsymbol{D}_{12}\boldsymbol{K})^T(\boldsymbol{C}_1+\boldsymbol{D}_{12}\boldsymbol{K}) < 0, \\ \qquad \boldsymbol{E}^T\boldsymbol{P} = \boldsymbol{P}^T\boldsymbol{E} \geq 0. \end{cases} \tag{4.30}$$

(ii) *There exists an output injection matrix $\boldsymbol{L}$ and a matrix $\boldsymbol{\Sigma}$ such that*

$$\begin{cases} (\boldsymbol{A}+\boldsymbol{L}\boldsymbol{C}_2)\boldsymbol{\Sigma}+\boldsymbol{\Sigma}^T(\boldsymbol{A}+\boldsymbol{L}\boldsymbol{C}_2)^T+\boldsymbol{\Sigma}\boldsymbol{C}_1^T\boldsymbol{C}_1\boldsymbol{\Sigma} \\ \qquad +(\boldsymbol{B}_1+\boldsymbol{L}\boldsymbol{D}_{21})(\boldsymbol{B}_1+\boldsymbol{L}\boldsymbol{D}_{21})^T < 0, \\ \qquad \boldsymbol{E}\boldsymbol{\Sigma} = \boldsymbol{\Sigma}^T\boldsymbol{E}^T \geq 0. \end{cases} \tag{4.31}$$

(iii) $\rho(\boldsymbol{P}\boldsymbol{\Sigma}) < \gamma^2$, *where $\rho(\bullet)$ denotes the spectral radius.* □

We can now prove the necessity part. Suppose that the $\mathcal{H}_\infty$ control problem is solvable. Then, from Lemma 4.8, Inequality (4.30) has a solution $\boldsymbol{P}$. Inequality (4.30) can be rewritten as

$$\begin{aligned} \mathrm{R}ic_1(\boldsymbol{P}) = {} & (\boldsymbol{A}-\boldsymbol{B}_2\boldsymbol{R}_1^{-1}\boldsymbol{D}_{12}^T\boldsymbol{C}_1)^T\boldsymbol{P}+\boldsymbol{P}^T(\boldsymbol{A}-\boldsymbol{B}_2\boldsymbol{R}_1^{-1}\boldsymbol{D}_{12}^T\boldsymbol{C}_1) \\ & +\boldsymbol{C}_1^T(\boldsymbol{I}-\boldsymbol{D}_{12}\boldsymbol{R}_1^{-1}\boldsymbol{D}_{12}^T)\boldsymbol{C}_1 \\ & +\boldsymbol{P}^T(\frac{1}{\gamma^2}\boldsymbol{B}_1\boldsymbol{B}_1^T-\boldsymbol{B}_2\boldsymbol{R}_1^{-1}\boldsymbol{B}_2^T)\boldsymbol{P} < 0. \end{aligned} \tag{4.32}$$

Now, Assumption (A4) implies that

$$\begin{bmatrix} -\mathrm{j}\omega\boldsymbol{E}+(\boldsymbol{A}-\boldsymbol{B}_2\boldsymbol{R}_1^{-1}\boldsymbol{D}_{12}^T\boldsymbol{C}_1) \\ \boldsymbol{C}_1^T(\boldsymbol{I}-\boldsymbol{D}_{12}\boldsymbol{R}_1^{-1}\boldsymbol{D}_{12}^T)\boldsymbol{C}_1 \end{bmatrix}$$

has full column rank for all $\omega \in \mathbb{R}$. Furthermore, we can show that $\{\boldsymbol{E}, (\boldsymbol{A} - \boldsymbol{B}_2\boldsymbol{R}_1^{-1}\boldsymbol{D}_{12}^T\boldsymbol{C}_1)\}$ is impulse-free. To see this, observe that

$$\begin{bmatrix} \boldsymbol{I} & -\boldsymbol{B}_2\boldsymbol{R}_1^{-1}\boldsymbol{D}_{12}^T \\ 0 & \boldsymbol{I} \end{bmatrix}\begin{bmatrix} -s\boldsymbol{E}+\boldsymbol{A} & \boldsymbol{B}_2 \\ \boldsymbol{C}_1 & \boldsymbol{D}_{12} \end{bmatrix} = \begin{bmatrix} -s\boldsymbol{E}+(\boldsymbol{A}-\boldsymbol{B}_2\boldsymbol{R}_1^{-1}\boldsymbol{D}_{12}^T\boldsymbol{C}_1) & 0 \\ \boldsymbol{C}_1 & \boldsymbol{D}_{12} \end{bmatrix}.$$

Applying Lemma 2.26 shows that there exists an admissible solution $\boldsymbol{X}_\infty$ satisfying $R_1(\boldsymbol{X}_\infty) = 0$ with $\boldsymbol{E}^T\boldsymbol{X}_\infty = \boldsymbol{X}_\infty^T\boldsymbol{E} \geq 0$. Similarly, the inequality (4.31) implies that there exists an admissible solution $\boldsymbol{Y}_\infty$ satisfying the GARE

$$\begin{aligned}\mathbf{R}ic_2(\boldsymbol{Y}) &= (\boldsymbol{A} - B_{-}1\boldsymbol{D}_{21}^T\boldsymbol{R}_2^{-1}\boldsymbol{C}_2)\boldsymbol{Y} + \boldsymbol{Y}^T(\boldsymbol{A} - \boldsymbol{B}_1\boldsymbol{D}_{21}^T\boldsymbol{R}_2^{-1}\boldsymbol{C}_2)^T \\ &+\boldsymbol{Y}^T(\frac{1}{\gamma^2}\boldsymbol{C}_1^T\boldsymbol{C}_1 - \boldsymbol{C}_2^T\boldsymbol{R}_2^{-1}\boldsymbol{C}_2)\boldsymbol{Y} \\ &+\boldsymbol{B}_1(\boldsymbol{I} - \boldsymbol{D}_{21}^T\boldsymbol{R}_2^{-1}\boldsymbol{D}_{21})\boldsymbol{B}_1^T = 0, \\ &\boldsymbol{E}\boldsymbol{Y} = \boldsymbol{Y}^T\boldsymbol{E}^T,\end{aligned}$$

with $\boldsymbol{E}\boldsymbol{Y}_\infty = \boldsymbol{Y}_\infty^T\boldsymbol{E}^T \geq 0$. Moreover, the spectral radius $\rho(\boldsymbol{Y}_\infty\boldsymbol{X}_\infty) < \gamma^2$. Define $\boldsymbol{Z}_\infty = (\boldsymbol{I} - \frac{1}{\gamma^2}\boldsymbol{Y}_\infty\boldsymbol{X}_\infty)^{-1}\boldsymbol{Y}_\infty = \boldsymbol{Y}_\infty(\boldsymbol{I} - \frac{1}{\gamma^2}\boldsymbol{X}_\infty\boldsymbol{Y}_\infty)^{-1}$. A little bit of algebra shows that $\boldsymbol{Z}_\infty$ is an admissible solution of Equation (4.26) with $\boldsymbol{E}\boldsymbol{Z}_\infty = \boldsymbol{Z}_\infty^T\boldsymbol{E}^T \geq 0$ **Q.E.D.**

Remark. For the proof of necessity, the use of GARI in the intermediate stage is needed. To avoid the former use of GARI, one might adopt the method proposed in [24] where the operator theory was used and thus the proofs given there were more involved.

We have given the necessary and sufficient conditions in the above theorem in terms of two coupled GAREs. It is possible to give an alternative set of necessary and sufficient conditions involving two uncoupled GAREs and a spectral radius condition, as given in [24] for systems in state-space model. The proof is quite standard and actually given in the necessity proof of the Theorem 4.7. This is summarized in the following statement.

Corollary 4.9. *Suppose that GARE $R_1(\boldsymbol{X}) = 0$ has an admissible solution $\boldsymbol{X}_\infty$ with $\boldsymbol{E}^T\boldsymbol{X}_\infty = \boldsymbol{X}_\infty^T\boldsymbol{E} \geq 0$. Then the condition (ii) of Theorem 4.7 holds if and only if the following conditions hold.*

1. *the GARE*

$$\begin{aligned}\mathbf{R}ic_2(\boldsymbol{Y}) &= (\boldsymbol{A} - \boldsymbol{B}_1\boldsymbol{D}_{21}^T\boldsymbol{R}_2^{-1}\boldsymbol{C}_2)\boldsymbol{Y} + \boldsymbol{Y}^T(\boldsymbol{A} - \boldsymbol{B}_1\boldsymbol{D}_{21}^T\boldsymbol{R}_2^{-1}\boldsymbol{C}_2)^T \\ &+\boldsymbol{Y}^T(\frac{1}{\gamma^2}\boldsymbol{C}_1^T\boldsymbol{C}_1 - \boldsymbol{C}_2^T\boldsymbol{R}_2^{-1}\boldsymbol{C}_2)\boldsymbol{Y} \\ &+\boldsymbol{B}_1(\boldsymbol{I} - \boldsymbol{D}_{21}^T\boldsymbol{R}_2^{-1}\boldsymbol{D}_{21})\boldsymbol{B}_1^T = 0, \\ &\boldsymbol{E}\boldsymbol{Y} = \boldsymbol{Y}^T\boldsymbol{E}^T,\end{aligned}$$

   has an admissible solution $\boldsymbol{Y}_\infty$ with $\boldsymbol{E}\boldsymbol{Y}_\infty = \boldsymbol{Y}_\infty^T\boldsymbol{E}^T \geq 0$,
2. *the spectral radius $\rho(\boldsymbol{Y}_\infty\boldsymbol{X}_\infty) < \gamma^2$.*

Moreover, when these conditions are satisfied, the matrices $\boldsymbol{X}_\infty$, $\boldsymbol{Y}_\infty$ and $\boldsymbol{Z}_\infty$ have the following relationship:

$$\boldsymbol{Z}_\infty = (\boldsymbol{I} - \frac{1}{\gamma^2}\boldsymbol{Y}_\infty\boldsymbol{X}_\infty)^{-1}\boldsymbol{Y}_\infty = \boldsymbol{Y}_\infty(\boldsymbol{I} - \frac{1}{\gamma^2}\boldsymbol{X}_\infty\boldsymbol{Y}_\infty)^{-1}.$$

□

In Theorem 4.7, we provide only one solution, namely the central controller. In what follows, we will give a complete derivation of state-space formulae for all controllers solving the standard $\mathcal{H}_\infty$ control problem for linear descriptor systems. The controllers obtained are parameterized by a fixed linear fractional transformation with a contractive, proper and stable free parameter. It may be considered to be the generalization of the parameterization of all $\mathcal{H}_\infty$ controllers given in [24] for systems in state-space model.

Theorem 4.10. *Suppose that Assumptions (A1) to (A8) are satisfied. Suppose also that Conditions (i) and (ii) of Theorem 4.7 hold. Then the set of all output feedback controllers that achieve closed-loop internal stability and render* $\|\boldsymbol{T_{zw}}\|_\infty < \gamma$ *can be parameterized as* $\boldsymbol{K} = \boldsymbol{\mathcal{LFT}}(\boldsymbol{M}, \boldsymbol{Q})$, *where*

$$\boldsymbol{M} = \left\{ \boldsymbol{E}, \left[\begin{array}{c|cc} \hat{\boldsymbol{A}}_0 & \hat{\boldsymbol{B}}_0 & \hat{\boldsymbol{B}}_1 \boldsymbol{R}_1^{-1} \boldsymbol{D}_{12}^T \\ \hline \hat{\boldsymbol{C}}_0 & 0 & \boldsymbol{R}_1^{-1} \boldsymbol{D}_{12}^T \\ -\boldsymbol{D}_{21}^T \boldsymbol{R}_2^{-1} (\boldsymbol{C}_2 + \boldsymbol{D}_{21} \boldsymbol{F}_1) & \boldsymbol{D}_{21}^T \boldsymbol{R}_2^{-1} & 0 \end{array} \right] \right\}, \tag{4.33}$$

and $\boldsymbol{Q} \in \mathcal{RH}^\infty$ *with* $\|\boldsymbol{Q}\|_\infty < \gamma$. $\hat{\boldsymbol{A}}_0$, $\hat{\boldsymbol{B}}_0$, $\hat{\boldsymbol{C}}_0$ *are defined as in Theorem 4.7,*

$$\begin{aligned} \hat{\boldsymbol{B}}_1 &\triangleq \boldsymbol{B}_2 - \boldsymbol{Z}_\infty^T \boldsymbol{F}_\infty^T, \\ \boldsymbol{F}_1 &\triangleq \frac{1}{\gamma^2} \boldsymbol{B}_1^T \boldsymbol{X}_\infty. \end{aligned} \tag{4.34}$$

□

Proof. We first make the following change of variabes:

$$\boldsymbol{q} \triangleq \boldsymbol{D}_{12}(\boldsymbol{u} - \boldsymbol{F}_\infty \boldsymbol{x}), \qquad \boldsymbol{r} \triangleq \boldsymbol{w} - \boldsymbol{F}_1 \boldsymbol{x}.$$

In terms of these new defined variables, the closed-loop system can be expressed as two interconnected subsystems below:

$$\hat{\boldsymbol{P}}(s) \triangleq \left\{ \begin{aligned} \boldsymbol{E}\dot{\boldsymbol{x}} &= (\boldsymbol{A} + \boldsymbol{B}_2 \boldsymbol{F}_\infty) x + \boldsymbol{B}_1 \boldsymbol{w} + \boldsymbol{B}_2 \boldsymbol{R}_1^{-1} \boldsymbol{D}_{12}^T \boldsymbol{q} \\ \boldsymbol{z} &= (\boldsymbol{C}_1 + \boldsymbol{D}_{12} \boldsymbol{F}_\infty) x + \boldsymbol{q} \\ \boldsymbol{r} &= -\boldsymbol{F}_1 \boldsymbol{x} + \boldsymbol{w}, \end{aligned} \right.$$

and

$$\hat{\boldsymbol{G}}(s) \triangleq \left\{ \begin{aligned} \boldsymbol{E}\dot{\boldsymbol{x}} &= (\boldsymbol{A} + \boldsymbol{B}_1 \boldsymbol{F}_1) \boldsymbol{x} + \boldsymbol{B}_1 \boldsymbol{r} + \boldsymbol{B}_2 \boldsymbol{u} \\ \boldsymbol{q} &= -\boldsymbol{D}_{12} \boldsymbol{F}_\infty \boldsymbol{x} + \boldsymbol{D}_{12} \boldsymbol{u} \\ \boldsymbol{y} &= (\boldsymbol{C}_2 + \boldsymbol{D}_{21} \boldsymbol{F}_1) \boldsymbol{x} + \boldsymbol{D}_{21} \boldsymbol{r}. \end{aligned} \right.$$

This is shown pictorically in Fig. 4.3.

It is straightforward but tedious to verify that $\boldsymbol{K}$ internally stabilizes $\boldsymbol{G}$ such that $\|\boldsymbol{T_{zw}}\|_\infty < \gamma$ if and only if $\boldsymbol{K}$ internally stabilizes $\hat{\boldsymbol{G}}$ such that $\|T_{\boldsymbol{qr}}\|_\infty < \gamma$. This is summarized in the following statement. Its proof only involves repeated use of the bounded real lemma for descriptor systems (Lemma 2.26), and is thus omitted.

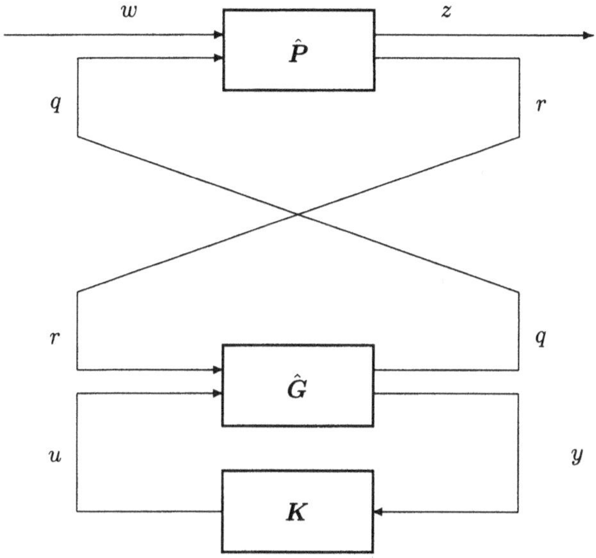

Fig. 4.3. Decomposition of the closed-loop system

Lemma 4.11. *Suppose that the GARE $R_1(\boldsymbol{X}) = 0$ has an admissible solution $\boldsymbol{X}_\infty$ with $\boldsymbol{E}^T\boldsymbol{X}_\infty = \boldsymbol{X}_\infty^T\boldsymbol{E} \geq 0$. Then $\boldsymbol{K}$ internally stabilizes $\boldsymbol{G}$ such that $\|\boldsymbol{T_{zw}}\|_\infty < \gamma$ if and only if $\boldsymbol{K}$ internally stabilizes $\hat{\boldsymbol{G}}$ such that $\|\boldsymbol{T_{qr}}\|_\infty < \gamma$.* □

Observe that $\boldsymbol{K}$ internally stabilizes $\hat{\boldsymbol{G}}$ and $\|T_{\boldsymbol{qr}}\|_\infty < \gamma$ if and only if $\boldsymbol{K}^T$ internally stabilizes $\hat{\boldsymbol{G}}^T$ and $\|\boldsymbol{T}_{\boldsymbol{qr}}^T\|_\infty < \gamma$. In order to find all controllers that internally stabilize $\hat{\boldsymbol{G}}^T$ and make $\|T_{\boldsymbol{qr}}^T\|_\infty < \gamma$, we introduce

$$
\begin{aligned}
\boldsymbol{G}_b = \hat{\boldsymbol{G}}^T &= \begin{cases} \boldsymbol{E}^T\dot{\tilde{\boldsymbol{x}}} = (\boldsymbol{A} + \frac{1}{\gamma^2}\boldsymbol{B}_1\boldsymbol{B}_1^T\boldsymbol{X}_\infty)^T\tilde{\boldsymbol{x}} - \boldsymbol{F}_\infty^T\boldsymbol{D}_{12}^T\tilde{\boldsymbol{r}} \\ \qquad +(\boldsymbol{C}_2 + \frac{1}{\gamma^2}\boldsymbol{D}_{21}\boldsymbol{B}_1^T\boldsymbol{X}_\infty)^T\tilde{\boldsymbol{u}} \\ \tilde{\boldsymbol{q}} = \boldsymbol{B}_1^T\tilde{\boldsymbol{x}} + \boldsymbol{D}_{21}^T\tilde{\boldsymbol{u}} \\ \tilde{\boldsymbol{y}} = \boldsymbol{B}_2^T\tilde{\boldsymbol{x}} + \boldsymbol{D}_{12}^T\tilde{\boldsymbol{r}}. \end{cases} \\
&\triangleq \begin{cases} \tilde{\boldsymbol{E}}\dot{\tilde{\boldsymbol{x}}} = \tilde{\boldsymbol{A}}\tilde{\boldsymbol{x}} + \tilde{\boldsymbol{B}}_1\tilde{\boldsymbol{r}} + \tilde{\boldsymbol{B}}_2\tilde{\boldsymbol{u}} \\ \tilde{\boldsymbol{q}} = \tilde{\boldsymbol{C}}_1\tilde{\boldsymbol{x}} + \tilde{\boldsymbol{D}}_{12}\tilde{\boldsymbol{u}} \\ \tilde{\boldsymbol{y}} = \tilde{\boldsymbol{C}}_2\tilde{\boldsymbol{x}} + \tilde{\boldsymbol{D}}_{21}\tilde{\boldsymbol{r}}. \end{cases}
\end{aligned}
$$

Now consider the GARE

$$
\begin{aligned}
\tilde{\mathbf{R}}ic_1(\tilde{X}) \triangleq = {} & (\tilde{\boldsymbol{A}} - \tilde{\boldsymbol{B}}_2\tilde{\boldsymbol{R}}_1^{-1}\tilde{\boldsymbol{D}}_{12}^T\tilde{\boldsymbol{C}}_1)^T\tilde{\boldsymbol{X}} + \tilde{\boldsymbol{X}}^T(\tilde{\boldsymbol{A}} - \tilde{\boldsymbol{B}}_2\tilde{\boldsymbol{R}}_1^{-1}\tilde{\boldsymbol{D}}_{12}^T\tilde{\boldsymbol{C}}_1) \\
& +\tilde{\boldsymbol{C}}_1^T(\boldsymbol{I} - \tilde{\boldsymbol{D}}_{12}\tilde{\boldsymbol{R}}_1^{-1}\tilde{\boldsymbol{D}}_{12}^T)\tilde{\boldsymbol{C}}_1 \\
& +\tilde{\boldsymbol{X}}^T(\tfrac{1}{\gamma^2}\tilde{\boldsymbol{B}}_1\tilde{\boldsymbol{B}}_1^T - \tilde{\boldsymbol{B}}_2\tilde{\boldsymbol{R}}_1^{-1}\tilde{\boldsymbol{B}}_2^T)\tilde{\boldsymbol{X}} = 0 \\
& \tilde{\boldsymbol{E}}^T\tilde{\boldsymbol{X}} = \tilde{\boldsymbol{X}}^T\tilde{\boldsymbol{E}},
\end{aligned}
$$

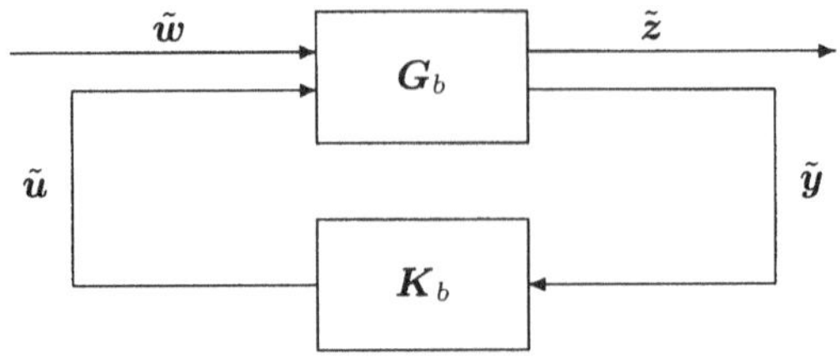

Fig. 4.4. Block diagram for $\mathcal{LFT}(G_b, K_b)$

where $\tilde{R}_1 \triangleq \tilde{D}_{12}^T \tilde{D}_{12}$. It is easy to verify that $\tilde{\mathbf{R}}ic_1(\tilde{X})$ is exactly $\mathbf{R}ic_2(\tilde{X})$, where $\mathbf{R}ic_2(\bullet)$ is as defined in Theorem 4.7. This shows that $\tilde{X} \triangleq Z_\infty$ is an admissible solution to the GARE $\tilde{\mathbf{R}}ic_1(X) = 0$. Denote by $\tilde{x}$, $\{\tilde{w}, \tilde{u}\}$ and $\{\tilde{z}, \tilde{y}\}$ the state, inputs and outputs of G_b, respectively, as shown in Fig. 4.4, and make another change of variables as follows:

$$\tilde{r} \triangleq \tilde{w} - \tilde{F}_1 \tilde{x}$$
$$\tilde{q} \triangleq \tilde{D}_{12}(\tilde{u} - \tilde{F}_\infty \tilde{x})$$

where

$$\tilde{F}_\infty \triangleq -\tilde{R}_1^{-1}(\tilde{B}_2^T \tilde{X} + \tilde{D}_{12}^T \tilde{C}_1)$$
$$\tilde{F}_1 \triangleq \frac{1}{\gamma^2} \tilde{B}_1^T \tilde{X}$$

Then the closed-loop system in Fig. 4.4 can be decomposed as two interconnected subsystems. This is shown pictorially in Fig. 4.5, where

$$P_b(s) = \left\{ \tilde{E}, \left[\begin{array}{c|cc} \tilde{A} + \tilde{B}_2 \tilde{F}_\infty & \tilde{B}_1 & \tilde{B}_2 \tilde{R}_1^{-1} \tilde{D}_{12}^T \\ \hline \tilde{C}_1 + \tilde{D}_{12} \tilde{F}_\infty & 0 & \tilde{D}_{12} \\ -\tilde{F}_1 & I & 0 \end{array} \right] \right\}$$

and

$$\hat{G}_b(s) = \left\{ \tilde{E}, \left[\begin{array}{c|cc} \tilde{A} + \tilde{B}_1 \tilde{F}_1 & \tilde{B}_1 & \tilde{B}_2 \\ \hline \tilde{F}_\infty & 0 & \tilde{D}_{12} \\ \tilde{C}_2 + \tilde{D}_{21} \tilde{F}_1 & \tilde{D}_{21} & 0 \end{array} \right] \right\} .$$

Again, applying Lemma 4.11 shows that K_b internally stabilizes G_b and $\|T_{qr}{}^T\| = \|T_{\tilde{z}\tilde{w}}\| < \gamma$ if and only if K_b internally stabilizes $\hat{G}_b$ and $\|T_{\tilde{q}\tilde{r}}\| < \gamma$.

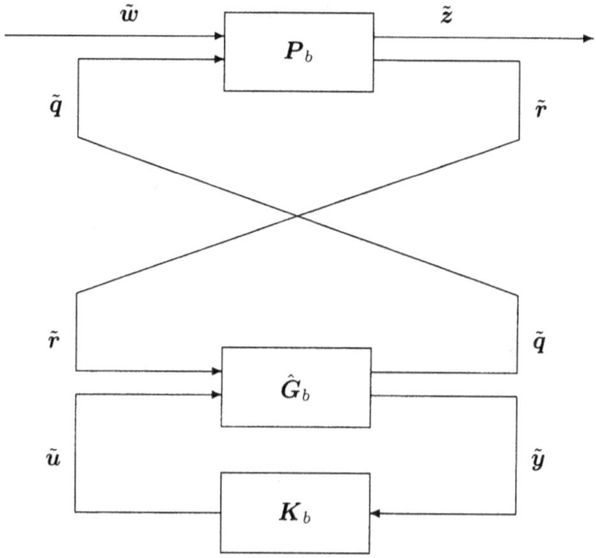

Fig. 4.5. Decomposition of $\mathcal{LFT}(G_b, K_b)$.

Now let $L \triangleq F_\infty^T = \hat{C}_0^T$, then simple algebra shows that $\{\tilde{E}, \tilde{A} + \tilde{B}_1\tilde{F}_1 + L(\tilde{C}_2 + \tilde{D}_{21}\tilde{F}_1)\} = \{E^T, (A + B_1F_1 + B_2F_\infty)^T\}$ which is admissible. Also note that if we let $F \triangleq -\hat{B}_0^T$, then it is straightforward to verify that $\{\tilde{E}, \tilde{A} + \tilde{B}_1\tilde{F}_1 + \tilde{B}_2F\} = \{E^T, (\tilde{A} - B_1D_{21}^T\tilde{C}_2)^T - (\tilde{C}_2^T\tilde{C}_2 - \frac{1}{\gamma^2}F_\infty^T F_\infty)Z_\infty\}$, which is also admissible since Z_∞ is an admissible solution to $R_2(Z) = 0$. Then, from Theorem 3.6, all internally stabilizing controllers for $\hat{G}_b$ can be parameterized as $K_b = \mathcal{LFT}(M_b, Q_b)$ with $Q_b \in \mathcal{RH}^\infty$, where

$$M_b = \left\{ \tilde{E}, \left[\begin{array}{c|cc} \tilde{A} + \tilde{B}_1\tilde{F}_1 - \tilde{B}_2\hat{B}_0^T + \hat{C}_0^T(\tilde{C}_2 + \tilde{D}_{21}\tilde{F}_1) & -\hat{C}_0^T & \tilde{B}_2 \\ \hline -\hat{B}_0^T & 0 & I \\ -(\tilde{C}_2 + \tilde{D}_{21}\tilde{F}_1) & I & 0 \end{array}\right] \right\}.$$

It is easy to see that $M_b(s)$ can be rewritten as

$$M_b = \left\{ E^T, \left[\begin{array}{c|cc} \hat{A}_0^T & -\hat{C}_0^T & (C_2 + D_{21}F_1)^T \\ \hline -\hat{B}_0^T & 0 & I \\ -\hat{B}_1^T & I & 0 \end{array}\right] \right\},$$

where $\hat{B}_1$ is as defined by (4.34).

Moreover, a little bit of algebra shows that

$$T_{\tilde{q}\tilde{r}} = \mathcal{LFT}(\hat{G}_b, \mathcal{LFT}(M_b, Q_b)) = D_{21}^T Q_b D_{12}^T.$$

So $\|T_{\tilde{q}\tilde{r}}\|_\infty < \gamma$ if and only if $\|D_{21}^T Q_b D_{12}^T\|_\infty < \gamma$. Thus all internally stabilizing controllers for $\hat{G}^T$ such that $\|T_{qr}^T\|_\infty < \gamma$ can be parameterized as $K_b = \mathcal{LFT}(M_b, Q_b)$, where $Q_b \in \mathcal{RH}^\infty$ with $\|D_{21}^T Q_b D_{12}^T\|_\infty < \gamma$. Consequently, all internally stabilizing controllers for $\hat{G}$ such that $\|T_{qr}\|_\infty < \gamma$ can be parameterized as $K = K_b^T = \mathcal{LFT}(M_b^T, \hat{Q})$ where $\hat{Q} \triangleq Q_b^T \in \mathcal{RH}^\infty$ with $\|D_{12}\hat{Q}D_{21}\|_\infty < \gamma$.

Now set $Q \triangleq D_{12}\hat{Q}D_{21}$. Then $Q \in \mathcal{RH}^\infty$ with $\|Q\|_\infty < \gamma$. It is easy to verify that $\mathcal{LFT}(M_b^T, \hat{Q}) = \mathcal{LFT}(M, Q)$. This completes the proof. **Q.E.D.**

Remark.

1. With $Q = 0$ we recover the controller (4.27) given in Theorem 4.7.
2. In the paper of Kawamoto and Katayama[57], a parameterization of all controllers solving the standard $\mathcal{H}_\infty$ control problem for descriptor systems has been provided. Nevertheless, only the result was presented in that paper, and no further details were discussed.

4.5 Converse Result-A Necessary Condition for the Nonlinear Problem

Lemma 4.2 establishes sufficient conditions for a DAE to have dissipative property. The conditions require the existence of a storage function $\mathsf{U}(x^e)$ that satisfies certain conditions. The conditions of this lemma cannot be checked directly on the data of the problem. Instead, one has to search for the auxiliary function. Faced with this searching problem, one question arises: Is there a function that satisfies the conditions of the lemma? In this section, we give necessary conditions for the existence of a storage function in terms of two Hamilton–Jacobi inequalities.

Suppose that the $\mathcal{H}_\infty$ control problem is solved by the output feedback controller Γ which has the following representation

$$\begin{aligned} \hat{E}\dot{\xi} &= \Phi(\xi, y), \\ u &= \Theta(\xi) \end{aligned} \tag{4.35}$$

and let U be a smooth function satisfying

$$\begin{aligned} \mathfrak{W}(x, \xi, w) = \begin{bmatrix} (U^x)^T & (U^\xi)^T \end{bmatrix} \begin{bmatrix} F(x, w, \Theta(\xi)) \\ \Phi(\xi, Y(x, w, \Theta(\xi))) \end{bmatrix} \\ + \|Z(x, w, \Theta(\xi))\|^2 - \gamma^2\|w\|^2 \le 0 \end{aligned} \tag{4.36}$$

for all $(\boldsymbol{x},\ \boldsymbol{\xi},\ \boldsymbol{w})$ in a neighborhood of (0, 0, 0). Here the function U satisfies $\dfrac{\partial \mathsf{U}}{\partial \boldsymbol{x}^e} = \boldsymbol{U}^T \boldsymbol{E}^e$ with $\boldsymbol{E}^{eT} \boldsymbol{U}_{\boldsymbol{x}^e} = \boldsymbol{U}_{\boldsymbol{x}^e}^T \boldsymbol{E}^e$, and $\boldsymbol{U} = \left[(\boldsymbol{U}^{\boldsymbol{x}})^T \quad (\boldsymbol{U}^{\boldsymbol{\xi}})^T \right]^T$ Consider the case that $\boldsymbol{\xi} \neq 0$ and $\boldsymbol{U}^{\boldsymbol{\xi}}(\boldsymbol{x},\ \boldsymbol{\xi}) \neq 0$. Since $\boldsymbol{\xi} \neq 0$, we have $\boldsymbol{\Phi} \neq 0$. Hence, from (4.36), we have

$$\inf_{\boldsymbol{\Phi},\, \boldsymbol{\Theta}} \max_{\boldsymbol{w}} \mathfrak{W}(\boldsymbol{x},\ \boldsymbol{\xi},\ \boldsymbol{w}) = -\infty,$$

because Inequality (4.36) contains a term linearly in $\boldsymbol{\Phi}$. Next, consider the case that $\boldsymbol{\xi} \neq 0$ but $\boldsymbol{U}^{\boldsymbol{\xi}}(\boldsymbol{x},\ \boldsymbol{\xi}) = 0$. Suppose that $\boldsymbol{U}_{\boldsymbol{\xi}}$ is nonsingular for every $(\boldsymbol{x},\ \boldsymbol{\xi})$ satisfying

$$\boldsymbol{U}^{\boldsymbol{\xi}}(\boldsymbol{x},\ \boldsymbol{\xi}) = 0. \tag{4.37}$$

Then, by the Implicit Function Theorem, Equation (4.37) has a differentiable solution $\boldsymbol{\xi} = \boldsymbol{\ell}(\boldsymbol{x})$ with $\boldsymbol{\ell}(0) = 0$. The previous statement is needed in the subsequent proof. We take it as a standing assumption.

(NA5) $\boldsymbol{U}^{\boldsymbol{\xi}}(\boldsymbol{x},\ \boldsymbol{\xi}) = 0$ if and only if $\boldsymbol{\xi} = \boldsymbol{\ell}(\boldsymbol{x})$ for some smooth function $\boldsymbol{\ell}(\bullet)$ with $\boldsymbol{\ell}(0) = 0$. Furthermore, $\boldsymbol{U}_{\boldsymbol{\xi}}(\boldsymbol{x},\ \boldsymbol{\xi})|_{\boldsymbol{\xi}=\boldsymbol{\ell}(\boldsymbol{x})}$ is nonsingular.

Define $\mathsf{V}(\boldsymbol{x}) \triangleq \mathsf{U}(\boldsymbol{x},\ \boldsymbol{\ell}(\boldsymbol{x}))$. Setting $\boldsymbol{\xi} = \boldsymbol{\ell}(\boldsymbol{x})$ in (4.8) yields

$$\boldsymbol{V}^T \boldsymbol{F}(\boldsymbol{x},\ \boldsymbol{w},\ \boldsymbol{\Theta}(\boldsymbol{\ell}(\boldsymbol{x}))) + \|\boldsymbol{Z}(\boldsymbol{x},\ \boldsymbol{w},\ \boldsymbol{\Theta}(\boldsymbol{\ell}(\boldsymbol{x})))\|^2 - \gamma^2 \|\boldsymbol{w}\|^2 \leq 0, \quad \forall \boldsymbol{w}. \tag{4.38}$$

This shows that

$$\inf_{0,\, \Theta(\boldsymbol{\xi})} \max_{\boldsymbol{w}} \mathfrak{W}(\boldsymbol{x},\ \boldsymbol{\xi},\ \boldsymbol{w}) = Y_1(\boldsymbol{x}),$$

where $\boldsymbol{\xi} = \boldsymbol{\ell}(\boldsymbol{x})$. Hence, the state feedback law $\boldsymbol{u} = \boldsymbol{\Theta}(\boldsymbol{\ell}(\boldsymbol{x}))$ solves the state feedback $\mathcal{H}_\infty$ control problem for $\mathcal{G}_P$. This shows that V is a solution of Y_1. A further necessary condition is obtained by restricting to the class of controller Γ which produces zero control input $\boldsymbol{u}$. Consider the Hamiltonian function $K_\gamma : \mathbb{R}^n \times \mathbb{R}^n \times \mathbb{R}^l \times \mathbb{R}^p \to \mathbb{R}$ defined as

$$K_\gamma(\boldsymbol{x}, \boldsymbol{p}, \boldsymbol{w}, \boldsymbol{y}) = \boldsymbol{p}^T \boldsymbol{F}(\boldsymbol{x}, \boldsymbol{w}, 0) - \boldsymbol{y}^T \boldsymbol{Y}(\boldsymbol{x}, \boldsymbol{w}, 0) + \|\boldsymbol{Z}(\boldsymbol{x}, \boldsymbol{w}, 0)\|^2 - \gamma^2 \|\boldsymbol{w}\|^2. \tag{4.39}$$

It is easy to verify that

$$\left(\frac{\partial^2 K_\gamma(\boldsymbol{x}, \boldsymbol{p}, \boldsymbol{w}, \boldsymbol{y})}{\partial \boldsymbol{w}^2} \right)_{(\boldsymbol{x}, \boldsymbol{p}, \boldsymbol{w}, \boldsymbol{y}) = (0,0,0,0)} = 2(\boldsymbol{D}_{11}^T \boldsymbol{D}_{11} - \gamma^2 \boldsymbol{I}).$$

This shows that there exists a smooth function $\hat{\boldsymbol{w}}(\boldsymbol{x}, \boldsymbol{p}, \boldsymbol{y})$ defined in a neighborhood of $(0, 0, 0)$ such that

$$\left(\frac{\partial K_\gamma(\boldsymbol{x}, \boldsymbol{p}, \boldsymbol{w}, \boldsymbol{y})}{\partial \boldsymbol{w}} \right)_{\boldsymbol{w} = \hat{\boldsymbol{w}}(\boldsymbol{x}, \boldsymbol{p}, \boldsymbol{y})} = 0, \qquad \hat{\boldsymbol{w}}(0, 0, 0) = 0.$$

Furthermore, it is also easy to check that

$$\left(\frac{\partial^2 K_\gamma(\boldsymbol{x},\boldsymbol{p},\hat{\boldsymbol{w}}(\boldsymbol{x},\boldsymbol{p},\boldsymbol{y}),\boldsymbol{y})}{\partial \boldsymbol{y}^2}\right)_{(\boldsymbol{x},\boldsymbol{p},\boldsymbol{y})=(0,0,0)} = \frac{1}{2}(\gamma^2 \boldsymbol{I} - \boldsymbol{D}_{11}^T\boldsymbol{D}_{11})^{-1}\boldsymbol{D}_{21}\boldsymbol{D}_{21}^T. \tag{4.40}$$

Thus, there exists a smooth function $\boldsymbol{y}^*(\boldsymbol{x},\boldsymbol{p})$ defined in a neighborhood of $(0,0)$ such that

$$\left(\frac{\partial K_\gamma(\boldsymbol{x},\boldsymbol{p},\hat{\boldsymbol{w}}(\boldsymbol{x},\boldsymbol{p},\boldsymbol{y}),\boldsymbol{y})}{\partial \boldsymbol{y}}\right)_{\boldsymbol{y}=\boldsymbol{y}^*(\boldsymbol{x},\boldsymbol{p})} = 0, \qquad \boldsymbol{y}^*(0,0)=0.$$

Set $\boldsymbol{w}^*(\boldsymbol{x},\boldsymbol{p}) = \hat{\boldsymbol{w}}(\boldsymbol{x},\boldsymbol{p},\boldsymbol{y}^*(\boldsymbol{x},\boldsymbol{p}))$. Then, we have

$$K_\gamma(\boldsymbol{x},\,\boldsymbol{p},\,\boldsymbol{w},\,\boldsymbol{y}) \le K_\gamma(\boldsymbol{x},\,\boldsymbol{p},\,\hat{\boldsymbol{w}}(\boldsymbol{x},\boldsymbol{p},\boldsymbol{y}),\,\boldsymbol{y}) \tag{4.41}$$

for all $(\boldsymbol{x},\boldsymbol{p},\boldsymbol{w},\boldsymbol{y})$ in a neighborhood of the origin and

$$K_\gamma(\boldsymbol{x},\,\boldsymbol{p},\,\hat{\boldsymbol{w}}(\boldsymbol{x},\boldsymbol{p},\boldsymbol{y}),\,\boldsymbol{y}) \ge K_\gamma(\boldsymbol{x},\,\boldsymbol{p},\,\boldsymbol{w}^*(\boldsymbol{x},\boldsymbol{p}),\,\boldsymbol{y}^*(\boldsymbol{x},\,\boldsymbol{p})) \tag{4.42}$$

for all $(\boldsymbol{x},\boldsymbol{p},\boldsymbol{y})$ in a neighborhood of the origin. We will show that it is necessary

$$K_\gamma(\boldsymbol{x},\boldsymbol{P}(\boldsymbol{x}),\boldsymbol{w}^*(\boldsymbol{x},\boldsymbol{P}(\boldsymbol{x})),\boldsymbol{y}^*(\boldsymbol{x},\boldsymbol{P}(\boldsymbol{x}))) \le 0 \tag{4.43}$$

for some storage function $\mathsf{P}(\boldsymbol{x})$ with $\dfrac{\partial \mathsf{P}}{\partial \boldsymbol{x}} = \boldsymbol{P}^T\boldsymbol{E}$. This is summarized in the following statement.

Theorem 4.12. *Consider System (4.1) and suppose Assumptions (NA1)-(NA5) hold. Suppose that the $\mathcal{H}_\infty$ control problem is solved by the output feedback controller (4.35). Suppose that there exists a smooth function $\mathsf{U}(\boldsymbol{x},\,\boldsymbol{\xi})$, which vanishes at $\boldsymbol{E}^e\boldsymbol{x}^e = 0$ and positive elsewhere with $(\boldsymbol{E}^e)^T\boldsymbol{U}_{\boldsymbol{x}^e} = \boldsymbol{U}_{\boldsymbol{x}^e}^T\boldsymbol{E}^e$, satisfying (4.8) for all $(\boldsymbol{x},\,\boldsymbol{\xi},\,\boldsymbol{w})$ in a neighborhood of $(0,\,0,\,0)$. Then the Hamilton–Jacobi inequalities*

$$Y_1(\boldsymbol{x}) \le 0$$

and

$$K_\gamma(\boldsymbol{x},\boldsymbol{P}(\boldsymbol{x}),\boldsymbol{w}^*(\boldsymbol{x},\boldsymbol{P}(\boldsymbol{x})),\boldsymbol{y}^*(\boldsymbol{x},\boldsymbol{P}(\boldsymbol{x}))) \le 0$$

have solutions $\mathsf{V}(\boldsymbol{x})$ and, respectively, $\mathsf{P}(\boldsymbol{x})$ (with $\dfrac{\partial P}{\partial \boldsymbol{x}} = \boldsymbol{P}^T\boldsymbol{E}$) given by

$$\mathsf{V}(\boldsymbol{x}) = \mathsf{U}(\boldsymbol{x},\,\boldsymbol{\ell}(\boldsymbol{x})) \ge 0$$

and, respectively,

$$\mathsf{P}(\boldsymbol{x}) = \mathsf{U}(\boldsymbol{x},\,0) \ge 0.$$

Furthermore,

$$\mathsf{Q}(\boldsymbol{x}) \triangleq \mathsf{P}(\boldsymbol{x}) - \mathsf{V}(\boldsymbol{x}) \ge 0$$

□

Proof. It is obvious that $\mathsf{V}(\boldsymbol{x})$ is a solution satisfying $Y_1(\boldsymbol{x}) \leq 0$ from our previous observation. It is claimed that $\mathsf{P}(\boldsymbol{x})$ is a solution of Inequality (4.43). To see this, setting $\boldsymbol{\xi} = 0$ in (4.8) yields

$$\boldsymbol{P}^T \boldsymbol{F}(\boldsymbol{x},\, \boldsymbol{w},\, 0) + (\boldsymbol{U}^{\boldsymbol{\xi}})^T(\boldsymbol{x}, 0)\boldsymbol{\Phi}(0, \boldsymbol{Y}(\boldsymbol{x}, \boldsymbol{w}, 0)) + \|\boldsymbol{Z}(\boldsymbol{x}, \boldsymbol{w}, 0)\|^2 - \gamma^2\|\boldsymbol{w}\|^2 \leq 0. \tag{4.44}$$

Let $\boldsymbol{U}^{\boldsymbol{\xi}}(\boldsymbol{x},\, 0)\boldsymbol{\Phi}(\boldsymbol{x},\, \boldsymbol{y}) = \boldsymbol{\Pi}^T(\boldsymbol{x},\, \boldsymbol{y})\boldsymbol{y}$, where $\boldsymbol{\Pi}(\boldsymbol{x},\, \boldsymbol{y})$ is a vector of smooth functions. This can always be done because the function $\boldsymbol{\Phi}(0,\, \boldsymbol{y})$ vanishes at $\boldsymbol{y} = 0$. Use $\boldsymbol{\Pi}^T(\boldsymbol{x},\, \boldsymbol{y})\boldsymbol{y}$ and choose $\boldsymbol{w} = \hat{\boldsymbol{w}}(\boldsymbol{x}, \boldsymbol{P}, \boldsymbol{y})$ in (4.8) to obtain

$$K_\gamma(\boldsymbol{x},\, \boldsymbol{P},\, \hat{\boldsymbol{w}}(\boldsymbol{x}, \boldsymbol{P}, \boldsymbol{y}),\, \boldsymbol{\Pi}(\boldsymbol{x},\, \boldsymbol{Y}((\boldsymbol{x},\, \hat{\boldsymbol{w}}(\boldsymbol{x}, \boldsymbol{P}, \boldsymbol{y})))) \leq 0. \tag{4.45}$$

Observe that the Hessian matrix of

$$\boldsymbol{y} - \boldsymbol{\Pi}(\boldsymbol{x},\, \boldsymbol{Y}((\boldsymbol{x},\, \hat{\boldsymbol{w}}(\boldsymbol{x}, \boldsymbol{P}, \boldsymbol{y})))$$

is nonsingular (from Assumption NA4 and Equation (4.40)). Hence, by the Implicit Function Theorem, there exists a unique solution, denoted by $\hat{\boldsymbol{y}}(\boldsymbol{x})$, satisfying

$$\boldsymbol{y}(\boldsymbol{x}) - \boldsymbol{\Pi}(\boldsymbol{x},\, \boldsymbol{Y}((\boldsymbol{x},\, \hat{\boldsymbol{w}}(\boldsymbol{x}, \boldsymbol{P}, \boldsymbol{y}(\boldsymbol{x}))) = 0, \qquad \boldsymbol{y}(0) = 0.$$

Set $\boldsymbol{y} = \hat{\boldsymbol{y}}(\boldsymbol{x})$ in (4.45) to obtain

$$K_\gamma(\boldsymbol{x},\, \boldsymbol{P}(\boldsymbol{x}),\, \hat{\boldsymbol{w}}(\boldsymbol{x},\, \boldsymbol{P}(\boldsymbol{x}),\, \hat{\boldsymbol{y}}(\boldsymbol{x})),\, \hat{\boldsymbol{y}}(\boldsymbol{x})) \leq 0.$$

This shows that

$$K_\gamma(\boldsymbol{x},\, \boldsymbol{P}(\boldsymbol{x}),\, \boldsymbol{w}^*(\boldsymbol{x}, \boldsymbol{P}(\boldsymbol{x})),\, \boldsymbol{y}^*(\boldsymbol{x},\, \boldsymbol{P}(\boldsymbol{x}))) \leq 0$$

from (4.42). In order to complete the proof, we have to show that $\mathsf{Q}(\boldsymbol{x}) \triangleq \mathsf{P}(\boldsymbol{x}) - \mathsf{V}(\boldsymbol{x}) \geq 0$. Note that the function $\mathsf{U}(\boldsymbol{x},\, \boldsymbol{\xi})$ has the following Taylor series expansion

$$\begin{aligned}\mathsf{U}(\boldsymbol{x},\, \boldsymbol{\xi}) &= \mathsf{U}(0,0) + \frac{\partial \mathsf{U}}{\partial \boldsymbol{x}}(0,0)\boldsymbol{x} + \frac{\partial \mathsf{U}}{\partial \boldsymbol{\xi}}(0,0)\boldsymbol{\xi} \\ &+ \begin{bmatrix}\boldsymbol{x}\\ \boldsymbol{\xi}\end{bmatrix}^T \begin{bmatrix} \frac{1}{2}\frac{\partial^2 \mathsf{U}}{\partial \boldsymbol{x}^2}(0,0) & \frac{1}{2}\frac{\partial^2 \mathsf{U}}{\partial \boldsymbol{x}\partial \boldsymbol{\xi}}(0,0) \\ \left(\frac{1}{2}\frac{\partial^2 \mathsf{U}}{\partial \boldsymbol{x}\partial \boldsymbol{\xi}}\right)^T(0,0) & \frac{1}{2}\frac{\partial^2 \mathsf{U}}{\partial \boldsymbol{\xi}^2}(0,0)\end{bmatrix} \begin{bmatrix}\boldsymbol{x}\\ \boldsymbol{\xi}\end{bmatrix} \\ &+ h.o.t.,\end{aligned} \tag{4.46}$$

where "*h.o.t.*" means high order terms. Moreover, we note that

$$\frac{\partial \mathsf{U}}{\partial \boldsymbol{x}}(\boldsymbol{x}, \boldsymbol{\xi}) = \boldsymbol{x}^T \frac{\partial^2 \mathsf{U}}{\partial \boldsymbol{x}^2}(0,0) + \boldsymbol{\xi}^T \left(\frac{\partial^2 \mathsf{U}}{\partial \boldsymbol{x}\partial \boldsymbol{\xi}}\right)^T(0,0) + (h.o.t.)_{\boldsymbol{x}}, \tag{4.47}$$

and

$$\frac{\partial^2 \mathrm{U}}{\partial \boldsymbol{x}^2}(\boldsymbol{x}, \boldsymbol{\xi}) = \frac{\partial^2 \mathrm{U}}{\partial \boldsymbol{x}^2}(0,0) + (h.o.t.)_{\boldsymbol{x}\boldsymbol{x}}. \tag{4.48}$$

Let $\boldsymbol{\xi} = \boldsymbol{\ell}(\boldsymbol{x})$ in (4.47) to obtain

$$\frac{\partial \mathrm{U}}{\partial \boldsymbol{x}}(\boldsymbol{x}, \boldsymbol{\ell}(\boldsymbol{x})) = \boldsymbol{x}^T \frac{\partial^2 \mathrm{U}}{\partial \boldsymbol{x}^2}(0,0) + \boldsymbol{\ell}^T(\boldsymbol{x}) \left(\frac{\partial^2 \mathrm{U}}{\partial \boldsymbol{x} \partial \boldsymbol{\xi}} \right)^T (0,0) + (h.o.t.)_{\boldsymbol{x}}|_{\boldsymbol{\xi}=\boldsymbol{\ell}(\boldsymbol{x})}.$$

Taking partial derivative with respect to $\boldsymbol{x}$ yields

$$\frac{\partial^2 \mathrm{U}}{\partial \boldsymbol{x}^2}(\boldsymbol{x}, \boldsymbol{\ell}(\boldsymbol{x})) = \frac{\partial^2 \mathrm{U}}{\partial \boldsymbol{x}^2}(0,0) + \frac{\partial^2 \mathrm{U}}{\partial \boldsymbol{x} \partial \boldsymbol{\xi}}(0,0)\boldsymbol{\ell}_{\boldsymbol{x}}(\boldsymbol{x}) + (h.o.t.)_{\boldsymbol{x}\boldsymbol{x}}|_{\boldsymbol{\xi}=\boldsymbol{\ell}(\boldsymbol{x})}.$$

Note that $\frac{\partial^2 \mathrm{U}}{\partial \boldsymbol{x}^2}(\boldsymbol{x}, \boldsymbol{\ell}(\boldsymbol{x})) = \frac{\partial^2 \mathrm{V}}{\partial \boldsymbol{x}^2}(\boldsymbol{x})$. Set $\boldsymbol{x} = 0$ in the previous relation to get

$$\frac{\partial^2 \mathrm{U}}{\partial \boldsymbol{x}^2}(0, \boldsymbol{\ell}(0)) = \frac{\partial^2 \mathrm{V}}{\partial \boldsymbol{x}^2}(0) - \frac{\partial^2 \mathrm{U}}{\partial \boldsymbol{x}^2}(0,0) + \frac{\partial^2 \mathrm{U}}{\partial \boldsymbol{x} \partial \boldsymbol{\xi}}(0,0)\boldsymbol{\ell}_{\boldsymbol{x}}(0).$$

Now, consider the following identity

$$\frac{\partial \mathrm{U}}{\partial \boldsymbol{\xi}}(\boldsymbol{x}, \boldsymbol{\xi}) = \boldsymbol{x}^T \frac{\partial^2 \mathrm{U}}{\partial \boldsymbol{x} \partial \boldsymbol{\xi}}(0,0) + \boldsymbol{\xi}^T \frac{\partial^2 \mathrm{U}}{\partial \boldsymbol{\xi}^2}(0,0) + (h.o.t.)_{\boldsymbol{\xi}}.$$

Set $\boldsymbol{\xi} = \boldsymbol{\ell}(\boldsymbol{x})$ in the previous identity to get

$$0 = \frac{\partial \mathrm{U}}{\partial \boldsymbol{\xi}}(\boldsymbol{x}, \boldsymbol{\ell}(\boldsymbol{x})) = \boldsymbol{x}^T \frac{\partial^2 \mathrm{U}}{\partial \boldsymbol{x} \partial \boldsymbol{\xi}}(0,0) + \boldsymbol{\ell}^T(\boldsymbol{x}) \frac{\partial^2 \mathrm{U}}{\partial \boldsymbol{\xi}^2}(0,0) + (h.o.t.)_{\boldsymbol{\xi}}|_{\boldsymbol{\xi}=\boldsymbol{\ell}(\boldsymbol{x})}.$$

Then, take the partial derivative of the above equation with respect to $\boldsymbol{x}$ to get

$$0 = \frac{\partial^2 \mathrm{U}}{\partial \boldsymbol{\xi} \partial \boldsymbol{x}}(\boldsymbol{x}, \boldsymbol{\ell}(\boldsymbol{x})) = \frac{\partial^2 \mathrm{U}^T}{\partial \boldsymbol{x} \partial \boldsymbol{\xi}}(0,0) + \frac{\partial^2 \mathrm{U}}{\partial \boldsymbol{\xi}^2}(0,0)\boldsymbol{\ell}_{\boldsymbol{x}}(\boldsymbol{x}) + (h.o.t.)_{\boldsymbol{x}\boldsymbol{\xi}}|_{\boldsymbol{\xi}=\boldsymbol{\ell}(\boldsymbol{x})}.$$

Setting $\boldsymbol{x} = 0$ into the above equation yields

$$0 = \frac{\partial^2 \mathrm{U}}{\partial \boldsymbol{\xi} \partial \boldsymbol{x}}(0,0) = \frac{\partial^2 \mathrm{U}^T}{\partial \boldsymbol{x} \partial \boldsymbol{\xi}}(0,0) + \frac{\partial^2 \mathrm{U}}{\partial \boldsymbol{\xi}^2}(0,0)\boldsymbol{\ell}_{\boldsymbol{x}}(0).$$

Next, observe that

$$\begin{aligned} \frac{\partial^2 \mathrm{Q}}{\partial \boldsymbol{x}^2}(0) &= \frac{\partial^2 \Pi}{\partial \boldsymbol{x}^2}(0) - \frac{\partial^2 \mathrm{V}}{\partial \boldsymbol{x}^2}(0) \\ &= \frac{\partial^2 \mathrm{U}}{\partial \boldsymbol{x}^2}(0,0) - (\frac{\partial^2 \mathrm{U}}{\partial \boldsymbol{x}^2}(0,0) + \frac{\partial^2 \mathrm{U}}{\partial \boldsymbol{x} \boldsymbol{\xi}}(0,0)\boldsymbol{\ell}_{\boldsymbol{x}}(0)) \qquad (4.49) \\ &= -\frac{\partial^2 \mathrm{U}}{\partial \boldsymbol{x} \boldsymbol{\xi}}(0,0)\boldsymbol{\ell}_{\boldsymbol{x}}(0) \\ &= \boldsymbol{\ell}_{\boldsymbol{x}}^T(0) \frac{\partial^2 \mathrm{U}}{\partial \boldsymbol{\xi}^2}(0,0)\boldsymbol{\ell}_{\boldsymbol{x}}(0) \geq 0. \end{aligned}$$

The last inequality holds by assumption NA5. This concludes that $\mathtt{Q}(\boldsymbol{x}) \geq 0$ by noting that $\mathtt{Q}(\boldsymbol{x})$ has the following Taylor series expansion.

$$\mathtt{Q}(\boldsymbol{x}) = \mathtt{Q}(0) + \frac{\partial \mathtt{Q}}{\partial \boldsymbol{x}}(0)\boldsymbol{x} + \frac{1}{2}\boldsymbol{x}^T \frac{\partial^2 \mathtt{Q}}{\partial \boldsymbol{x}^2}(0)\boldsymbol{x} + h.o.t. \geq 0.$$

It is nonnegative around the origin because it vanishes at the origin together with its first order derivative, and its second order derivative is positive by equation (4.49). This completes the proof. **Q.E.D.**

Notes and References

1. The contents of this chapter are largely based on Wang *et al.*[116], [117], and Yung *et al.*[129].
2. For general information concerning the $\mathcal{H}_\infty$ control problem for linear systems, one can consult, for example, Francis[30], Green and Limebeer[37], and Zhou[140]. The state-space solution to the standard linear suboptimal $\mathcal{H}_\infty$ control problem is due to Doyle *et al.*[24]. See also e.g. Petersen *et al.*[93], Sampei *et al.*[99], and Zhou[139]. The complete characterization of optimal controllers for the general setup can be found in Glover *et al.*[36].
3. Solutions to $\mathcal{H}_\infty$ control problem for linear descriptor systems were first given in Takaba *et al.* [111]. They dealt with the problem using a J-spectral factorization approach and only the sufficient condition for solutions to exist was given. Recently, Masubuchi *et al.*[82] have considered a similar problem by using a matrix inequalities approach. A parameterization of all the output feedback $\mathcal{H}_\infty$ controllers for linear descriptor systems can be found in [57]. See also [129]. For the differential game approach to linear $\mathcal{H}_\infty$ control theory see the excellent text book [8] by Basar and Bernhard.
4. The solution to the nonlinear suboptimal $\mathcal{H}_\infty$ control problem was given in van der Schaft[104], Isidori and Astolfi[47][48], Ball *et al.*[6]. $\mathcal{H}_\infty$ control for general nonlinear systems (not necessarily affine in the inputs and disturbances) was treated in Isidori and Kang [49], Ball *et al.*[6], Van der Schaft[105], and Yung *et al.*[135]. The parameterization of a class of $\mathcal{H}_\infty$ controllers for nonlinear systems can be found in, for example, [74], [130], [134], [135].

5

Balanced Realization

In this chapter, we introduce balancing for stable nonlinear descriptor systems. The approach used here is an extension of balancing for conventional systems. The classical ideas is extended naturally to the descriptor systems. That is, we will consider the past input and output energy of the descriptor systems. More precisely, the controllability function is defined as the past input energy and the observability function as the future output energy. Under certain sufficient conditions, these functions may be transformed into a form that gives us a measure of importance of a state component. A nonlinear descriptor system with its controllability and observability functions in this form is called a balanced representation.

5.1 Balanced Realization

Consider a nonlinear descriptor system of the form

$$\begin{aligned} \dot{\boldsymbol{x}}_1 &= \boldsymbol{f}_1(\boldsymbol{x}_1, \boldsymbol{x}_2) + \boldsymbol{g}_1(\boldsymbol{x}_1, \boldsymbol{x}_2)\boldsymbol{u}, \quad \boldsymbol{u} \in \mathbb{U} \subset \mathrm{I\!R}^m \\ 0 &= \boldsymbol{f}_2(\boldsymbol{x}_1, \boldsymbol{x}_2) + \boldsymbol{g}_2(\boldsymbol{x}_1, \boldsymbol{x}_2)\boldsymbol{u} \\ \boldsymbol{y} &= \boldsymbol{h}(\boldsymbol{x}_1, \boldsymbol{x}_2), \boldsymbol{y} \in \mathbb{Y} \subset \mathrm{I\!R}^p, \end{aligned} \tag{5.1}$$

or in a more compact form

$$\begin{aligned} \boldsymbol{E}\dot{\boldsymbol{x}} &= \boldsymbol{f}(\boldsymbol{x}) + \boldsymbol{g}(\boldsymbol{x})\boldsymbol{u} \\ \boldsymbol{y} &= \boldsymbol{h}(\boldsymbol{x}), \end{aligned} \tag{5.2}$$

where $\boldsymbol{x} = (\boldsymbol{x}_1, \boldsymbol{x}_2)$ are local coordinates for a state-space manifold $\mathcal{X}$. Note that system (5.1) is linear in $\boldsymbol{u}$. The controllability and observability functions, L_c and L_o, respectively, of the system (5.2) are defined as follows.

Definition 5.1. *Consider (5.2). The controllability function of (5.2) at $\boldsymbol{x}_0 \in \Im m\boldsymbol{E}$ is the minimum amount of control energy required to reach the state $\boldsymbol{x}_0$, which can be defined as*

$$\mathrm{L}_c(\boldsymbol{x}_0) \triangleq \min_{\substack{\boldsymbol{u}\in\mathcal{L}_2(-\infty,0)\\ \boldsymbol{x}(-\infty)=0,\ \boldsymbol{x}(0)=\boldsymbol{x}_0}} \frac{1}{2}\int_{\infty}^{0} \|\boldsymbol{u}(t)\|^2 \mathrm{d}t. \tag{5.3}$$

On the other hand, the observability function of (5.2) at $\boldsymbol{x}_0 \in \Im m\boldsymbol{E}$ is the amount of output energy generated by the initial state $\boldsymbol{x}_0$ with $\boldsymbol{u}(t) \equiv 0$.

$$\mathrm{L}_o(\boldsymbol{x}_0) \triangleq \frac{1}{2}\int_0^{\infty} \|\boldsymbol{y}(t)\|^2 \mathrm{d}t, \quad \boldsymbol{x}(0) = \boldsymbol{x}_0 \in \Im m\boldsymbol{E}, \quad \boldsymbol{u}(t) \equiv 0,\ 0 \le t \le \infty. \tag{5.4}$$

□

Note that the previous definition does not assume *a priori* asymptotical stability and index one property of the system (5.2). Hence these functions defined above do not necessarily exist, i.e., are not necessarily finite. For example, if $\boldsymbol{x}_0$ cannot be reached from 0 backward in time, then by convention (See [102]) L_c is defined to be infinite. On the other hand, if the system has higher index, then the output $\boldsymbol{y}$, in general, will involve differentiation of the input $\boldsymbol{u}$ and $\boldsymbol{x}$. In this case, L_o may be infinite even though the system is asymptotically stable. We shall give some conditions under which L_c and L_o exist. Those conditions are not only sufficient, the reverse implications are also true. This is summarized in the following statement.

Theorem 5.2. *Consider (5.2). Assume that the system $\boldsymbol{E}\dot{\boldsymbol{x}} = \boldsymbol{f}(\boldsymbol{x})$ is asymptotically stable with index one. Suppose that there exists a smooth function $\tilde{\mathrm{L}}_o(\boldsymbol{x})$ with $\dfrac{\partial\tilde{\mathrm{L}}_o}{\partial\boldsymbol{x}}(\boldsymbol{x}) = \tilde{\boldsymbol{L}}_o^T(\boldsymbol{x})\boldsymbol{E}$ satisfying the following equation*

$$\tilde{\boldsymbol{L}}_o^T(\boldsymbol{x})\boldsymbol{f}(\boldsymbol{x}) + \frac{1}{2}\boldsymbol{h}^T(\boldsymbol{x})\boldsymbol{h}(\boldsymbol{x}) = 0, \quad \tilde{\mathrm{L}}_o(0) = 0 \tag{5.5}$$

for all $\boldsymbol{x}_0 \in \mathcal{X}$. Then L_o exists and the function restricted to $\Im m\boldsymbol{E}$ is the unique smooth solution of (5.5). Furthermore, suppose that there exists a smooth function $\tilde{\mathrm{L}}_c(\boldsymbol{x})$ with $\dfrac{\partial\tilde{\mathrm{L}}_c}{\partial\boldsymbol{x}}(\boldsymbol{x}) = \tilde{\boldsymbol{L}}_c(\boldsymbol{x})\boldsymbol{E}$ satisfying the following equation

$$\tilde{\boldsymbol{L}}_c(\boldsymbol{x})\boldsymbol{f}(\boldsymbol{x}) + \frac{1}{2}\tilde{\boldsymbol{L}}_c(\boldsymbol{x})\boldsymbol{g}(\boldsymbol{x})\boldsymbol{g}^T(\boldsymbol{x})\tilde{\boldsymbol{L}}_c^T(\boldsymbol{x}) = 0, \quad \tilde{\mathrm{L}}(0) = 0 \tag{5.6}$$

for all $\boldsymbol{x}_0 \in \mathcal{X}$ such that the system

$$\boldsymbol{E}\dot{\boldsymbol{x}} = -\left(\boldsymbol{f}(\boldsymbol{x}) + \boldsymbol{g}(\boldsymbol{x})\boldsymbol{g}^T(\boldsymbol{x})\tilde{\boldsymbol{L}}_c^T(\boldsymbol{x})\right)$$

is asymptotically stable with index one on $\mathcal{X}$. Then L_c exists and the function restricted to $\Im m\boldsymbol{E}$ is the unique smooth solution of (5.6) such that $\boldsymbol{E}\dot{\boldsymbol{x}} = -\left(\boldsymbol{f}(\boldsymbol{x}) + \boldsymbol{g}(\boldsymbol{x})\boldsymbol{g}^T(\boldsymbol{x})\boldsymbol{L}_c^T(\boldsymbol{x})\right)$ is asymptotically stable with index one.

Conversely, assume that the system $\boldsymbol{E}\dot{\boldsymbol{x}} = \boldsymbol{f}(\boldsymbol{x})$ is asymptotically stable with index one on a neighborhood $\mathcal{X}$ of the origin. Suppose that $\mathrm{L}_o(\boldsymbol{x})$ exists and *is smooth on* $\Im m\boldsymbol{E} \subset \mathcal{X}$*. Then for all* $\boldsymbol{x} \in \Im m\boldsymbol{E}$*,* $\mathrm{L}_o(\boldsymbol{x})$ *is the unique smooth solution of*

$$\boldsymbol{L}_o^T(\boldsymbol{x})\boldsymbol{f}(\boldsymbol{x}) + \frac{1}{2}\boldsymbol{h}^T(\boldsymbol{x})\boldsymbol{h}(\boldsymbol{x}) = 0, \quad \mathrm{L}_o(0) = 0. \tag{5.7}$$

It is unique in the sense that, if $\tilde{\mathrm{L}}_o(\boldsymbol{x})$ *is any other solution, then* $\mathrm{L}_o(\boldsymbol{x}) = \tilde{\mathrm{L}}_o(x)$ *for all* $\boldsymbol{x} \in \Im m\boldsymbol{E}$*. Moreover, suppose that* $\mathrm{L}_c(\boldsymbol{x})$ *exists and is smooth on* $\Im m\boldsymbol{E} \subset \mathcal{X}$*. Then for all* $\boldsymbol{x} \in \Im m\boldsymbol{E}$*,* $\mathrm{L}_c(\boldsymbol{x})$ *is the unique solution of*

$$\boldsymbol{L}_c(\boldsymbol{x})\boldsymbol{f}(\boldsymbol{x}) + \frac{1}{2}\boldsymbol{L}_c(\boldsymbol{x})\boldsymbol{g}(\boldsymbol{x})\boldsymbol{g}^T(\boldsymbol{x})\boldsymbol{L}_c^T(\boldsymbol{x}) = 0, \quad \mathrm{L}(0) = 0 \tag{5.8}$$

such that $\boldsymbol{E}\dot{\boldsymbol{x}} = -\left(\boldsymbol{f}(\boldsymbol{x}) + \boldsymbol{g}(\boldsymbol{x})\boldsymbol{g}^T(\boldsymbol{x})\boldsymbol{L}_c^T(\boldsymbol{x})\right)$ *is asymptotically stable with index one on* $\mathcal{X}$*.* □

Proof.

Sufficiency. Suppose that (5.5) has a smooth solution $\tilde{\mathrm{L}}_o(\boldsymbol{x})$ on $\mathcal{X}$. Hence, by definition we have, from (5.5), the following identities.

$$\begin{aligned}
\mathrm{L}_o(\boldsymbol{x}_0) &= \frac{1}{2}\int_0^\infty \|\boldsymbol{y}(t)\|^2 \mathrm{d}t = \frac{1}{2}\int_0^\infty \boldsymbol{h}^T(\boldsymbol{x})\boldsymbol{h}(\boldsymbol{x})\mathrm{d}t \\
&= -\int_0^\infty \tilde{\boldsymbol{L}}_o^T(\boldsymbol{x}(t))\boldsymbol{f}(\boldsymbol{x}(t))\mathrm{d}t = -\int_0^\infty \frac{\mathrm{d}}{\mathrm{d}t}\tilde{\mathrm{L}}_o(\boldsymbol{x}(t))\mathrm{d}t \\
&= -\tilde{\mathrm{L}}_o(\boldsymbol{x}(\infty)) + \tilde{\mathrm{L}}_o(\boldsymbol{x}(0)) = \tilde{\mathrm{L}}_o(\boldsymbol{x}_0), \qquad \forall \boldsymbol{x}_0 \in \Im m\boldsymbol{E}.
\end{aligned}$$

The last equality holds because of the asymptotical stability of the system. This shows that $\mathrm{L}_o(\boldsymbol{x}_0)$ is the unique solution of (5.5) for all $\boldsymbol{x}_0 \in \Im m\boldsymbol{E}$. Next, assume that Equation (5.6) has a smooth solution $\tilde{\mathrm{L}}_c(\boldsymbol{x})$ on $\mathcal{X}$ with $\frac{\partial}{\partial \boldsymbol{x}}\tilde{\mathrm{L}}_c(\boldsymbol{x}) = \tilde{\boldsymbol{L}}_c(\boldsymbol{x})$. Then, by constructing an input $\boldsymbol{u}(t)$ such that $\boldsymbol{x}(0) = \boldsymbol{x}_0 \in \Im m\boldsymbol{E}$ and $\boldsymbol{x}(-\infty) = 0$ (See Definition 5.1), we have

$$\begin{aligned}
\frac{\mathrm{d}}{\mathrm{d}t}\tilde{\mathrm{L}}_c(\boldsymbol{x}) &= \tilde{\boldsymbol{L}}_c(\boldsymbol{x})\boldsymbol{E}\dot{\boldsymbol{x}} = \tilde{\boldsymbol{L}}_c(\boldsymbol{x})\left(\boldsymbol{f}(\boldsymbol{x}) + \boldsymbol{g}(\boldsymbol{x})\boldsymbol{u}\right) \\
&= -\frac{1}{2}\tilde{\boldsymbol{L}}_c(\boldsymbol{x})\boldsymbol{g}(\boldsymbol{x})\boldsymbol{g}^T(\boldsymbol{x})\tilde{\boldsymbol{L}}_c^T(\boldsymbol{x}) + \tilde{\boldsymbol{L}}_c(\boldsymbol{x})\boldsymbol{g}(\boldsymbol{x})\boldsymbol{u} \\
&= \frac{1}{2}\|\boldsymbol{u}\|^2 - \frac{1}{2}\|\boldsymbol{u} - \boldsymbol{g}^T(\boldsymbol{x})\tilde{\boldsymbol{L}}_c^T(\boldsymbol{x})\|^2.
\end{aligned}$$

With the above identity in hand, we then have

$$\begin{aligned}
\tilde{\mathrm{L}}_c(\boldsymbol{x}_0) &- \int_\infty^0 \frac{\mathrm{d}}{\mathrm{d}t}\tilde{\mathrm{L}}_c(\boldsymbol{x}(t))\mathrm{d}t \\
&= \frac{1}{2}\int_\infty^0 \|\boldsymbol{u}(t)\|^2 \mathrm{d}t - \frac{1}{2}\int_\infty^0 \|\boldsymbol{u}(t) - \boldsymbol{g}^T(\boldsymbol{x})\tilde{\boldsymbol{L}}_c^T(\boldsymbol{x})\|^2 \mathrm{d}t \\
&\le \frac{1}{2}\int_\infty^0 \|\boldsymbol{u}\|\mathrm{d}t, \qquad \forall \boldsymbol{x}_0 \in \Im m\boldsymbol{E}.
\end{aligned}$$

This shows that $\frac{1}{2}\int_{\infty}^{0}\|\boldsymbol{u}\|\mathrm{d}t$ is bounded from below by $\tilde{\mathrm{L}}_c(\boldsymbol{x}_0)$. In view of the previous inequality, It is obvious that the lower bound is achieved by the selection of input $\boldsymbol{u} = \boldsymbol{g}^T(\boldsymbol{x})\boldsymbol{L}_c^T(\boldsymbol{x})$, i.e., $\tilde{\mathrm{L}}_c(\boldsymbol{x}_0)$ is equal to the lower bound. Note that the resulting closed-loop system by feeding back the input $\boldsymbol{u} = \boldsymbol{g}^T(\boldsymbol{x})\boldsymbol{L}_c^T(\boldsymbol{x})$ is such that $\boldsymbol{x}(-\infty) = 0$ by hypothesis. Therefore, for all $\boldsymbol{x}_0 \in \Im m\boldsymbol{E} \subset \mathcal{X}$, we have

$$\mathrm{L}_c(\boldsymbol{x}_0) = \min_{\substack{\boldsymbol{u}\in\mathcal{L}_2(-\infty,0)\\ \boldsymbol{x}(-\infty)=0,\ \boldsymbol{x}(0)=\boldsymbol{x}_0}} \frac{1}{2}\int_{\infty}^{0}\|\boldsymbol{u}(t)\|^2\mathrm{d}t = \tilde{\mathrm{L}}_c(\boldsymbol{x}_0).$$

Necessity. We now move on to the reverse implication of this theorem. First, assume that $\mathrm{L}_o(\boldsymbol{x})$ exists and is smooth on $\Im m\boldsymbol{E}$. Note that by definition $\mathrm{L}_o(0) = 0$. Upon differentiating $\mathrm{L}_o(\boldsymbol{x})$ with respect to t gives (Recall that $\mathrm{L}_o(\boldsymbol{x}_0) \triangleq \frac{1}{2}\int_0^{\infty}\|\boldsymbol{y}(t)\|^2\mathrm{d}t,\quad \boldsymbol{x}(0) = \boldsymbol{x}_0 \in \Im m\boldsymbol{E},\quad \boldsymbol{u}(t) \equiv 0,\ 0 \le t \le \infty$)

$$\frac{\mathrm{d}}{\mathrm{d}t}\mathrm{L}_o(\boldsymbol{x}(t)) = -\frac{1}{2}\boldsymbol{h}^T(\boldsymbol{x}(t))\boldsymbol{h}(\boldsymbol{x}(t)),$$

which implies that

$$\boldsymbol{L}_o^T(\boldsymbol{x})\boldsymbol{f}(\boldsymbol{x}) + \frac{1}{2}\boldsymbol{h}^T(\boldsymbol{x})\boldsymbol{h}(\boldsymbol{x}) = 0.$$

This shows that $\mathrm{L}_o(\boldsymbol{x})$ is a solution of (5.7). The uniqueness is rather obvious since it has aleady been shown in the proof of the sufficiency part. Next, suppose that $\mathrm{L}_c(\boldsymbol{x}(t))$ exists and is smooth on $\Im m\boldsymbol{E}$. The rest of the proof follows the same line of [102](of course, in descriptor systems context now). Since, by hypothesis, the function $\mathrm{L}_c(\boldsymbol{x})$ exists, this implies that the optimal control problem

$$\mathrm{L}_c(\boldsymbol{x}(t)) = \min_{\substack{\boldsymbol{u}\in\mathcal{L}_2(-\infty,0)\\ \boldsymbol{x}(-\infty)=0,\ \boldsymbol{x}(0)=\boldsymbol{x}_0}} \frac{1}{2}\int_{\infty}^{t}\|\boldsymbol{u}(\tau)\|^2\mathrm{d}\tau$$

has a solution with the optimal control law $\boldsymbol{u}^*(t)$. Differentiating the previous equation with respect to t yields

$$\begin{aligned}\frac{\mathrm{d}}{\mathrm{d}t}\mathrm{L}_c(\boldsymbol{x}(t)) &= \frac{1}{2}\|\boldsymbol{u}^*(t)\|^2\\ \Rightarrow \boldsymbol{L}_c(\boldsymbol{x}(t))\left(\boldsymbol{f}(\boldsymbol{x}(t)) + \boldsymbol{g}(\boldsymbol{x}(t))\boldsymbol{u}^*(t)\right) &- \tfrac{1}{2}\|\boldsymbol{u}^*(t)\|^2.\end{aligned}$$

Then, completing the square gives

$$\boldsymbol{L}_c(\boldsymbol{x})\boldsymbol{f}(\boldsymbol{x}) + \frac{1}{2}\boldsymbol{L}_c(\boldsymbol{x})\boldsymbol{g}(\boldsymbol{x})\boldsymbol{g}^T(\boldsymbol{x})\boldsymbol{L}_c(\boldsymbol{x}) - \frac{1}{2}\|\boldsymbol{u}^* - \boldsymbol{g}^T(\boldsymbol{x})\boldsymbol{L}_c(\boldsymbol{x})\|^2 = 0. \quad (5.9)$$

Let $\boldsymbol{u}(\bullet)$ be any continuous admissible control law which steers the state from $\boldsymbol{x}(t)$ at time t to $\boldsymbol{x}(-\infty) = 0$. Next, construct a control input of the following form:

$$\tilde{\boldsymbol{u}}(\tau) \triangleq \begin{cases} \boldsymbol{u}(\tau), & \text{for } t-\delta \leq \tau \leq t \\ \boldsymbol{u}^*(\tau), & \text{for } -\infty < \tau < t-\delta, \end{cases}$$

and define a performance index

$$J(\tilde{\boldsymbol{u}}(\bullet)) \triangleq \frac{1}{2}\int_{\infty}^{t} \|\tilde{\boldsymbol{u}}(\tau)\|^2 \mathrm{d}\tau = \frac{1}{2}\int_{\infty}^{t-\delta} \|\tilde{\boldsymbol{u}}(\tau)\|^2 \mathrm{d}\tau + \frac{1}{2}\int_{t-\delta}^{t} \|\tilde{\boldsymbol{u}}(\tau)\|^2 \mathrm{d}\tau.$$

$J(\tilde{\boldsymbol{u}}(\bullet))$ can be rewritten as follows (by the definition of $\mathrm{L}_c(\boldsymbol{x})$).

$$J(\tilde{\boldsymbol{u}}(\bullet)) = \mathrm{L}_c(\boldsymbol{x}(t-\delta)) + \frac{1}{2}\delta \cdot \|\boldsymbol{u}(t)\|^2 + o(\delta),$$

where

$$\lim_{\delta\to 0} \frac{o(\delta)}{\delta} = 0.$$

Moreover, by the smoothness of $\mathrm{L}_c(\boldsymbol{x})$ we have

$$\mathrm{L}_c(\boldsymbol{x}(t)) = \mathrm{L}_c(\boldsymbol{x}(t-\delta)) + \delta \cdot (\boldsymbol{L}_c(\boldsymbol{x}(t))[\boldsymbol{f}(\boldsymbol{x}(t)) + \boldsymbol{g}(\boldsymbol{x}(t))\boldsymbol{u}^*(t)]) + o(\delta).$$

Because of $J(\tilde{\boldsymbol{u}}(\bullet)) \geq \mathrm{L}_c(\boldsymbol{x}(\bullet))$ it is concluded that

$$\boldsymbol{L}_c(\boldsymbol{x}(t))\,(\boldsymbol{f}(\boldsymbol{x}(t)) + \boldsymbol{g}(\boldsymbol{x}(t))\boldsymbol{u}^*(t)) - \frac{1}{2}\|\boldsymbol{u}(t)\|^2 \leq 0.$$

Therefore

$$\boldsymbol{L}_c(\boldsymbol{x})\boldsymbol{f}(\boldsymbol{x}) + \frac{1}{2}\boldsymbol{L}_c(\boldsymbol{x})\boldsymbol{g}(\boldsymbol{x})\boldsymbol{g}^T(\boldsymbol{x})\boldsymbol{L}_c(\boldsymbol{x}) - \frac{1}{2}\|\boldsymbol{u} - \boldsymbol{g}^T(\boldsymbol{x})\boldsymbol{L}_c(\boldsymbol{x})\|^2 \leq 0. \quad (5.10)$$

Observe that, from (5.9), the maximum of (5.10) is attained (which is equal to zero) by the selection $\boldsymbol{u}(t) = \boldsymbol{g}^T(\boldsymbol{x}(t))\boldsymbol{L}_c^T(\boldsymbol{x}(t))$. This shows that $\boldsymbol{u}^*(t) = \boldsymbol{g}^T(\boldsymbol{x}(t))\boldsymbol{L}_c^T(\boldsymbol{x}(t))$ is the optimal control. Since by definition $\mathrm{L}_c(0) = 0$, $\mathrm{L}_c(\boldsymbol{x}(t))$ satisfies Equation (5.8) such that

$$\boldsymbol{E}\dot{\boldsymbol{x}} = -\left(\boldsymbol{f}(\boldsymbol{x}) + \boldsymbol{g}(\boldsymbol{x})\boldsymbol{g}^T(\boldsymbol{x})\boldsymbol{L}_c^T(\boldsymbol{x})\right)$$

is asymptotically stable with index one. The uniqueness of the solution follows directly from the proof of the sufficiency part. **Q.E.D.**

We still haven't examined the definiteness of the previous two functions. To do this, we need the following definition.

Definition 5.3. *The system (5.1) is reachable from $\boldsymbol{x}_0$ if for any $\bar{\boldsymbol{x}} \in \Im m\boldsymbol{E}$ there exists a $\bar{t} \geq 0$, and input $\boldsymbol{u}$ such that $\bar{\boldsymbol{x}} = \boldsymbol{x}(\bar{t}, 0, \boldsymbol{x}_0, \boldsymbol{u})$. On the other hand, the system (5.1) is said to be zero-state observable if $\boldsymbol{u}(t) \equiv 0$ and $\boldsymbol{y}(t) \equiv 0$ implies that $\boldsymbol{x}_1(t) \equiv 0$; or in other words, for all $\boldsymbol{x} \in \mathcal{X}$, if $\boldsymbol{h}(\boldsymbol{x}(t, 0, \boldsymbol{x}, 0)) = 0$ for all $t \geq 0$, then $\boldsymbol{x}(t, 0, \boldsymbol{x}, 0)) = 0$ for all $t \geq 0$.* □

Theorem 5.4. *Consider (5.2). Suppose that $\boldsymbol{E}\dot{\boldsymbol{x}} = \boldsymbol{f}(\boldsymbol{x})$ is asymptotically stable with index one on $\mathcal{X}$. Then we have the following.*

1. *Suppose that (5.6) has a solution $\mathrm{L}_c(\boldsymbol{x})$ on $\mathcal{X}$. Then $\mathrm{L}_c(\boldsymbol{x}_0) > 0$ for $\boldsymbol{x}_0 \in \Im m\boldsymbol{E} \subset \mathcal{X}$, $\boldsymbol{x}_0 \neq 0$, if and only if the system*
$$\boldsymbol{E}\dot{\boldsymbol{x}} = -\left(\boldsymbol{f}(\boldsymbol{x}) + \boldsymbol{g}(\boldsymbol{x})\boldsymbol{g}^T(\boldsymbol{x})\boldsymbol{L}^T(\boldsymbol{x})\right)$$
is asymptotically stable with index one on $\mathcal{X}$.
2. *If the system (5.6) is zero-state observable and $\mathrm{L}_o(\boldsymbol{x})$ exists and is smooth on $\mathcal{X}$, then $\mathrm{L}_o(\boldsymbol{x}_0) > 0$ for all $\boldsymbol{x}_0 \in \Im m\boldsymbol{E}$, $\boldsymbol{x}_0 \neq 0$.*

□

The characterization of the observability and controllability functions of a nonlinear descriptor system as given above will turn out to be useful for obtaining a balanced representation of the corresponding system.

For the remainder of this section, we consider System (5.2) with controllability and observability functions L_c and L_o, respectively, and make the following standing assumptions.

A1 $\boldsymbol{E}\dot{\boldsymbol{x}} = \boldsymbol{f}(\boldsymbol{x})$ is asymptotically stable with index one on $\mathcal{X}$.
A2 The systme is zero-state observable on $\mathcal{X}$.
A3 L_c and L_o exist and are smooth on $\mathcal{X}$.
A4 $\boldsymbol{E}\frac{\partial}{\partial \boldsymbol{x}}\boldsymbol{L}_c(0) = \frac{\partial}{\partial \boldsymbol{x}}\boldsymbol{L}_c^T(0)\boldsymbol{E}^T \geq 0$ and $\boldsymbol{E}^T\frac{\partial}{\partial \boldsymbol{x}}\boldsymbol{L}_o(0) = \frac{\partial}{\partial \boldsymbol{x}}\boldsymbol{L}_o^T(0)\boldsymbol{E} \geq 0.$
Furthermore, the two operators
$$\boldsymbol{E}\frac{\partial}{\partial \boldsymbol{x}}\boldsymbol{L}_c(0) \quad \text{and} \quad \boldsymbol{E}^T\frac{\partial}{\partial \boldsymbol{x}}\boldsymbol{L}_o(0)$$
restricted to $\Im m\boldsymbol{E}$ are, in fact, positive definite.

The following two lemmas are the first result for the balanced representation. Their proofs are pretty much similar to the corresponding ones given in [102].

Lemma 5.5. *Consider System (5.2) together with its controllability and observability functions L_c and L_o. There exists a coordinate transformation $\boldsymbol{x} = \boldsymbol{\phi}(\bar{\boldsymbol{x}})$ with $\boldsymbol{\phi}(0) = 0$ such that in the new coordinates $\bar{\boldsymbol{x}} = \boldsymbol{\phi}^{-1}(\boldsymbol{x})$ the function $\mathrm{L}_c(\boldsymbol{x})$ has the following form*
$$\mathrm{L}_c(\boldsymbol{\phi}(\bar{\boldsymbol{x}})) = \frac{1}{2}\bar{\boldsymbol{x}}^T\begin{bmatrix}\boldsymbol{I} & 0\\ 0 & 0\end{bmatrix}\bar{\boldsymbol{x}}. \tag{5.11}$$
Furthermore, in the new coordinates $\bar{\boldsymbol{x}} = \boldsymbol{\phi}^{-1}(\boldsymbol{x})$, we can write $\mathrm{L}_o(\boldsymbol{x})$ in the form
$$\mathrm{L}_o(\boldsymbol{\phi}(\bar{\boldsymbol{x}})) = \frac{1}{2}\bar{\boldsymbol{x}}^T\boldsymbol{M}(\bar{\boldsymbol{x}})\bar{\boldsymbol{x}}, \quad \text{where } \boldsymbol{M}(0) = \boldsymbol{E}^T\frac{\partial \boldsymbol{L}_o}{\partial \boldsymbol{x}}(0). \tag{5.12}$$
Here $\boldsymbol{M}(\bar{\boldsymbol{x}})$ is an $n \times n$ symmetric matrix such that its entries are smooth functions of $\bar{\boldsymbol{x}}$. □

Lemma 5.6. *Consider System (5.2). Suppose the condition of Lemma 5.5 holds. Then, locally on a neighborhood $\bar{\mathcal{X}}$ of the origin, there exists a coordinate transformation $\boldsymbol{x} = \boldsymbol{\psi}(\boldsymbol{z})$ with $\boldsymbol{\psi}(0) = 0$ such that, in the new coordinates $\boldsymbol{z} = \boldsymbol{\psi}^{-1}(\boldsymbol{x}) \in \mathcal{Z} \subset \Im m\boldsymbol{E}$, the function L_c has the following form*

$$\bar{\mathrm{L}}_c(\boldsymbol{z}) \triangleq \mathrm{L}_c(\boldsymbol{\psi}(\boldsymbol{z})) = \frac{1}{2}\boldsymbol{z}^T \begin{bmatrix} \boldsymbol{I} & 0 \\ 0 & 0 \end{bmatrix} \boldsymbol{z}, \tag{5.13}$$

and the functions $\mathrm{L}_o(\boldsymbol{x})$ has the form of

$$\bar{\mathrm{L}}_o(\boldsymbol{z}) \triangleq \mathrm{L}_o(\boldsymbol{\psi}(\boldsymbol{z})) = \frac{1}{2}\boldsymbol{z}^T \begin{bmatrix} \gamma(\boldsymbol{z}) & 0 \\ 0 & 0 \end{bmatrix} \boldsymbol{z}, \text{ with } \gamma(\boldsymbol{z}) \triangleq \begin{bmatrix} \tau_1(\boldsymbol{z}) & & 0 \\ & \ddots & \\ 0 & & \tau_r(\boldsymbol{z}) \end{bmatrix} \tag{5.14}$$

where $\tau_1(\boldsymbol{z}) \geq \tau_2(\boldsymbol{z}) \geq \ldots \geq \tau_r(\boldsymbol{z})$ are smooth functions of $\boldsymbol{z}$, and are called the singular value functions of the system. □

With the previous two lemmas in hand, we are ready to find a balanced representation for the nonlinear descriptor system (5.2). Observe that $\boldsymbol{M}(\bar{\boldsymbol{x}})$ can be written as follows (because $\boldsymbol{M}(\bar{x})$ is symmetric).

$$\boldsymbol{M}(\bar{\boldsymbol{x}}) = \boldsymbol{T}(\bar{\boldsymbol{x}})\Lambda(\bar{\boldsymbol{x}})\boldsymbol{T}^T(\bar{\boldsymbol{x}}), \quad \text{where } \Lambda(\bar{\boldsymbol{x}}) = \begin{bmatrix} \lambda_1(\bar{\boldsymbol{x}}) & & 0 \\ & \ddots & \\ 0 & & \lambda_r(\bar{\boldsymbol{x}}) \end{bmatrix}$$

with $\lambda_1(\bar{\boldsymbol{x}}) \leq \ldots \leq \lambda_r(\bar{\boldsymbol{x}})$. Here $\lambda_i(\bar{\boldsymbol{x}})$, $i = 1, \ldots, r$ are the eigenvalues of $\boldsymbol{M}(\bar{\boldsymbol{x}})$, and $\boldsymbol{T}(\bar{\boldsymbol{x}})$ is the associated orthogonal matrix consisting of the normalized eigenvectors, i.e., $\boldsymbol{T}^T(\bar{\boldsymbol{x}})\boldsymbol{T}(\bar{\boldsymbol{x}}) = \boldsymbol{I}$, for all $\bar{\boldsymbol{x}}$ on a neighborhood of the origin. Therefore, the function $\mathrm{L}_o(\boldsymbol{\phi}(\bar{\boldsymbol{x}}))$ can be rewritten as

$$\mathrm{L}_o(\boldsymbol{\phi}(\bar{\boldsymbol{x}})) = \frac{1}{2}\bar{\boldsymbol{x}}^T\boldsymbol{T}(\bar{\boldsymbol{x}})\Lambda(\bar{\boldsymbol{x}})\boldsymbol{T}^T(\bar{\boldsymbol{x}})\bar{\boldsymbol{x}},$$

where $\bar{\boldsymbol{x}}$ belongs to a neighborhood of the origin. In order to yield a diagonalized form of the observability function, define a new coordinate transformation $\boldsymbol{z} = \boldsymbol{v}(\bar{\boldsymbol{x}}) \triangleq \boldsymbol{T}^T(\bar{\boldsymbol{x}})\bar{\boldsymbol{x}}$ to get

$$\mathrm{L}_o(\boldsymbol{\phi}(\boldsymbol{v}^{-1}(\boldsymbol{z}))) = \frac{1}{2}\boldsymbol{z}^T\Lambda(\boldsymbol{v}^{-1}(\boldsymbol{z}))\boldsymbol{z}.$$

In this case,

$$\mathrm{L}_c(\boldsymbol{\phi}(\boldsymbol{v}^{-1}(\boldsymbol{z}))) = \frac{1}{2}\boldsymbol{z}^T\boldsymbol{T}^T(\bar{\boldsymbol{x}})\boldsymbol{T}(\bar{\boldsymbol{x}})\boldsymbol{z} = \frac{1}{2}\boldsymbol{z}^T\boldsymbol{z}.$$

As a matter of fact, the functions $\tau_i(\boldsymbol{z})$ and $\lambda_i(\boldsymbol{v}^{-1}(\boldsymbol{z}))$, and the coordinate transformations $\boldsymbol{\phi}$ and $\boldsymbol{\psi}$ have the following relationship.

$$\tau_i(\boldsymbol{z}) = \lambda_i(\boldsymbol{v}^{-1}(\boldsymbol{z})), \quad i = 1, \ldots, r,$$
$$\boldsymbol{\phi} = \boldsymbol{\psi} \circ \boldsymbol{v}^{-1}.$$

Here "$\circ$" denotes the composition of functions. To achieve the balanced realization, we construct a last smooth coordinate transformation of the form

$$\bar{\boldsymbol{z}} = \eta(\boldsymbol{z}) \triangleq [\eta_1(\boldsymbol{z}_1) \ \ldots \ \eta_r(\boldsymbol{z}_r)\ 0 \ \ldots \ 0]^T,$$

where $\bar{\boldsymbol{z}}_i = \eta_i(\boldsymbol{z}_i) \triangleq \tau_i(0, \ldots, 0, \boldsymbol{z}_i, 0, \ldots, 0)^{\frac{1}{4}} \boldsymbol{z}_i$, $i = 1, \ldots, r$. Define $\bar{\mathrm{L}}_c(\bar{\boldsymbol{z}}) \triangleq \bar{\mathrm{L}}_c(\eta^{-1}(\bar{\boldsymbol{z}}))$ and $\bar{\mathrm{L}}_o(\bar{\boldsymbol{z}}) \triangleq \bar{\mathrm{L}}_o(\eta^{-1}(\bar{\boldsymbol{z}}))$. Under these new coordinates, Equations (5.13) and (5.14) can be put in the following form.

$$\bar{\mathrm{L}}_c(\bar{\boldsymbol{z}}) = \frac{1}{2}\bar{\boldsymbol{z}}^T \begin{bmatrix} \boldsymbol{\Sigma}(\bar{\boldsymbol{z}}) & 0 \\ 0 & 0 \end{bmatrix} \bar{\boldsymbol{z}}, \text{ with } \boldsymbol{\Sigma}(\boldsymbol{z}) \triangleq \begin{bmatrix} \sigma_1(\bar{\boldsymbol{z}}_1)^{-1} & & 0 \\ & \ddots & \\ 0 & & \sigma_r(\bar{\boldsymbol{z}}_r)^{-1} \end{bmatrix} \tag{5.15}$$

and

$$\bar{\mathrm{L}}_o(\bar{\boldsymbol{z}}) = \frac{1}{2}\bar{\boldsymbol{z}}^T \begin{bmatrix} \boldsymbol{\Sigma}(\bar{\boldsymbol{z}})\gamma(\eta^{-1}(\bar{\boldsymbol{z}})) & 0 \\ 0 & 0 \end{bmatrix} \bar{\boldsymbol{z}}, \tag{5.16}$$

respectively. In the above equations, the functions $\sigma_i(\bar{\boldsymbol{z}}_i)$s are defined as follows.

$$\sigma_i(\bar{\boldsymbol{z}}_i) = \tau_i(0, \ldots, 0, \eta_i^{-1}(\bar{\boldsymbol{z}}_i), 0, \ldots, 0)^{\frac{1}{2}}, \quad \text{for } i = 1, \ldots, r.$$

Right now, we arrive at the point that, after the coordinate transformation $\boldsymbol{x} = \chi(\bar{\boldsymbol{z}}) := \boldsymbol{\phi}(\eta^{-1}(\bar{\boldsymbol{z}}))$, the system is transformed to

$$\begin{aligned} \boldsymbol{E}\dot{\bar{\boldsymbol{z}}} &= \bar{\boldsymbol{f}}(\bar{\boldsymbol{z}}) + \bar{\boldsymbol{g}}(\bar{\boldsymbol{z}})\boldsymbol{u}, \\ \boldsymbol{y} &= \bar{\boldsymbol{h}}(\bar{\boldsymbol{z}}). \end{aligned} \tag{5.17}$$

with the controllability and observability functions given by (5.15) and (5.16), respectively. In case that the matrix $\boldsymbol{E}$ is nonsingualr, the system (5.17) is said to be balanced (See for example [102]). To sum up the discussion given above, we mention the following defintion.

Definition 5.7. *Consider the nonlinear descriptor system (5.17). The system is said to be balancd if its controllability and observability functions are of the following form:*

$$\bar{\mathrm{L}}_c(\bar{\boldsymbol{z}}) = \frac{1}{2}\bar{\boldsymbol{z}}^T \begin{bmatrix} \boldsymbol{\Sigma}(\bar{\boldsymbol{z}}) & 0 \\ 0 & 0 \end{bmatrix} \bar{\boldsymbol{z}}, \text{ with } \boldsymbol{\Sigma}(\boldsymbol{z}) \triangleq \begin{bmatrix} \sigma_1(\bar{\boldsymbol{z}}_1)^{-1} & & 0 \\ & \ddots & \\ 0 & & \sigma_r(\bar{\boldsymbol{z}}_r)^{-1} \end{bmatrix}$$

and

$$\bar{\mathrm{L}}_o(\bar{\boldsymbol{z}}) = \frac{1}{2}\bar{\boldsymbol{z}}^T \begin{bmatrix} \boldsymbol{\Sigma}(\bar{\boldsymbol{z}})\gamma(\eta^{-1}(\bar{\boldsymbol{z}})) & 0 \\ 0 & 0 \end{bmatrix} \bar{\boldsymbol{z}}, \text{ with } \gamma(\boldsymbol{z}) \triangleq \begin{bmatrix} \tau_1(\boldsymbol{z}) & & 0 \\ & \ddots & \\ 0 & & \tau_r(\boldsymbol{z}) \end{bmatrix},$$

respectively. □

For conventional nonlinear systems, the interpretation of the balanced realization is that the states $\bar{z}_1, \ldots, \bar{z}_r$ are increasingly less important in terms of the input and output energy. From our previous derivation, it is clearly seen that this is also true for nonlinear descriptor systems.

Let us now examine the linearizations of (5.2) and (5.17). Suppose that (5.2) and (5.17) have linearizations, locally around the equilibrium point and $\boldsymbol{u} = 0$, given by

$$\begin{aligned} \boldsymbol{E}\dot{\hat{\boldsymbol{x}}} &= \boldsymbol{A}\hat{\boldsymbol{x}} + \boldsymbol{B}\boldsymbol{u}, \\ \boldsymbol{y} &= \boldsymbol{C}\hat{\boldsymbol{x}}, \end{aligned} \tag{5.18}$$

and

$$\begin{aligned} \boldsymbol{E}\dot{\hat{\boldsymbol{z}}} &= \hat{\boldsymbol{A}}\hat{\boldsymbol{z}} + \hat{\boldsymbol{B}}\boldsymbol{u}, \\ \boldsymbol{y} &= \hat{\boldsymbol{C}}\hat{\boldsymbol{z}}, \end{aligned} \tag{5.19}$$

respectively, where

$$\begin{aligned} \boldsymbol{A} &= \frac{\partial \boldsymbol{f}}{\partial \boldsymbol{x}}(0), \ \boldsymbol{B} = \boldsymbol{g}(0), \ \text{and } \boldsymbol{C} = \frac{\partial \boldsymbol{h}}{\partial \boldsymbol{x}}(0), \\ \hat{\boldsymbol{A}} &= \frac{\partial \boldsymbol{f}}{\partial \bar{\boldsymbol{z}}}(0), \ \hat{\boldsymbol{B}} = \bar{\boldsymbol{g}}(0), \ \text{and } \hat{C} = \frac{\partial \boldsymbol{h}}{\partial \bar{\boldsymbol{z}}}(0). \end{aligned}$$

Assume that the pair $\{\boldsymbol{E}, \boldsymbol{A}\}$ is stable and impulse-free. The system (5.18), respectively (5.19), have controllability and observability functions given by $\mathtt{M}_c(\hat{\boldsymbol{x}})$ and $\mathtt{M}_o(\hat{\boldsymbol{x}})$, respectively $\hat{\mathtt{M}}_c(\hat{\boldsymbol{z}})$ and $\hat{\mathtt{M}}_o(\hat{\boldsymbol{z}})$. It is plain to check that the controllability and observability functions of the linearized systems have the following form.

$$\begin{aligned} \mathtt{M}_c(\hat{\boldsymbol{x}}) &= \frac{1}{2}\hat{\boldsymbol{x}}^T \boldsymbol{E}^T \frac{\partial \boldsymbol{L}_c}{\partial \boldsymbol{x}}(0)\hat{\boldsymbol{x}}, \quad \mathtt{M}_o(\hat{\boldsymbol{x}}) = \tfrac{1}{2}\hat{\boldsymbol{x}}^T \boldsymbol{E}\tfrac{\partial \boldsymbol{L}_o}{\partial \boldsymbol{x}}(0)\hat{\boldsymbol{x}}, \\ \hat{\mathtt{M}}_c(\hat{\boldsymbol{z}}) &= \frac{1}{2}\hat{\boldsymbol{z}}^T E^T \frac{\partial \hat{\boldsymbol{L}}_c}{\partial \bar{\boldsymbol{z}}}(0)\hat{\boldsymbol{z}}, \quad \hat{\mathtt{M}}_o(\hat{\boldsymbol{z}}) = \tfrac{1}{2}\hat{\boldsymbol{z}}^T \boldsymbol{E}\tfrac{\partial \hat{\boldsymbol{L}}_o}{\partial \bar{\boldsymbol{z}}}(0)\hat{\boldsymbol{z}}. \end{aligned}$$

Define $\boldsymbol{S} \triangleq \frac{\partial \boldsymbol{x}}{\partial \bar{\boldsymbol{z}}}(0)$. Then, it is easily verified that System (5.18) is transformed into (5.19) by the coordinate transformation $\hat{\boldsymbol{x}} = \boldsymbol{S}\hat{\boldsymbol{z}}$, so that $\hat{\boldsymbol{A}} = \boldsymbol{S}^{-1}\boldsymbol{A}\boldsymbol{S}$, $\hat{\boldsymbol{B}} = \boldsymbol{S}^{-1}\boldsymbol{B}$ and $\hat{\boldsymbol{C}} = \boldsymbol{C}\boldsymbol{S}$. Furthermore,

$$\boldsymbol{E}^T \frac{\partial \hat{\boldsymbol{L}}_c}{\partial \bar{\boldsymbol{z}}}(0) = \boldsymbol{S}^T \boldsymbol{E}^T \frac{\partial \boldsymbol{L}_c}{\partial \boldsymbol{x}}(0)\boldsymbol{S}, \qquad \boldsymbol{E}^T \frac{\partial \hat{\boldsymbol{L}}_o}{\partial \bar{\boldsymbol{z}}}(0) = \boldsymbol{S}^T \boldsymbol{E}^T \frac{\partial \boldsymbol{L}_o}{\partial \boldsymbol{x}}(0)\boldsymbol{S}.$$

We shall make some further investigation on the linearized system (5.18) here. Consider the linear descriptor system (5.18). From now on, we suppose that the system (5.18) is minimal (i.e., both finite and infinite modes are controllable and observable). First, partiton the system in the following form.

$$\begin{aligned} \begin{bmatrix} \boldsymbol{I} & 0 \\ 0 & 0 \end{bmatrix} \begin{bmatrix} \dot{\boldsymbol{x}}_1 \\ \dot{\boldsymbol{x}}_2 \end{bmatrix} &= \begin{bmatrix} \boldsymbol{A}_{11} & \boldsymbol{A}_{12} \\ \boldsymbol{A}_{21} & \boldsymbol{A}_{22} \end{bmatrix} \begin{bmatrix} \boldsymbol{x}_1 \\ \boldsymbol{x}_2 \end{bmatrix} + \begin{bmatrix} \boldsymbol{B}_1 \\ \boldsymbol{B}_2 \end{bmatrix} \boldsymbol{u}, \\ \boldsymbol{y} &= \begin{bmatrix} \boldsymbol{C}_1 & \boldsymbol{C}_2 \end{bmatrix} \begin{bmatrix} \boldsymbol{x}_1 \\ \boldsymbol{x}_2 \end{bmatrix}. \end{aligned}$$

Since the pair $\{\boldsymbol{E},\ \boldsymbol{A}\}$ is assumed to be stable and impulse-free, the algebraic part of the system can be solved for $\boldsymbol{x}_2$ to get

$$\begin{aligned}\boldsymbol{A}_0 &\triangleq \boldsymbol{A}_{11} - \boldsymbol{A}_{12}\boldsymbol{A}_{22}^{-1}\boldsymbol{A}_{21},\\ \boldsymbol{B}_0 &\triangleq \boldsymbol{B}_1 - \boldsymbol{A}_{12}\boldsymbol{A}_{22}^{-1}\boldsymbol{B}_2,\\ \boldsymbol{C}_0 &\triangleq \boldsymbol{C}_1 - \boldsymbol{C}_2\boldsymbol{A}_{22}^{-1}\boldsymbol{A}_{21}.\end{aligned}$$

The following two theorems are certain generalizations of existing results.

Theorem 5.8. *Consider the system (5.18). Suppose that the realization* $\{\boldsymbol{E}, \boldsymbol{A}, \boldsymbol{B}, \boldsymbol{C}\}$ *is minimal. Then*

$$\mathrm{L}_c(\boldsymbol{x}_0) = \frac{1}{2}\boldsymbol{x}_0^T \boldsymbol{E}\boldsymbol{W}_c^{-1}\boldsymbol{x}_0 = \frac{1}{2}\boldsymbol{x}_0^T \begin{bmatrix} \boldsymbol{W}_{c0}^{-1} & 0 \\ 0 & 0 \end{bmatrix} \boldsymbol{x}_0$$

and

$$\mathrm{L}_o(\boldsymbol{x}_0) = \frac{1}{2}\boldsymbol{x}_0^T \boldsymbol{E}^T\boldsymbol{W}_o\boldsymbol{x}_0 = \frac{1}{2}\boldsymbol{x}_0^T \begin{bmatrix} \boldsymbol{W}_{o0} & 0 \\ 0 & 0 \end{bmatrix} \boldsymbol{x}_0,$$

where

$$\boldsymbol{W}_{c0} = \int_0^\infty \exp^{\boldsymbol{A}_0 t}\, \boldsymbol{B}_0\boldsymbol{B}_0^T \exp^{\boldsymbol{A}_0^T t}\, \mathrm{d}t$$

is the controllability Gramian and

$$\boldsymbol{W}_{o0} = \int_0^\infty \exp^{\boldsymbol{A}_0^T t}\, \boldsymbol{C}_0^T\boldsymbol{C}_0 \exp^{\boldsymbol{A}_0 t}\, \mathrm{d}t$$

is the observability Gramian. Furthermore, $\boldsymbol{W}_c$ *and* $\boldsymbol{W}_o$ *are solutions to the generalized Lyapunov equations*

$$\begin{cases} \boldsymbol{A}\boldsymbol{W}_c + \boldsymbol{W}_c^T\boldsymbol{A}^T = -\boldsymbol{B}\boldsymbol{B}^T \\ \boldsymbol{E}\boldsymbol{W}_c = \boldsymbol{W}_c^T\boldsymbol{E}^T \geq 0 \end{cases} \tag{5.20}$$

and

$$\begin{cases} \boldsymbol{A}^T\boldsymbol{W}_o + \boldsymbol{W}_c^T\boldsymbol{A} = -\boldsymbol{C}^T\boldsymbol{C} \\ \boldsymbol{E}^T\boldsymbol{W}_o = \boldsymbol{W}_o^T\boldsymbol{E} \geq 0, \end{cases} \tag{5.21}$$

respectively. □

Theorem 5.9. *The eigenvalues of* $\boldsymbol{W}_{o0}\boldsymbol{W}_{c0}$ *are invariant under similarity transformation. Furthermore, there exists a transformation matrix that transforms the system (5.18) into the balanced form (5.19) with*

$$\mathcal{W} := \boldsymbol{W}_{o0} = \boldsymbol{W}_{c0} = \begin{bmatrix} \sigma_1 & & 0 \\ & \ddots & \\ 0 & & \sigma_r \end{bmatrix}. \tag{5.22}$$

Here $\sigma_1 \geq \sigma_2 \geq \ldots \geq \sigma_r > 0$ *denote the square roots of the eigenvalues of* $\boldsymbol{W}_{o0}\boldsymbol{W}_{c0}$ *and are called the Hankel singular values.* □

Remark. The largest Hankel singular value and the *Hankel norm* of linear descriptor systems are eventually identical. Denote by $\boldsymbol{G}(s) = \boldsymbol{C}(s\boldsymbol{E}-\boldsymbol{A})^{-1}\boldsymbol{B}$ the transfer function matrix of the system (5.18). Then, following the obvious notation, we can obtain the next identity.

$$\begin{aligned}\|\boldsymbol{G}(s)\|_H^2 &= \max_{\substack{\boldsymbol{x} \in \Im m\, \boldsymbol{E} \\ \boldsymbol{x} \neq 0}} \frac{\mathrm{L}_o(\boldsymbol{x})}{\mathrm{L}_c(\boldsymbol{x})} = \max_{\substack{\boldsymbol{x} \in \Im m\, \boldsymbol{E} \\ \boldsymbol{x} \neq 0}} \frac{\boldsymbol{x}^T \boldsymbol{W}_o \boldsymbol{x}}{\boldsymbol{x}^T \boldsymbol{W}_c^{-1} \boldsymbol{x}} \\ &= \max_{\substack{\bar{\boldsymbol{x}}_1 \in \mathbb{R}^r \\ \bar{\boldsymbol{x}}_1 \neq 0}} \frac{\bar{\boldsymbol{x}}_1^T \mathcal{W}^2 \bar{\boldsymbol{x}}_1}{\bar{\boldsymbol{x}}_1^T \bar{\boldsymbol{x}}_1} = \sigma_1^2\end{aligned}$$

Note that the Hankel norm of a linear system is defined as the $\mathcal{L}_2$-gain from past inputs to future outputs. However, we cannot achieve the same conclusion -the largest Hankel singular value equal to Hankel norm- for nonlinear descriptor systems. To verify the aforementioned statement, let's consider the system (5.17) and define $\tau_1^{\max} \triangleq \max_{\boldsymbol{z} \in \Im m\, E} \tau_1(\boldsymbol{z})$. Then,

$$\begin{aligned}\max_{\substack{\bar{\boldsymbol{z}} \in \Im m\, E \\ \bar{\boldsymbol{z}} \neq 0}} \frac{\bar{\mathrm{L}}_o(\bar{\boldsymbol{z}})}{\bar{\mathrm{L}}_c(\bar{\boldsymbol{z}})} &= \max_{\substack{\boldsymbol{z} \in \Im m\, E \\ \boldsymbol{z} \neq 0}} \frac{\bar{\mathrm{L}}_o(\boldsymbol{z})}{\bar{\mathrm{L}}_c(\boldsymbol{z})} = \max_{\substack{\boldsymbol{z} \in \Im m\, E \\ \boldsymbol{z} \neq 0}} \frac{\sum_{i=1}^r \tau_i(\boldsymbol{z}) \boldsymbol{z}_i^2}{\sum_{i=1}^r \boldsymbol{z}_i^2} \\ &\leq \max_{\substack{\boldsymbol{z} \in \Im m\, E \\ \boldsymbol{z} \neq 0}} \frac{\tau_1^{\max} \boldsymbol{z}_1^2 + \sum_{i=1}^r \tau_i(\boldsymbol{z}) \boldsymbol{z}_i^2}{\sum_{i=1}^r \boldsymbol{z}_i^2} = \tau_1^{\max}.\end{aligned}$$

It is clearly seen that the equality is not necessarily true. This shows that $\tau_1^{\max}$, in general, is merely an upper bound for the $\mathcal{L}_2$-gain from the past inputs to future outputs.

Two other representations, that may be obtained from (5.19) by coordinate transformations

$$\boldsymbol{x} = \begin{bmatrix} \mathcal{W}^{-\frac{1}{2}} & 0 \\ 0 & \boldsymbol{I} \end{bmatrix} \bar{\boldsymbol{x}}, \text{ and } \boldsymbol{x} = \begin{bmatrix} \mathcal{W}^{\frac{1}{2}} & 0 \\ 0 & \boldsymbol{I} \end{bmatrix} \bar{\boldsymbol{x}},$$

respectively, follows easily from the above theorem.

Definition 5.10. [84] *A state-space representation is an input-normal representation if $\boldsymbol{W}_{c0} = \boldsymbol{I}$ and $\boldsymbol{W}_{o0} = \mathcal{W}^2$, where $\mathcal{W}$ is given by (5.22). Furthermore, it is an output-normal representation if $\boldsymbol{W}_{c0} = \mathcal{W}^2$ and $\boldsymbol{W}_{o0} = \boldsymbol{I}$.*

□

5.2 Model Reduction via Balanced Truncation

Let's first exaime the model reduction for stable linear descriptor systems. Consider the linear descriptor system (5.18) together with its balanced representation. (5.19). Assume that the Hankel singular values of the system are decreasingly ordered so that $\sigma_1 \geq \sigma_2 \geq \ldots \geq \sigma_r$. Suppose $\sigma_k \gg \sigma_{k+1}$ for some $k < r$. Then, by a classical result, the balanced realization implies that those states corresponding to the singular values of $\sigma_{k+1}, \ldots, \sigma_r$ are less controllable and less observable than those states corresponding to the first k singular values. In other words, the state components $\sigma_{k+1}, \ldots, \sigma_r$ are far less important and may be removed to reduce the number of states of the system without losing much information about the system. Partitoning the system (5.19) in a corresponding way yields

$$\begin{bmatrix} \boldsymbol{I} & 0 & 0 \\ 0 & \boldsymbol{I} & 0 \\ 0 & 0 & 0 \end{bmatrix} \begin{bmatrix} \dot{\hat{\boldsymbol{z}}}_1^s \\ \dot{\hat{\boldsymbol{z}}}_2^s \\ \dot{\hat{\boldsymbol{z}}}^f \end{bmatrix} = \begin{bmatrix} \hat{\boldsymbol{A}}_{11}^s & \hat{\boldsymbol{A}}_{12}^s & \hat{\boldsymbol{A}}_{121} \\ \hat{\boldsymbol{A}}_{21}^s & \hat{\boldsymbol{A}}_{22}^s & \hat{\boldsymbol{A}}_{122} \\ \hat{\boldsymbol{A}}_{211} & \hat{\boldsymbol{A}}_{212} & \hat{\boldsymbol{A}}^f \end{bmatrix} \begin{bmatrix} \hat{\boldsymbol{z}}_1^s \\ \hat{\boldsymbol{z}}_2^s \\ \hat{\boldsymbol{z}}^f \end{bmatrix} + \begin{bmatrix} \hat{\boldsymbol{B}}_1^s \\ \hat{\boldsymbol{B}}_2^s \\ \hat{\boldsymbol{B}}^f \end{bmatrix} \boldsymbol{u}$$

$$\boldsymbol{y} = \begin{bmatrix} \hat{\boldsymbol{C}}_1^s & \hat{\boldsymbol{C}}_2^s & \hat{\boldsymbol{C}}^f \end{bmatrix} \begin{bmatrix} \hat{\boldsymbol{z}}_1^s \\ \hat{\boldsymbol{z}}_2^s \\ \hat{\boldsymbol{z}}^f \end{bmatrix},$$

and

$$\mathcal{W} = \begin{bmatrix} \mathcal{W}_1 & 0 \\ 0 & \mathcal{W}_2 \end{bmatrix}$$

Theorem 5.11. *Consider System (5.19). Assume the realization*

$$\{\boldsymbol{E}, \hat{\boldsymbol{A}}, \hat{\boldsymbol{B}}, \hat{\boldsymbol{C}}\}$$

is minimal. Suppose that the pair $\{\boldsymbol{E}, \hat{\boldsymbol{A}}\}$ is stable and impulse-free. Then the following two subsystems

$$\begin{bmatrix} \boldsymbol{I} & 0 \\ 0 & 0 \end{bmatrix} \begin{bmatrix} \dot{\hat{\boldsymbol{z}}}_1^s \\ \dot{\hat{\boldsymbol{z}}}^f \end{bmatrix} = \begin{bmatrix} \hat{\boldsymbol{A}}_{11}^s & \hat{\boldsymbol{A}}_{121} \\ \hat{\boldsymbol{A}}_{211} & \hat{\boldsymbol{A}}^f \end{bmatrix} \begin{bmatrix} \hat{\boldsymbol{z}}_1^s \\ \hat{\boldsymbol{z}}^f \end{bmatrix} + \begin{bmatrix} \hat{\boldsymbol{B}}_1^s \\ \hat{\boldsymbol{B}}^f \end{bmatrix} \boldsymbol{u}$$
$$\boldsymbol{y} = \begin{bmatrix} \hat{\boldsymbol{C}}_1^s & \hat{\boldsymbol{C}}^f \end{bmatrix} \begin{bmatrix} \hat{\boldsymbol{z}}_1^s \\ \hat{\boldsymbol{z}}^f \end{bmatrix}, \tag{5.23}$$

and

$$\begin{bmatrix} \boldsymbol{I} & 0 \\ 0 & 0 \end{bmatrix} \begin{bmatrix} \dot{\hat{\boldsymbol{z}}}_2^s \\ \dot{\hat{\boldsymbol{z}}}^f \end{bmatrix} = \begin{bmatrix} \hat{\boldsymbol{A}}_{22}^s & \hat{\boldsymbol{A}}_{122} \\ \hat{\boldsymbol{A}}_{212} & \hat{\boldsymbol{A}}^f \end{bmatrix} \begin{bmatrix} \hat{\boldsymbol{z}}_2^s \\ \hat{\boldsymbol{z}}^f \end{bmatrix} + \begin{bmatrix} \hat{\boldsymbol{B}}_2^s \\ \hat{\boldsymbol{B}}^f \end{bmatrix} \boldsymbol{u}$$
$$\boldsymbol{y} = \begin{bmatrix} \hat{\boldsymbol{C}}_2^s & \hat{\boldsymbol{C}}^f \end{bmatrix} \begin{bmatrix} \hat{\boldsymbol{z}}_2^s \\ \hat{\boldsymbol{z}}^f \end{bmatrix}, \tag{5.24}$$

are in balanced form, and their controllability and observability Gramians are equal to $\mathcal{W}_1$ and $\mathcal{W}_2$, respectively. Furthermore, assume that $\mathcal{W}_1$ and $\mathcal{W}_2$ have no diagonal elements in common. Then both systems are stable and impulse-free. □

For nonlinear descriptor systems, we can use the same kind of reasoning to truncate the less important states. Consider the system (5.2) together with its balanced representation (5.17). Suppose that $\tau_k(\boldsymbol{z}) > \tau_{k+1}(\boldsymbol{z})$ (or equivalently,

$\sigma(\bar{z}_k)^{-1}\tau_k(\eta^{-1}(\bar{z})) > \sigma(\bar{z}_{k+1})^{-1}\tau_{k+1}(\eta^{-1}(\bar{z})))$ for some $k < r$. Then, from the system energy point of view, the state variable $\bar{z}_k$ is more important than the state variable $\bar{z}_{k+1}$. Partitioning the system in a corresponding way yields

$$\begin{bmatrix} I & 0 & 0 \\ 0 & I & 0 \\ 0 & 0 & 0 \end{bmatrix} \begin{bmatrix} \dot{\bar{z}}_1^s \\ \dot{\bar{z}}_2^s \\ \dot{\bar{z}}^f \end{bmatrix} = \begin{bmatrix} \bar{f}_1^s(\bar{z}_1^s, \bar{z}_2^s, \bar{z}^f) \\ \bar{f}_2^s(\bar{z}_1^s, \bar{z}_2^s, \bar{z}^f) \\ \bar{f}^f(\bar{z}_1^s, \bar{z}_2^s, \bar{z}^f) \end{bmatrix} + \begin{bmatrix} \bar{g}_1^s(\bar{z}_1^s, \bar{z}_2^s, \bar{z}^f) \\ \bar{g}_2^s(\bar{z}_1^s, \bar{z}_2^s, \bar{z}^f) \\ \bar{g}^f(\bar{z}_1^s, \bar{z}_2^s, \bar{z}^f) \end{bmatrix} u$$
$$y = \bar{h}(\bar{z}_1^s, \bar{z}_2^s, \bar{z}^f)$$

where $\bar{z}_1^s = [\bar{z}_1 \ldots \bar{z}_k]^T$, $\bar{z}_2^s = [\bar{z}_{k+1} \ldots \bar{z}_r]^T$. To truncate the less important states, we simply set $\bar{z}_2^s = 0$. From (5.15), it is clear that $\dfrac{\partial \bar{L}_c}{\partial \bar{z}_1^s}(\bar{z}_1^s, 0, \bar{z}^f) = 0$ for all $\bar{z} \in \Im m E$, since $\sigma_1(\bar{z}_1), \ldots \sigma_k(\bar{z}_k)$ do not depend on $\bar{z}_2^s$. However, in view of equation (5.16), we cannot draw the same conclusion for $\dfrac{\partial \bar{L}_o}{\partial \bar{z}_1^s}(\bar{z}_1^s, 0, \bar{z}^f) = 0$. The Hamilton–Jacobi equations (5.7) and (5.8) of the system (5.17) with $\bar{z}_2^s = 0$ become

$$\begin{aligned} &\left[(\bar{L}_o)_1^s(\bar{z}_1^s, 0, \bar{z}^f) \; (\bar{L}_o)_2^s(\bar{z}_1^s, 0, \bar{z}^f) \; (\bar{L}_o)^f(\bar{z}_1^s, 0, \bar{z}^f) \right] \begin{bmatrix} \bar{f}_1^s(\bar{z}_1^s, 0, \bar{z}^f) \\ \bar{f}_2^s(\bar{z}_1^s, 0, \bar{z}^f) \\ \bar{f}^f(\bar{z}_1^s, 0, \bar{z}^f) \end{bmatrix} \\ &+ \tfrac{1}{2}\bar{h}^T(\bar{z}_1^s, 0, \bar{z}^f)\bar{h}(\bar{z}_1^s, 0, \bar{z}^f) = 0 \end{aligned} \tag{5.25}$$

$$\begin{aligned} &\left[(\bar{L}_c)_1^s(\bar{z}_1^s, 0, \bar{z}^f) \; (\bar{L}_c)^f(\bar{z}_1^s, 0, \bar{z}^f) \right] \begin{bmatrix} \bar{f}_1^s(\bar{z}_1^s, 0, \bar{z}^f) \\ \bar{f}^f(\bar{z}_1^s, 0, \bar{z}^f) \end{bmatrix} \\ &+ \tfrac{1}{2} \begin{bmatrix} \left(\bar{L}_c^T\right)_1^s(\bar{z}_1^s, 0, \bar{z}^f) \\ \left(\bar{L}_c^T\right)^f(\bar{z}_1^s, 0, \bar{z}^f) \end{bmatrix}^T \begin{bmatrix} \bar{g}_1^s(\bar{z}_1^s, 0, \bar{z}^f) \\ \bar{g}^f(\bar{z}_1^s, 0, \bar{z}^f) \end{bmatrix} \begin{bmatrix} \bar{g}_1^s(\bar{z}_1^s, 0, \bar{z}^f) \\ \bar{g}^f(\bar{z}_1^s, 0, \bar{z}^f) \end{bmatrix}^T \cdot \\ &\begin{bmatrix} \left(\bar{L}_c^T\right)_1^s(\bar{z}_1^s, 0, \bar{z}^f) \\ \left(\bar{L}_c^T\right)^f(\bar{z}_1^s, 0, \bar{z}^f) \end{bmatrix} = 0 \end{aligned} \tag{5.26}$$

Upon defining

$$\tilde{z} = \begin{bmatrix} \bar{z}_1^s \\ \bar{z}^f \end{bmatrix}, \qquad \tilde{E} = \begin{bmatrix} I & 0 \\ 0 & 0 \end{bmatrix},$$
$$\tilde{f}(\tilde{z}) = \begin{bmatrix} \bar{f}_1^s(\bar{z}_1^s, 0, \bar{z}^f) \\ \bar{f}^f(\bar{z}_1^s, 0, \bar{z}^f) \end{bmatrix}, \tilde{g}(\tilde{x}) = \begin{bmatrix} \bar{g}_1^s(\bar{z}_1^s, 0, \bar{z}^f) \\ \bar{g}^f(\bar{z}_1^s, 0, \bar{z}^f) \end{bmatrix},$$

and

$$\tilde{h}(\tilde{z}) = \bar{h}(\bar{z}_1^s, 0, \bar{z}^f)$$

for all $\tilde{z} \in \tilde{\mathcal{Z}}$, where $\tilde{\mathcal{Z}} \triangleq \{\tilde{z} = [(\bar{z}_1^s)^T \; (\bar{z}^f)^T]^T | [(\bar{z}_1^s)^T \; 0 \; (\bar{z}^f)^T]^T \in \mathcal{Z}\}$, we then obtain the following reduced order system.

$$\begin{aligned} \tilde{E}\dot{\tilde{z}} &= \tilde{f}(\tilde{z}) + \tilde{g}(\tilde{z})u, \\ \tilde{y} &= \tilde{h}(\tilde{z}). \end{aligned} \tag{5.27}$$

It follows immediately from (5.25) that the controllability function of the reduced system, $\tilde{\mathrm{L}}_c(\tilde{z})$, is equal to $\bar{\mathrm{L}}_c(\bar{z}_1^s,\, 0,\, \bar{z}^f)$. For the observability function of the reduced system (5.27), we have the following lemma.

Lemma 5.12. *Consider (5.27) and (5.25). Suppose that*

$$\bar{L}_{o2}^s(\bar{z}_1^s, 0, \bar{z}^f)\bar{f}_2^s(\bar{z}_1^s, 0, \bar{z}^f) = 0$$

for $(\bar{z}^s,\, \bar{z}^f) \in \mathcal{Z}$. *Then the observability function of the reduced system (5.27) is equal to* $\bar{\mathrm{L}}_o(\bar{z}_1^s,\, 0,\, \bar{z}^f)$. *Moreover, the reduced system (5.27) is in balanced form with singular value function given by*

$$\tau_1(\bar{z}_1^s, 0) \geq \ldots \geq \tau_k(\bar{z}_1^s, 0), \quad \text{for } (\bar{z}_1^s, 0) = \eta^{-1}(\bar{z}_1^s, 0),$$

where η *is defined as in the previous section.* □

Proof. The proof of the lemma is fairly straightforward, and is thus omitted.

With the help of the linearized system and Theorem 5.11, we can prove the asymptotical stability of the reduced nonlinear system.

Theorem 5.13. *The subsystems of the balanced system (5.17), namely system (5.27) and the following system*

$$\begin{aligned} \begin{bmatrix} I & 0 \\ 0 & 0 \end{bmatrix} \begin{bmatrix} \dot{\bar{z}}_2^s \\ \dot{\bar{z}}^f \end{bmatrix} &= \begin{bmatrix} \bar{f}_2^s(0, \bar{z}_2^s, \bar{z}^f) \\ \bar{f}^f(0, \bar{z}_2^s, \bar{z}^f) \end{bmatrix} + \begin{bmatrix} \bar{g}_2^s(0, \bar{z}_2^s, \bar{z}^f) \\ \bar{g}^f(0, \bar{z}_2^s, \bar{z}^f) \end{bmatrix} u \\ \tilde{y}_2 &= \bar{h}(0, \bar{z}_2^s, \bar{z}^f), \end{aligned} \tag{5.28}$$

are both locally asymptotically stable with index one. □

Proof. From Theorem 5.11, it is concluded that both the linearizations of the subsystems (5.27) and (5.28), namely Systems (5.18) and (5.19), are stable and impulse-free. This in turn implies that both Systems (5.27) and (5.28) are asymptotically stable with index one by the Lyapunov's direct method.
Q.E.D.

To conclude this section, we mention the following theorem.

Theorem 5.14. *Let* $G(s)$ *and* $G_r(s)$ *denote the transfer function matrices of the systems (5.18) and (5.23), respectively. Then the following inequality holds*

$$\|G(s) - G_r(s)\|_H \leq \|G(s) - G_r(s)\|_\infty \leq 2(\sigma_{k+1} + \ldots + \sigma_r).$$

□

5.3 Network Synthesis

In this section, we will apply the balanced realization scheme to the positive real network synthesis problem. We will first find a balanced state-space representation for a given positive real descriptor system. This may be called *ESPR-balanced realization.* A formal definition of a synthesis question may be given as follows[2]: Given certain ideal elements which may be interconnected according to certain interconnection laws, what systems can be realized in this way, and, for a given system, what elements should be used and how should they be interconnected? Inspired by the above definition, we will focus our attention on the synthesizing problem in which a network can be realized by means of some ideal elements, like resistors, capacitors, inductors and transformers, but no gyrators.

Throughout this section, we consider the system

$$\begin{aligned} E\dot{x} &= Ax + Bu \\ y &= Cx + Du \\ Ex(0^-) &= Ex_0, \quad G(s) \triangleq D + C(sE - A)^{-1}B, \end{aligned} \tag{5.29}$$

together with its Weierstrass from representation

$$E = \begin{bmatrix} I & 0 \\ 0 & 0 \end{bmatrix}, \quad A = \begin{bmatrix} A_1 & 0 \\ 0 & -I \end{bmatrix}, \quad B = \begin{bmatrix} B_1 \\ B_2 \end{bmatrix}, \text{and} \quad C = \begin{bmatrix} C_1 & C_2 \end{bmatrix}. \tag{5.30}$$

The following proposition connects the GARE with the generalized linear matrix inequality(GLMI).

Proposition 5.15. *Suppose that* $\Delta \triangleq D + D^T > 0$. *Then GARE*

$$\begin{cases} A^T X + X^T A + (X^T B - C^T)\Delta^{-1}(X^T B - C^T)^T = 0, \\ E^T X = X^T E \end{cases}, \tag{5.31}$$

has an admissible solution if and only if there exists a matrix P *satisfying the GLMI:*

$$\begin{bmatrix} -A^T P - P^T A & P^T B - C^T \\ B^T P - C & (D + D^T) \end{bmatrix} > 0, \qquad E^T P = P^T E \geq 0. \tag{5.32}$$

□

Proof. The proof follows by identifying that the existence of a solution to the GLMI is equivalent to the existence of a solution to the following generalized algebraic Riccati inequality:

$$\begin{cases} A^T P + P^T A + (P^T B - C^T)\Delta^{-1}(P^T B - C^T)^T < 0, \\ E^T P = P^T E \geq 0. \end{cases}$$

The later is equivalent to the existence of admissible solutions to the GARE (5.31). **Q.E.D.**

Remark 1. If both the finite and infinite modes of the system (5.29) is controllable and observable , then, without loss of generality, we can suppose that P is a symmetric block diagonal matrix with the partition coinciding with the Weierstrass form.

Before proceeding to the synthesis problem, we should first introduce the notion of reciprocity.

Definition 5.16. *Consider System (5.29) with transfer function matrix $G(s)$. Then $G(s)$ is said to be reciprocal if $G(s) = G^T(s)$.* □

The above definition is the most simple and straightforward one. For more details, see [28][121]. The notion of reciprocity has special importance in classical dynamics and circuit theory(e.g. thermodynamics, electric circuit synthesis...) especially in a discussion of the time reversibility of dynamical systems. In fact, a time-reversible system must be reciprocal. Since our goal is network synthesis, we will not go further into this notion; some basic ideas will be enough. The following lemma which converts the frequency domain condition of reciprocity to an algebraic constraint, is useful in the synthesis problem.

Lemma 5.17. (Reciprocity)
Suppose that the realization of System (5.29) is minimal (i.e., both finite and infinite modes are controllable and observable). Suppose that $\{E, A\}$ is impulse-free and $D = D^T$. Then $G(s)$ is reciprocal if and only if there exists a nonsingular matrix P satisfying the following conditions:

$$\begin{aligned} P^T A &= A^T P, \\ E^T P &= P^T E, \\ P^T B &= -C^T. \end{aligned} \tag{5.33}$$

□

Proof.
Sufficiency. It is easy to verify that $G(s) = G^T(s)$ by direct substituting (5.33) into the state space representation of $G(s)$.

Necessity.
Suppose that $G(s) = G^T(s)$. By setting $s = \infty$, we immediately have

$$C_2 B_2 = B_2^T C_2^T \tag{5.34}$$

since, by hypothesis, $D = D^T$. Observe that (5.34) holds if and only if

$$C_2(sI + I)^{-1} B_2 = B_2^T (sI + I)^{-1} C_2^T$$

holds for all s; i.e., $G_{22} \triangleq C_2(sI+I)^{-1}B_2$ is symmetric. Then, by the standard results for linear time-invariant reciprocal systems [2], this implies that there exists a symmetric matrix P_{22} such that

$$\begin{aligned} P_{22}(-I) &= (-I)^T P_{22}, \\ P_{22}B_2 &= -C_2^T. \end{aligned} \tag{5.35}$$

Next, the symmetry of $G(s)$ implies that

$$\begin{aligned} &\begin{bmatrix} C_1 \; C_2 \end{bmatrix} \begin{bmatrix} (sI - A_1)^{-1} & 0 \\ 0 & I \end{bmatrix} \begin{bmatrix} B_1 \\ B_2 \end{bmatrix} \\ &= \begin{bmatrix} B_1^T \; B_2^T \end{bmatrix} \begin{bmatrix} (sI - A_1)^{-T} & 0 \\ 0 & I \end{bmatrix} \begin{bmatrix} C_1^T \\ C_2^T \end{bmatrix}. \end{aligned}$$

This, in turn, implies that $C_1(sI - A_1)^{-1}B_1 = B_1^T(sI - A_1)^{-T}C_1^T$. Since the realization is assumed to be minimal, this, again by the standard results for linear time-invariant reciprocal systems, implies that there exists a matrix P_{11} such that

$$\begin{aligned} P_{11}A_1 &= A_1^T P_{11}, \\ P_{11}B_1 &= -C_1^T. \end{aligned} \tag{5.36}$$

Combining (5.35) and (5.36), and setting $P = \begin{bmatrix} P_{11} & 0 \\ 0 & P_{22} \end{bmatrix}$, we have

$$\begin{aligned} P^T A &= A^T P, \\ E^T P &= P^T E, \\ P^T B &= -C^T. \end{aligned}$$

Note that in these particular coordinate bases P is symmetric, but, in general, P does not need to be symmetric, we only require that $E^T P$ be symmetric. Therefore the superscript "T" is needed in (5.33). **Q.E.D.**

Consider again system (5.29), and suppose that it is ESPR and reciprocal. Then, from Lemma 5.17, there exists a nonsingular matrix P such that (5.33) is satisfied. Note that, without loss of generality, we can assume that P is symmetry and block-diagonal. As long as P is symmetric, we can always find three matrices[2][32], Σ, V, and H, where V is orthogonal, H is symmetric positive definite and Σ is a signature matrix(i.e., a diagonal matrix with diagonal entries either $+1$ or -1) such that

$$P = HV\Sigma V^T = V\Sigma V^T H. \tag{5.37}$$

Furthermore, $V\Sigma V^T$ commutes with $H^{\frac{1}{2}}$, the unique positive definite square root of H. In fact, some further observations can be made: these matrices are all in block diagonal form since P is a block diagonal matrix.

On the other hand, since $G(s)$ is ESPR, there exists a nonsingular symmetric block diagonal matrix $\hat{P}$ such that (5.32) is satisfied. Define a equivalent transformation:

$$\begin{aligned}\{MEN, MAN, MB, N^{-1}C, D\} &= \{\hat{P}^T E, \hat{P}^T A, \hat{P}^T B, C, D\} \\ &\triangleq \{\bar{E}, \bar{A}, \bar{B}, \bar{C}, \bar{D}\},\end{aligned}$$

where $M \triangleq \hat{P}^T$ and $N \triangleq I$. In particular, the transformed system will satisfy the following simplified GLMI:

$$\begin{bmatrix} -\bar{A} - \bar{A}^T & \bar{B} - \bar{C}^T \\ \bar{B}^T - \bar{C} & (\bar{D} + \bar{D}^T) \end{bmatrix} > 0, \qquad \bar{E} = \bar{E}^T \geq 0. \tag{5.38}$$

In fact, some further simplification can be made. Observe that

$$\bar{E} = \begin{bmatrix} P_{11} & 0 \\ 0 & 0 \end{bmatrix},$$

where P_{11}[1] is a positive definite matrix. Again, define a new transformation matrix

$$T_1 \triangleq \begin{bmatrix} J & 0 \\ 0 & I \end{bmatrix},$$

where J is the unique positive definite square root of P_{22}. Then the resulting equivalent system,

$$\{\bar{E}_1, \bar{A}_1, \bar{B}_1, \bar{C}_1, \bar{D}_1\} \triangleq \{T_1^{-1}\bar{E}T_1^{-1}, T_1^{-1}\bar{A}T_1^{-1}, T_1^{-1}\bar{B}, \bar{C}T_1^{-1}, \bar{D}\},$$

also satisfys Inequality (5.38) but with $\bar{E}_1 = \begin{bmatrix} I & 0 \\ 0 & 0 \end{bmatrix}$.

In order to simplify our notation, we will use the same notation as in (5.38) but, without loss of generality, assume that

$$\bar{E} = \begin{bmatrix} I & 0 \\ 0 & 0 \end{bmatrix}.$$

With Equation (5.37) and Inequality (5.38) in hand, we have the following very important lemma.

Lemma 5.18. (Realizability: Without Gyrator)
Let $\{\bar{E}, \bar{A}, \bar{B}, \bar{C}, \bar{D}\}$ be a minimal realization of a ESPR reciprocal $G(s)$ satisfying the special GLMI (5.38) with $\bar{D} = \bar{D}^T$. Let V, H, and Σ be defined as in (5.37). Perform an equivalent transformation on $G(s)$ to get

[1] Note that we have used the same symbol in a different context, but one should not identify these symbols as being the same.

$$\{E_0, A_0, B_0, C_0, D_0\} = \{T\bar{E}T^{-1}, T\bar{A}T^{-1}, T\bar{B}, \bar{C}T^{-1}, \bar{D}\}$$

where the matrix T is given by

$$T = V^T H^{\frac{1}{2}}. \tag{5.39}$$

Then, the matrix

$$M_0 = \begin{bmatrix} -A_0 & B_0 \\ -C_0 & D \end{bmatrix}$$

will satisfy the following properties:

$$M_0 + M_0^T \geq 0, \tag{5.40}$$

$$E_0 = E_0^T \geq 0, \tag{5.41}$$

$$\operatorname{diag}[\Sigma, I]\, M_0 = M_0^T \operatorname{diag}[\Sigma, I]\,. \tag{5.42}$$

□

Proof.
The proof is fairly straightforward; by some simple direct calculations we can easily obtain (5.40) and (5.42). It follows that $E_0 \geq 0$ from the fact that T is block-diagonal. **Q.E.D.**

Remark 2. A great more deal can be said about (5.40), (5.41) and (5.40).
(i) A network is realizable without gyrators if and only if (5.40) and (5.42) hold. Each of the -1's in the signature matrix Σ corresponds to a capacitor in the realization, and each of the $+1$'s corresponds to an inductor.

(ii) The impedance matrix sE_0 is simply synthesizable, for the nonnegativity of $\bar{E}$ guarantees that a matrix L exists with n rows and $l = \text{rank}E_0$ columns such that

$$E_0 = L^T L.$$

Then unit inductors(or capacitors) terminating the secondary ports of a multiport transformer of turns-ratio matrix L yields a synthesis of sE_0.

Based on Lemma 5.18 and the above remarks, a passive reciprocal synthesis is readily obtained. The constant hybrid matrix M_0 is synthesizable with a network of multiport ideal transformers[9] and positive resistors only. A synthesis of the prescribed $G(s)$ then follows by appropriately terminating this network in sE_0, where sE_0 is simply realizable as described in Remark 2. (ii) (see Fig. 5.1).

Remark 3.
(i) The transfer function matrix $G(s)$ is, in general, a hybrid matrix. Therefore, we use V^i, the excitation varibles, and V^r, the response variables instead of currents or voltages.

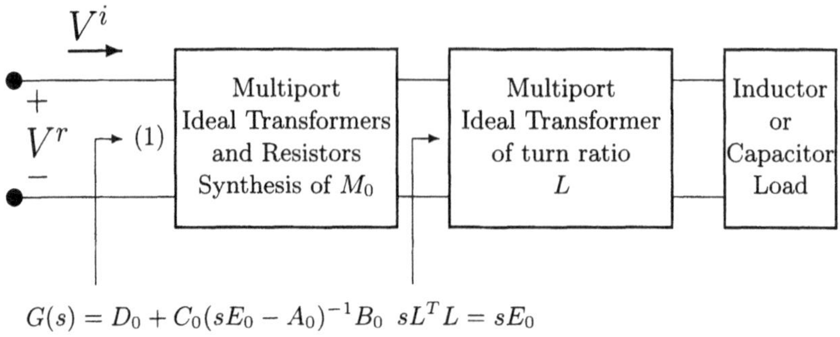

Fig. 5.1. Reciprocal ESPR network synthesis

(ii) The connection between M_0 and sE_0 is called a cascade-load connection. Suppose $M_0 = \begin{bmatrix} M_{11} & M_{12} \\ M_{21} & M_{22} \end{bmatrix}$. Then, the transfer function matrix looking into port **(1)**(see Fig. 5.1) can be obtained as follows.

$$G(s) \triangleq \frac{V^r}{V^i} = M_{22} - M_{21}(sE_0 - M_{11})^{-1}M_{12}.$$

(iii) It is obvious that the previous synthesis is, in fact, in a output-normal form.

Notes and Reference

Balanced realization first appeared with the work of Mullis and Roberts[86], who were interested in realizations of digital filters that are optimal with respect to round-off errors in the state update calculations. The balanced truncation that was used as a method to reduce the order of a stable linear was then initiated by Moore[84]. The twice-the-sum-of-the-tail infinity norm error bound for balanced truncation is due to Enns[27]. He also proposed a frequency weighting method, in order to have great flexibility and to reach more satisfactory results. The twice-the-sum-of-the-tail error bound was also proved by Glover[34] using a different approach. The LQG balancing was introduced by Jonckheere and Silverman[52][53], while the $\mathcal{H}_\infty$ balancing for linear systems being introduced by Mustafa and Glover [88][89][90], to overcome the drawback of the original balancing method, because it only applied to stable systems and emphasized the open-loop input-output characteristics of the systems. Most of the results for linear systems were then extended to the nonlinear systems by Scherpen in a series of paper [100][101][103]. See also[102].

6

Some Further Topics

6.1 An Algebraic Approach to Reduced-Order Controller Design

As we have shown in Chapter 4, the full-order $\mathcal{H}_\infty$ controllers can be constructed from two GAREs for linear descriptor systems or two Hamilton–Jacobi inequalities for nonlinear descriptor systems. The controllers thus obtained have a state dimension not less than that of the generalized plant. Since the generalized plant is built from the physical plant and some weighting functions that are used to reflect performance and robustness requirements, the order of the genralized plant may be very high. Under the circumstances, the full-order controlers may be of limited use in practical applications. On the other hand, since the plant is in descriptor form, the controllers solving the corresponding problem are, in general, also in descriptor form. The above discussions sum up to motivate the work of constructing reduced-order controllers that are in conventional state-space form, so that they can be implemented more easily. The present section is largely based on Wang *et al.* [118]

Consider, as before, a nonlinear descriptor system described by a set of differential-algebraic equations, which is reproduced here for the sake of clarity.

$$\boldsymbol{\Sigma} \triangleq \begin{cases} \boldsymbol{E}\dot{\boldsymbol{x}} = \boldsymbol{F}(\boldsymbol{x},\, \boldsymbol{w},\, \boldsymbol{u}),\ \boldsymbol{w} \in \mathbb{W} \subset \mathbb{R}^l,\ \ \boldsymbol{u} \in \mathbb{U} \subset \mathbb{R}^m, \\ \boldsymbol{z} = \boldsymbol{Z}(\boldsymbol{x},\, \boldsymbol{w},\, \boldsymbol{u}),\ \boldsymbol{z} \in \mathbb{Z} \subset \mathbb{R}^s, \\ \boldsymbol{y} = \boldsymbol{Y}(\boldsymbol{x},\, \boldsymbol{w},\, \boldsymbol{u}),\ \boldsymbol{y} \in \mathbb{Y} \subset \mathbb{R}^p, \end{cases} \tag{6.1}$$

where $\boldsymbol{x} \in \mathcal{X}$. Here $\boldsymbol{u}$ stands for the vector of control inputs, $\boldsymbol{w}$ is the exogenous input (disturbances to-be-rejected or signals to-be-tracked), $\boldsymbol{y}$ is the measured output, and finally $\boldsymbol{z}$ denotes the to-be-controlled outputs (tracking errors, cost variables). It is assumed, as before, that $\boldsymbol{F}(0,\, 0,\, 0) = 0,\ \boldsymbol{Z}(0,\, 0,\, 0) = 0$ and $\boldsymbol{Y}(0,\, 0,\, 0) = 0$. Here we again restrict ourselves to the consideration of systems satisfying the similar standing assumptions made in Chapter 4. For convenience, we reproduce them in the following.

(NA1) The matrix $\boldsymbol{D}_{12}$ has rank m and the matrix $\boldsymbol{D}_{11}^T\boldsymbol{D}_{11}-\gamma^2\boldsymbol{I}$ is negative definite, where $\boldsymbol{D}_{12} = (\frac{\partial \boldsymbol{Z}}{\partial \boldsymbol{u}})_{(\boldsymbol{x},\boldsymbol{w},\boldsymbol{u})=(0,0,0)}$ and $\boldsymbol{D}_{11} = (\frac{\partial \boldsymbol{Z}}{\partial \boldsymbol{w}})_{(\boldsymbol{x},\boldsymbol{w},\boldsymbol{u})=(0,0,0)}$.

(NA2) Any bounded trajectory $\boldsymbol{x}(t)$ of the system $\boldsymbol{E}\dot{\boldsymbol{x}}(t) = F(\boldsymbol{x}(t),\, 0,\, \boldsymbol{u}(t))$ satisfying $\boldsymbol{Z}(\boldsymbol{x}(t),\, 0,\, \boldsymbol{u}(t)) = 0$ for all $t \geq 0$ is such that $\lim_{t\to\infty} \boldsymbol{x}(t) = 0$.

(NA3) The matrix pencil

$$\begin{bmatrix} \boldsymbol{A} - \mathrm{j}\omega \boldsymbol{E} & \boldsymbol{B}_2 \\ \boldsymbol{C}_1 & \boldsymbol{D}_{12} \end{bmatrix}$$

has full column rank for all $\omega \in \mathbb{R}\cup\{\infty\}$, where $\boldsymbol{A} = (\frac{\partial \boldsymbol{F}}{\partial \boldsymbol{x}})_{(\boldsymbol{x},\boldsymbol{w},\boldsymbol{u})=(0,0,0)}$, $\boldsymbol{B}_2 = (\frac{\partial \boldsymbol{F}}{\partial \boldsymbol{u}})_{(\boldsymbol{x},\boldsymbol{w},\boldsymbol{u})=(0,0,0)}$, and $\boldsymbol{C}_1 = (\frac{\partial \boldsymbol{Z}}{\partial \boldsymbol{x}})_{(\boldsymbol{x},\boldsymbol{w},\boldsymbol{u})=(0,0,0)}$.

(NA4) The matrix $\boldsymbol{D}_{21} = (\frac{\partial \boldsymbol{Y}}{\partial \boldsymbol{w}})_{(\boldsymbol{x},\boldsymbol{w},\boldsymbol{u})=(0,0,0)}$ has rank p.

Consider the following reduced-order controller, which is in a conventional state-space form.

$$\begin{aligned} \dot{\boldsymbol{\zeta}}(t) &= \hat{\boldsymbol{F}}_r(\boldsymbol{\zeta}(t)) + \hat{\boldsymbol{G}}_r(\boldsymbol{\zeta}(t))\boldsymbol{y}(t), \\ \boldsymbol{u}(t) &= \hat{\boldsymbol{H}}_r(\boldsymbol{\zeta}(t)), \end{aligned} \tag{6.2}$$

where $\boldsymbol{\zeta} \in \mathbb{R}^r$ is defined on a neighborhood of the origin. Assume that $\hat{\boldsymbol{F}}_r(0) = 0$ and $\hat{\boldsymbol{H}}_r(0) = 0$. Our aim is to find a reduced-order output feedback controller of the form (6.2), such that the resulting closed-loop system has a locally asymptotically stable equilibrium point at the origin with index one, and has a $\mathcal{L}_2$ gain less than or equal to a prescribed attenuation level γ. Recall that a full-order controller and a family of parameterized output feedback controllers are provided in Chapter 4, Theorem 4.3 and Theorem 4.6, respectively. Let's first make a change of variables defined as follows.

$$\hat{\boldsymbol{\zeta}} \triangleq -\boldsymbol{\phi}(\boldsymbol{x}) + \boldsymbol{\zeta}, \tag{6.3}$$

where $\boldsymbol{\phi}(\bullet) : \mathcal{X} \to \mathbb{R}^r$ is a smooth function locally defined around the equilibrium point with $\boldsymbol{\phi}(0) = 0$. Then, in terms of the new state variables, the closed-loop system can be written as:

$$\begin{aligned} \boldsymbol{E}^e\dot{\boldsymbol{x}}^e &= \boldsymbol{F}^e(\boldsymbol{x}^e,\, \boldsymbol{w}) \\ \boldsymbol{z} &= \boldsymbol{H}^e(\boldsymbol{x}^e), \end{aligned} \tag{6.4}$$

where $\boldsymbol{x}^e = col(\boldsymbol{x}, \hat{\boldsymbol{\zeta}})$, $\boldsymbol{E}^e = \begin{bmatrix} \boldsymbol{E} & 0 \\ 0 & \boldsymbol{I} \end{bmatrix}$,

$$\boldsymbol{F}^e(\boldsymbol{x}^e,\, \boldsymbol{w}) = \left[\begin{array}{c} \boldsymbol{F}(\boldsymbol{x},\, \boldsymbol{w},\, \hat{\boldsymbol{H}}_r(\hat{\boldsymbol{\zeta}} + \boldsymbol{\phi}(\boldsymbol{x}))) \\ \hline -\dot{\boldsymbol{\phi}}(\boldsymbol{x})\boldsymbol{F}(\boldsymbol{x},\, \boldsymbol{w},\, \hat{\boldsymbol{H}}_r(\hat{\boldsymbol{\zeta}} + \boldsymbol{\phi}(\boldsymbol{x})) + \hat{\boldsymbol{F}}_r(\hat{\boldsymbol{\zeta}} + \boldsymbol{\phi}(\boldsymbol{x})) \\ +\hat{\boldsymbol{G}}_r(\hat{\boldsymbol{\zeta}} + \boldsymbol{\phi}(\boldsymbol{x}))\boldsymbol{Y}(\boldsymbol{x},\, \boldsymbol{w},\, \hat{\boldsymbol{H}}_r(\hat{\boldsymbol{\zeta}} + \boldsymbol{\phi}(\boldsymbol{x}))) \end{array} \right],$$

and

$$\boldsymbol{H}^e(\boldsymbol{x}^e) = \boldsymbol{H}(\boldsymbol{x}, \boldsymbol{w}, \hat{\boldsymbol{H}}_r(\hat{\zeta} + \boldsymbol{\phi}(\boldsymbol{x}))).$$

Recall that the problem of choosing a control law (6.2) in such a way that the $\mathcal{L}_2$-gain of the closed-loop system (6.4) from the exogeneous input $\boldsymbol{w}$ to the penalty output $\boldsymbol{z}$ is less than of equal to γ can be viewed as a game problem of rendering the Hamiltonian function $\mathfrak{H} : \mathbb{R}^{n+r} \times \mathbb{R}^{n+r} \times \mathbb{R}^l \to \mathbb{R}$, defined as

$$\mathfrak{H}(\boldsymbol{x}^e, \boldsymbol{p}, \boldsymbol{w}) \triangleq \boldsymbol{p}^T \boldsymbol{F}^e(\boldsymbol{x}^e, \boldsymbol{w}) + \|\boldsymbol{H}^e(\boldsymbol{x}^e)\|^2 - \gamma^2 \|\boldsymbol{w}\|^2,$$

nonpositive for all $\boldsymbol{x}^e$ and all $(\boldsymbol{p}, \boldsymbol{w})$; or, in other words, we shall try to render the closed-loop system locally dissipative with respect to the supply rate $\gamma^2\|\boldsymbol{w}\|^2 - \|\boldsymbol{z}\|^2$. Clearly, it suffices to show that there exists a smooth nonnegative function $\mathtt{W}(\boldsymbol{x}^e)$ with $\dfrac{\partial \mathtt{W}}{\partial \boldsymbol{x}^e} = \boldsymbol{W}^T \boldsymbol{E}^e$ and $\boldsymbol{E}^{eT} \boldsymbol{W}_{\boldsymbol{x}} = \boldsymbol{W}_{\boldsymbol{x}}^T \boldsymbol{E}^e$ satisfying

$$\boldsymbol{W}^T \boldsymbol{F}^e(\boldsymbol{x}^e, \boldsymbol{w}) + \|\boldsymbol{H}^e(\boldsymbol{x}^e, \boldsymbol{w})\|^2 - \gamma^2 \|\boldsymbol{w}\|^2 \leq 0, \qquad \text{for all } \boldsymbol{w}, \tag{6.5}$$

such that the closed-loop system is locally asymptotically stable with index one. In this regard, we have the following preliminary result.

Lemma 6.1. *Consider (6.1), (6.2) and (6.4). Suppose that Assumptions (NA1) to (NA4) hold. Suppose also that there exists a smooth nonnegative function* $\mathtt{W}(\boldsymbol{x}^e)$ *locally defined on a neighborhood of the origin in* $\mathbb{R}^{n+r}$, *vanishing at* $\boldsymbol{x}^e \in \Im m \boldsymbol{E}^e$, *that satisfies* $\dfrac{\partial \mathtt{W}}{\partial \boldsymbol{x}^e} = \boldsymbol{W}^T \boldsymbol{E}^e$ *and* $\boldsymbol{E}^{eT} \boldsymbol{W}_{\boldsymbol{x}} = \boldsymbol{W}_{\boldsymbol{x}}^T \boldsymbol{E}^e$. *The function* $\mathtt{W}(\boldsymbol{x}^e)$ *thus defined is such that the Hamilton–Jacobi inequality (dissipation inequality) (6.5) is negative for all nonzero* x^e. *Then the controller (6.2) locally internally stabilizes the resulting closed-loop system (6.4) and renders its* $\mathcal{L}_2$ *gain less than or equal to* γ. *Moreover, the closed-loop system has index one.* □

Proof. The proof that the closed-loop system (6.4) has $\mathcal{L}_2$ gain less than or equal to γ is easy. We need only to show that the closed-loop system is locally asymptotically stable with index one. Regarding this, let's consider the candidate Lyapunov function $\mathtt{W}(\boldsymbol{x}^e)$. Then, it is easy to see that, along any trajectory of the closed-loop system, the following Hamilton–Jacobi inequality holds.

$$\mathfrak{H}(\boldsymbol{x}^e, \boldsymbol{W}^T, w) = \frac{\mathrm{d}\mathtt{W}}{\mathrm{d}t} + \|\boldsymbol{z}\|^2 - \gamma^2 \|\boldsymbol{w}\|^2 < 0. \tag{6.6}$$

Setting $\boldsymbol{w} = 0$ in the above inequality yields

$$\frac{\mathrm{d}\mathtt{W}}{\mathrm{d}t} = -\|\boldsymbol{z}\|^2 < 0,$$

i.e., $\dfrac{\mathrm{d}\mathtt{W}}{\mathrm{d}t}$ is negative definite near $\boldsymbol{x}^e = 0$. This proves that the equilibrium point $\boldsymbol{x}^e = 0$ of the closed-loop system is locally asymptotically stable with index one. **Q.E.D.**

By Assumption (NA1), it is easy toe see that, in a neighborhood of the point $(\boldsymbol{x}^e, \boldsymbol{p}, \boldsymbol{w}) = (0,0,0,0)$, the function $\mathfrak{H}(\boldsymbol{x}^e,\, \boldsymbol{p},\, \boldsymbol{w})$ has a unique local maximal point in $\boldsymbol{w}$ for each $(\boldsymbol{x}^e, \boldsymbol{p})$. More precisely, there exists a unique smooth function $\boldsymbol{w}^*(\boldsymbol{x}, \boldsymbol{p})$, defined in a neighborhood of $(0,0)$, satisfying

$$\frac{\partial \mathfrak{H}}{\partial \boldsymbol{w}}(\boldsymbol{x}^e,\, \boldsymbol{p},\, \boldsymbol{w}^*(\boldsymbol{x},\, \boldsymbol{p})) = 0$$

with $\boldsymbol{w}^*(0,0) = 0$, and such that

$$\mathfrak{H}(\boldsymbol{x}^e, \boldsymbol{p}, \boldsymbol{w}) \leq \mathfrak{H}(\boldsymbol{x}^e, \boldsymbol{p}, \boldsymbol{w}^*(\boldsymbol{x},\, \boldsymbol{p}))$$

for each $(\boldsymbol{x}^e, \boldsymbol{p})$ in a neighborhood of $(\boldsymbol{x}^e, \boldsymbol{p}, \boldsymbol{w}) = (0,0,0)$. Set

$$\mathfrak{H}_*(\boldsymbol{x}^e,\, \boldsymbol{W}^T) := \mathfrak{H}(\boldsymbol{x}^e,\, \boldsymbol{W}^T,\, \boldsymbol{w}^*(\boldsymbol{x}, \boldsymbol{W}^T)).$$

The function thus constructed offers an intermediate result.

Proposition 6.2. *Consider (6.1), (6.2) and (6.4). Suppose that Assumptions (NA1) to (NA4) hold. Suppose also that there exists a smooth nonnegative function $\mathtt{W}(\boldsymbol{x}^e)$ locally defined on a neighborhood of the origin in $\mathbb{R}^{n+r}$, vanishing at $\boldsymbol{x}^e \in \Im m \boldsymbol{E}^e$, that satisfies $\dfrac{\partial \mathtt{W}}{\partial \boldsymbol{x}^e} = \boldsymbol{W}^T \boldsymbol{E}^e$ and $\boldsymbol{E}^{eT} \boldsymbol{W}_{\boldsymbol{x}} = \boldsymbol{W}_{\boldsymbol{x}}^T \boldsymbol{E}^e$. The function $\mathtt{W}(\boldsymbol{x}^e)$ thus defined is such that*

$$\mathfrak{H}_*(\boldsymbol{x}^e,\, \boldsymbol{W}^T) := \mathfrak{H}(\boldsymbol{x}^e,\, \boldsymbol{W}^T,\, w^*(\boldsymbol{x}, \boldsymbol{W}^T))$$

is negative definite for all nonzero $\boldsymbol{x}^e$. Then the controller (6.2) locally internally stabilizes the resulting closed-loop system (6.4) and renders its $\mathcal{L}_2$ gain less than or equal to γ. Moreover, the closed-loop system has index one. □

The remaining problem now is how to choose the to-be-determined parameters $\hat{\boldsymbol{F}}_r(\bullet)$, $\hat{\boldsymbol{G}}_r(\bullet)$, $\hat{\boldsymbol{H}}_r(\bullet)$ and the change-of-the-variables mapping $\phi(\bullet)$ so that the conditions in Lemma 6.1 can be met. One further assumption is needed. Denote by $\mathcal{N} \subset \mathbb{R}^r$ a locally r-dimensional smooth manifold containing the origin.

(NA5) The inverse mapping of $\phi(\bullet) : \mathcal{X} \to \mathbb{R}^r$, denoted by $\phi^{-1}(\bullet) : \mathcal{N} \to \mathbb{R}^n$, is an locally *univalent immersion*, i.e., it is an immersion and is injective. Note that a mapping $F : N \to M$ is said to be an immersion if $\mathrm{rank}(F)=\dim(N)$.

It is worth noting that, under assumption **(NA5)**, $\mathcal{N}$ can be identified with $\phi(\mathcal{X})$. The following theorem provides a sufficient condition.

Theorem 6.3. *Consider (6.1), (6.2) and (6.4). Suppose that Assumptions (NA1) to (NA5) hold, and that Hypotheses (H1) of Lemma 4.1 and (H2) of Theorem 4.3 are satisfied. Let $\phi(\boldsymbol{x}) : \mathcal{X} \to \mathbb{R}^r$ be a smooth function,* *locally defined on a neighborhood of the origin* $\boldsymbol{x} = 0$ *in* $\mathbb{R}^n$ *with* $\phi(0) = 0$ *and* $\boldsymbol{\Phi}(0)\boldsymbol{\Phi}^T(0) = I$. *Here the function* $\boldsymbol{\Phi}(\boldsymbol{x})$ *is defined by* $\dfrac{\partial \phi}{\partial \boldsymbol{x}}(\boldsymbol{x}) = \boldsymbol{\Phi}(\boldsymbol{x})\boldsymbol{E}$. *Moreover, let* $\mathsf{U}(\bullet)$ *be a smooth positive definite function, locally defined on a neighborhood of the origin in* $\mathbb{R}^r$, *which satisfies* $\mathsf{U}(\zeta) = \mathsf{Q}(\phi^{-1}(\zeta))$ *(*$\mathsf{Q}(\boldsymbol{x})$ *is defined as in Theorem 4.3). Choose* $\hat{\boldsymbol{F}}_r(\bullet)$, $\hat{\boldsymbol{G}}_r(\bullet)$ *and* $\hat{\boldsymbol{H}}_r(\bullet)$ *in such a way that the following algebraic equations hold*

$$\boldsymbol{\Phi}(\boldsymbol{x})^T \hat{\boldsymbol{F}}_r(\phi(\boldsymbol{x})) = \boldsymbol{F}_k(\boldsymbol{x}), \tag{6.7}$$

$$\boldsymbol{\Phi}(\boldsymbol{x})^T \hat{\boldsymbol{G}}_r(\phi(\boldsymbol{x})) = \boldsymbol{G}_k(\boldsymbol{x}), \tag{6.8}$$

and

$$\hat{\boldsymbol{H}}_r(\phi(\boldsymbol{x})) = \boldsymbol{H}_k(\boldsymbol{x}). \tag{6.9}$$

Then the $\mathcal{H}_\infty$ *output feedback control problem for the nonlinear descriptor system (6.1) is solved by the r-th order normal controller (6.2).* □

To prove the above theorem, the following lemma is useful. See Lin and Byrnes[71].

Lemma 6.4. *Consider a function of the form*

$$\varphi(x, e) \;=\; e^T P(x) e + R(x, e)$$

where $x \in \mathbb{R}^n$, $e \in \mathbb{R}^r$, *and* $\varphi : \mathbb{R}^n \times \mathbb{R}^r \to \mathbb{R}$. *Suppose* $P(0)$ *is an* $r \times r$ *positive definite matrix and* $P(x)$ *is* C^0. *Suppose also that* $R(x, e)$ *satisfying*

$$R(x, 0) = 0, \quad \frac{\partial R}{\partial e}(x, 0) = 0, \quad \frac{\partial^2 R}{\partial e^2}(x, 0) = 0,$$

is C^k $(k \geq 2)$ *with respect to e. Then there exists a real constant* $c > 0$ *such that* $\varphi(x, e)$ *is positive definite for all* $x \in \mathbb{B}_c(x) \triangleq \{x : \|x\| < c\}$ *and all* $e \in \mathbb{B}_c(e) - \{0\}$. □

Proof of Theorem 6.3. Set $\mathsf{W}(\boldsymbol{x}^e) = \mathsf{V}(\boldsymbol{x}) + \mathsf{U}(\hat{\zeta})$, which is nonnegative by construction. Then, by using (6.7), (6.8) and (6.9), it is straightforward but tedious to veryfied that the function $\mathfrak{H}^*(\boldsymbol{x}^e) := \mathfrak{H}^*(\boldsymbol{x}, \hat{\zeta})$ satisfies $\mathfrak{H}(\boldsymbol{x}, 0) = Y_1(\boldsymbol{x})$, $\mathfrak{H}^*_{\hat{\zeta}}(\boldsymbol{x}, 0) = 0$, and $\mathfrak{H}^*_{\hat{\zeta}\hat{\zeta}} = \boldsymbol{\Phi}(0) Y_{2\boldsymbol{x}\boldsymbol{x}}(0) \boldsymbol{\Phi}^T(0)$. Hence, by the Taylor series expansion theorem, we have

$$\mathfrak{H}^*(\boldsymbol{x}, \hat{\zeta}) = Y_1(\boldsymbol{x}) + \frac{1}{2}\hat{\zeta}^T \mathfrak{H}^*_{\hat{\zeta}\hat{\zeta}} \hat{\zeta} + \text{h.o.t.},$$

locally around $\hat{\zeta} = 0$, where "h.o.t." denotes the high order terms. Since that $Y_{2\boldsymbol{x}\boldsymbol{x}}(0)$ is negative definite under Hypothesis (H2) and $Y_1(\boldsymbol{x})$ is a negative definite function by Hypothesis (H1). This implies that, by Lemma 6.4, $\mathfrak{H}^*(\boldsymbol{x}, \hat{\zeta})$ is negative for all nonzero $\boldsymbol{x}^e$, hence the proof is completed. **Q.E.D.**

The following statement gives the linear version of the previous theorem.

Proposition 6.5. *Assume that the descriptor system $\boldsymbol{\Sigma}$ is described by the set of linear differential-algebraic equations of the form*

$$\begin{aligned} \boldsymbol{E}\dot{\boldsymbol{x}} &= \boldsymbol{A}\boldsymbol{x} + \boldsymbol{B}_1\boldsymbol{w} + \boldsymbol{B}_2\boldsymbol{u} \\ \boldsymbol{z} &= \boldsymbol{C}_1\boldsymbol{x} + \boldsymbol{D}_{12}\boldsymbol{u} \\ \boldsymbol{y} &= \boldsymbol{C}_2\boldsymbol{x} + \boldsymbol{D}_{21}\boldsymbol{w}. \end{aligned} \tag{6.10}$$

Suppose that the matrices $\boldsymbol{R}_1 \triangleq \boldsymbol{D}_{12}^T\boldsymbol{D}_{12}$ and $\boldsymbol{R}_2 \triangleq \boldsymbol{D}_{21}\boldsymbol{D}_{21}^T$ are nonsingular. Suppose also that the following hypotheses are satisfied.

(L1) *There exists a matrix $\boldsymbol{X}$ satisfying the generalized algebraic Riccati inequality (GARI)*

$$\begin{aligned} \mathrm{R}ic_{\boldsymbol{X}}(\boldsymbol{X}) = (&\boldsymbol{A} - \boldsymbol{B}_2\boldsymbol{R}_1^{-1}\boldsymbol{D}_{12}^T\boldsymbol{C}_1)^T\boldsymbol{X} + \boldsymbol{X}^T(\boldsymbol{A} - \boldsymbol{B}_2\boldsymbol{R}_1^{-1}\boldsymbol{D}_{12}^T\boldsymbol{C}_1) \\ &+\boldsymbol{C}_1^T(\boldsymbol{I} - \boldsymbol{D}_{12}\boldsymbol{R}_1^{-1}\boldsymbol{D}_{12}^T)\boldsymbol{C}_1 \\ &+\boldsymbol{X}^T(\frac{1}{\gamma^2}\boldsymbol{B}_1\boldsymbol{B}_1^T - \boldsymbol{B}_2\boldsymbol{R}_1^{-1}\boldsymbol{B}_2^T)\boldsymbol{X} < 0, \\ &\boldsymbol{E}^T\boldsymbol{X} = \boldsymbol{X}^T\boldsymbol{E} \end{aligned}$$

with $\boldsymbol{E}^T\boldsymbol{X} = \boldsymbol{X}^T\boldsymbol{E} \geq 0$.

(L2) *There exists a matrix $\boldsymbol{S}$ satisfying the GARI*

$$\begin{aligned} \mathrm{R}ic_{\boldsymbol{S}} = (&\tilde{\boldsymbol{A}} - \boldsymbol{B}_1\boldsymbol{D}_{21}^T\boldsymbol{R}_2^{-1}\tilde{\boldsymbol{C}}_2)^T\boldsymbol{S} + \boldsymbol{S}^T(\tilde{\boldsymbol{A}} - \boldsymbol{B}_1\boldsymbol{D}_{21}^T\boldsymbol{R}_2^{-1}\tilde{\boldsymbol{C}}_2) \\ &+\tilde{\boldsymbol{F}}_\infty^T\boldsymbol{R}_1\tilde{\boldsymbol{F}}_\infty - \gamma^2\tilde{\boldsymbol{C}}_2^T\boldsymbol{R}_2^{-1} \\ &+\frac{1}{\gamma^2}\boldsymbol{S}^T\boldsymbol{B}_1(\boldsymbol{I} - \boldsymbol{D}_{21}^T\boldsymbol{R}_2^{-1}\boldsymbol{D}_{21})\boldsymbol{B}_1^T\boldsymbol{S} < 0, \\ &\boldsymbol{E}^T\boldsymbol{S} = \boldsymbol{S}^T\boldsymbol{E} \end{aligned}$$

with $\boldsymbol{E}^T\boldsymbol{S} = \boldsymbol{S}^T\boldsymbol{E} \geq 0$, where

$$\tilde{\boldsymbol{A}} = \boldsymbol{A} + \frac{1}{\gamma^2}\boldsymbol{B}_1\boldsymbol{B}_1^T\boldsymbol{X},$$

$$\tilde{\boldsymbol{C}}_2 = \boldsymbol{C}_2 + \frac{1}{\gamma^2}\boldsymbol{D}_{21}^T\boldsymbol{B}_1^T\boldsymbol{X},$$

and

$$\tilde{\boldsymbol{F}}_\infty = -\boldsymbol{R}_1^{-1}(\boldsymbol{B}_2^T\boldsymbol{X} + \boldsymbol{D}_{21}^T\boldsymbol{C}_1).$$

Furthermore, suppose that $\boldsymbol{V}$ is an $r \times n$ matrix which satisfies $\boldsymbol{V}\boldsymbol{V}^T = \boldsymbol{I}$. Then the r-th order controller

$$\begin{aligned} \dot{\zeta}(t) &= \hat{\boldsymbol{F}}_r\zeta(t) + \hat{\boldsymbol{G}}_r\boldsymbol{y}(t) \\ \boldsymbol{u}(t) &= \hat{\boldsymbol{H}}_r\zeta(t) \end{aligned}$$

solves the $\mathcal{H}_\infty$ control problem via output feedback, where

$$\begin{aligned} \boldsymbol{V}^T \hat{\boldsymbol{F}}_r \boldsymbol{V} \boldsymbol{E} &= (\tilde{\boldsymbol{A}} + \boldsymbol{B}_2 \tilde{\boldsymbol{F}}_\infty - \tilde{\boldsymbol{G}} \tilde{\boldsymbol{C}}_2), & (6.11) \\ \boldsymbol{V}^T \hat{\boldsymbol{G}}_r &= \tilde{\boldsymbol{G}}, & (6.12) \end{aligned}$$

and

$$\hat{\boldsymbol{H}}_r \boldsymbol{V} \boldsymbol{E} = \tilde{\boldsymbol{F}}_\infty. \tag{6.13}$$

Here the matrix $\tilde{\boldsymbol{G}}$ is defined by

$$\tilde{\boldsymbol{G}} \triangleq (\gamma^2 \boldsymbol{S}^{-1} \tilde{\boldsymbol{C}}_2^T + \boldsymbol{B}_1 \boldsymbol{D}_{21}^T) \boldsymbol{R}_2^{-1}.$$

□

Proof. Set $\phi(\boldsymbol{x}) := \boldsymbol{V}\boldsymbol{E}\boldsymbol{x}$. Then, simple algebraic calculation shows that the hypotheses in Theorem 6.3 are indeed satisfied by assuming (L1) and (L2). Moreover, Equations (6.7)- (6.9) are exactly reduced to (6.11)-(6.13), respectively. **Q.E.D.**

6.1.1 GARE Approach to the Linear Problem

In this subsection, we will take a different approach to find the reduced-order controller for linear descriptor systems.The approach taken is mainly based on the solutions of two generalized algebraic Riccati equations (GARE), while exploiting the structure of the deflating subspace of the pencil $\{\boldsymbol{E}, \boldsymbol{W}_\infty\}$, where $\boldsymbol{W}_\infty$ is an admissible solution to a GARE that is the descriptor systems counterpart of a certain ARE (algebraic Riccati equation) first developed by Petersen *et al.*[93]. This approach has the advantage that, by proper selection of the bases of the deflating space and suitable assumptions, the reduced-order controller may be in a normal form, namely the $\boldsymbol{E}$-matrix of the controller is nonsingular. We should first review some necessary materials concerning deflating spaces of matrix pencils. Let us first begin with the following definition.

Definition 6.6. *Let $s\boldsymbol{E}-\boldsymbol{A}$ be a regular pencil with $\boldsymbol{E}, \boldsymbol{A} \in \mathbb{R}^{n\times n}$. The linear space $\mathcal{V}$ is called a deflating subspace of $s\boldsymbol{E}-\boldsymbol{A}$ if* $\dim(\boldsymbol{E}\mathcal{V} + \boldsymbol{A}\mathcal{V}) = \dim\mathcal{V}$. □

The following two theorems are taken from Lancaster and Rodman[63].

Theorem 6.7. *Let $s\boldsymbol{E}-\boldsymbol{A}$ be a regular pencil with* $\det(\lambda_0 \boldsymbol{E} - \boldsymbol{A}) \neq 0$. *Then a subspace $\mathcal{S}$ is invariant under $(\lambda_0\boldsymbol{E}-\boldsymbol{A})^{-1}\boldsymbol{E}$ if and only if $\mathcal{S}$ is deflating for the pair $\{\boldsymbol{E}, \boldsymbol{A}\}$.* □

In the case that $\boldsymbol{E} = \boldsymbol{I}$ then $\mathcal{S}$ is $\boldsymbol{A}$-invariant, and so this concept generalizes that of an invariant subspace.

Theorem 6.8. *Let $\lambda\boldsymbol{E}-\boldsymbol{A}$ be a regular matrix pencil. If $\mathcal{S}$ is an r-dimensional deflating subspace for $\{\boldsymbol{E}, \boldsymbol{A}\}$ then there are nonsingular matrices $\boldsymbol{T}_1$ and $\boldsymbol{T}_2$ for which*

$$\boldsymbol{T}_1 \boldsymbol{E} \boldsymbol{T}_2 = \begin{bmatrix} \boldsymbol{E}_1 & \boldsymbol{E}_2 \\ 0 & \boldsymbol{E}_3 \end{bmatrix}, \quad \boldsymbol{T}_1 \boldsymbol{A} \boldsymbol{T}_2 = \begin{bmatrix} \boldsymbol{A}_1 & \boldsymbol{A}_2 \\ 0 & \boldsymbol{A}_3 \end{bmatrix}, \tag{6.14}$$

where $\boldsymbol{E}_1$, $\boldsymbol{A}_1$ are of size $r \times r$, and $\mathcal{S}$ is the span of the first r columns of $\boldsymbol{T}_2$. Conversely, if (6.14) holds for nonsingular matrices $\boldsymbol{T}_1$ and $\boldsymbol{T}_2$, and $\boldsymbol{V}_1$ is the leading $n \times r$ partition of $\boldsymbol{T}_2$ then $\Im m \boldsymbol{V}_1$ is deflating for $\{\boldsymbol{E}, \boldsymbol{A}\}$. □

We are now ready to state the main result of this subsection. The state-space formula of the reduced controller is based on the central controller given in Theorem 4.7. Henceforth, we suppose the conditions (i) and (ii) of Theorem 4.7 hold. In this case, according to the proof of Theorem 4.7, the following GARE

$$\begin{array}{l} \mathbf{R}ic_4(\boldsymbol{W}) = \boldsymbol{A}_0^T \boldsymbol{W} + \boldsymbol{W}^T \boldsymbol{A}_0 + \frac{1}{\gamma^2} \boldsymbol{W}^T \boldsymbol{B}_0 \boldsymbol{B}_0^T \boldsymbol{W} \\ + \boldsymbol{C}_0^T \boldsymbol{C}_0 = 0, \ \boldsymbol{W}^T \boldsymbol{E} = \boldsymbol{E}^T \boldsymbol{W} \end{array} \tag{6.15}$$

has also an admissible solution $\boldsymbol{W}_\infty$ with $\boldsymbol{W}_\infty^T \boldsymbol{E} = \boldsymbol{E}^T \boldsymbol{W}_\infty \geq 0$, where

$$\begin{aligned} \boldsymbol{A}_0 &= \tilde{\boldsymbol{A}} - \boldsymbol{B}_1 \boldsymbol{D}_{21}^T \boldsymbol{R}_2^{-1} \tilde{\boldsymbol{C}}_2 - \boldsymbol{Z}_\infty^T \tilde{\boldsymbol{C}}_2^T \boldsymbol{R}_2^{-1} \tilde{\boldsymbol{C}}_2 \\ \boldsymbol{B}_0 &= \tilde{\boldsymbol{B}}_1 - \boldsymbol{Z}_\infty^T \tilde{\boldsymbol{C}}_2^T \boldsymbol{R}_2^{-1} \boldsymbol{D}_{21} \\ \boldsymbol{C}_0 &= -\boldsymbol{R}_1^{-\frac{1}{2}} (\boldsymbol{B}_2^T \boldsymbol{X}_\infty + \boldsymbol{D}_{21}^T \boldsymbol{C}_1) = \boldsymbol{R}_1^{\frac{1}{2}} \boldsymbol{F}_\infty \end{aligned}$$

Without loss of generality, assume that the pair $\{\boldsymbol{E}, \boldsymbol{W}_\infty\}$ has the following form.

$$\{\boldsymbol{E}, \boldsymbol{W}_\infty\} = \left\{ \begin{bmatrix} \boldsymbol{I} & 0 \\ 0 & 0 \end{bmatrix}, \begin{bmatrix} \boldsymbol{W}_\infty^{11} & 0 \\ \boldsymbol{W}_\infty^{21} & \boldsymbol{W}_\infty^{22} \end{bmatrix} \right\},$$

where the matrices $\boldsymbol{W}_\infty^{11}$ and $\boldsymbol{W}_\infty^{22}$ are symmetric. For technical reasons, we assume that $\{\boldsymbol{E}, \boldsymbol{W}_\infty\}$ is regular and impulse-free. Let $\Im m \boldsymbol{E}^T \boldsymbol{W}_\infty$ be an $\hat{r}$-dimensional vector space, i.e., $\boldsymbol{W}_\infty^{11}$ is of rank $\hat{r}$. Then, we can choose a matrix $\hat{\boldsymbol{V}}$ such that its columns form an orthonormal basis of $\Im m \boldsymbol{W}_\infty^{11}$, and a matrix $\hat{\boldsymbol{U}}$ such that its columns forms an orthonormal basis of Ker$\boldsymbol{W}_\infty^{11}$. Under these circumstances, it can be shown that

$$\begin{bmatrix} \hat{\boldsymbol{V}}^T \\ \hat{\boldsymbol{U}}^T \end{bmatrix} \boldsymbol{W}_\infty^{11} [\hat{\boldsymbol{V}} \ \hat{\boldsymbol{U}}] = \begin{bmatrix} \hat{\boldsymbol{W}} & 0 \\ 0 & 0 \end{bmatrix}$$

where the matrix $\hat{\boldsymbol{W}} = \hat{\boldsymbol{V}}^T \boldsymbol{W}_\infty^{11} \hat{\boldsymbol{V}} \geq 0$. As a consequence of the orthogonality of the matrix $\begin{bmatrix} \hat{\boldsymbol{V}} \ \hat{\boldsymbol{U}} \end{bmatrix}$, we have

$$\boldsymbol{W}_\infty^{11} = [\hat{\boldsymbol{V}} \ \hat{\boldsymbol{U}}] \begin{bmatrix} \hat{\boldsymbol{W}} & 0 \\ 0 & 0 \end{bmatrix} \begin{bmatrix} \hat{\boldsymbol{V}}^T \\ \hat{\boldsymbol{U}}^T \end{bmatrix} = \hat{\boldsymbol{V}} \hat{\boldsymbol{W}} \hat{\boldsymbol{V}}^T$$

Next, in a similar way, let $\tilde{\boldsymbol{V}}$ be an orthonormal basis matrix of $\Im m \boldsymbol{W}_\infty^{22}$. Define the following two matrices

$$\boldsymbol{V} = \begin{bmatrix} [\hat{\boldsymbol{V}} \ \hat{\boldsymbol{U}}] & 0 \\ 0 & \tilde{\boldsymbol{V}} \end{bmatrix}, \quad \boldsymbol{Q} = \begin{bmatrix} \boldsymbol{I} & 0 \\ -(\boldsymbol{W}_\infty^{22})^{-1} \boldsymbol{W}_\infty^{21} & \boldsymbol{I} \end{bmatrix},$$

where the inverse of $\boldsymbol{W}_\infty^{22}$ exists provided that $\{\boldsymbol{E}, \boldsymbol{W}_\infty\}$ is impulse-free. It is therefore not difficult to see

$$\boldsymbol{V}^T\boldsymbol{E}\boldsymbol{Q}\boldsymbol{V} = \begin{bmatrix} \boldsymbol{I} & 0 \\ 0 & 0 \end{bmatrix}, \quad \boldsymbol{V}^T\boldsymbol{W}_\infty\boldsymbol{Q}\boldsymbol{V} = \begin{bmatrix} \begin{bmatrix} \hat{\boldsymbol{W}} & 0 \\ 0 & 0 \end{bmatrix} & 0 \\ 0 & \tilde{\boldsymbol{W}} \end{bmatrix}.$$

Here the actual form of $\tilde{\boldsymbol{W}}$ is not important. It is then straightforward to show that the following identity holds.

$$\boldsymbol{V}_1^T\boldsymbol{W}_\infty = \begin{bmatrix} \hat{\boldsymbol{W}} & 0 \\ 0 & \tilde{\boldsymbol{W}} \end{bmatrix} \boldsymbol{V}_2 \triangleq \boldsymbol{W}_2\boldsymbol{V}_2,$$

where

$$\boldsymbol{V}_1 = \begin{bmatrix} \hat{\boldsymbol{V}} & 0 \\ 0 & \tilde{\boldsymbol{V}} \end{bmatrix}, \text{ and } \boldsymbol{V}_2 = \boldsymbol{V}_1^T\boldsymbol{Q}^{-1}$$

Motivated by the work of Yung[136], we will now claim that $\Im m \begin{bmatrix} \hat{\boldsymbol{V}} \\ 0 \end{bmatrix}$ is a deflating subspace for $\{\boldsymbol{E}^T, \hat{\boldsymbol{A}}_0^T\}$. Since the pair $\{\boldsymbol{E}, \boldsymbol{A}_0\}$ is regular and impulse-free, it is assumed without loss of generality that they have the following form

$$\{\boldsymbol{E}, \boldsymbol{A}_0\} = \left\{ \begin{bmatrix} \boldsymbol{I} & 0 \\ 0 & 0 \end{bmatrix}, \begin{bmatrix} \boldsymbol{A}_1 & 0 \\ 0 & \boldsymbol{I} \end{bmatrix} \right\}. \tag{6.16}$$

Lemma 6.9. *Let the pair $\{\boldsymbol{E}, \boldsymbol{W}_\infty\}$ be regular and impulse-free. Then the equations*

$$\boldsymbol{E}_{\hat{r}}\boldsymbol{V}_1^T\boldsymbol{Q}^{-1} = \boldsymbol{V}_1^T\boldsymbol{E}, \text{ and } \boldsymbol{F}\boldsymbol{V}_1^T\boldsymbol{Q}^{-1} = \boldsymbol{V}_1^T\hat{\boldsymbol{A}}_0 \tag{6.17}$$

has a solution $\boldsymbol{F}$, where $\boldsymbol{E}_{\hat{r}} = \begin{bmatrix} \boldsymbol{I}_{\hat{r}} & 0 \\ 0 & 0 \end{bmatrix}$. □

Proof. Let $\boldsymbol{B}_0$ and $\boldsymbol{C}_0$ be partitioned as follows

$$\boldsymbol{B}_0 = \begin{bmatrix} \boldsymbol{B}_{01} \\ \boldsymbol{B}_{02} \end{bmatrix}, \ \boldsymbol{C}_0 = \begin{bmatrix} \boldsymbol{C}_{01} & \boldsymbol{C}_{02} \end{bmatrix},$$

where the partitions are compatible with (6.16).

Observe that GARE $Ric_3(\boldsymbol{W}_\infty)$ can be written as

$$\begin{aligned} Ric_3(\boldsymbol{W}_\infty) &= \boldsymbol{A}_0^T\boldsymbol{W}_\infty + \boldsymbol{W}_\infty^T\boldsymbol{A}_0 + \tfrac{1}{\gamma^2}\boldsymbol{W}_\infty^T\boldsymbol{B}_0\boldsymbol{B}_0^T\boldsymbol{W}_\infty \\ &\quad + \boldsymbol{C}_0^T\boldsymbol{C}_0 \\ &= \begin{bmatrix} (1,1) & (1,2) \\ (2,1) & (2,2) \end{bmatrix}, \end{aligned}$$

where the $(1,1)$-block is an ARE given by

$$\begin{aligned}&(\boldsymbol{A}_1+\frac{1}{\gamma^2}\boldsymbol{B}_{01}\boldsymbol{B}_{02}^T\boldsymbol{W}_\infty^{21})^T\boldsymbol{W}_\infty^{11}+\boldsymbol{W}_\infty^{11}(\boldsymbol{A}_1+\frac{1}{\gamma^2}\boldsymbol{B}_{01}\boldsymbol{B}_{02}^T\boldsymbol{W}_\infty^{21})\\&+\boldsymbol{C}_{01}^T\boldsymbol{C}_{01}+\frac{1}{\gamma^2}(\boldsymbol{W}_\infty^{21})^T\boldsymbol{B}_{02}\boldsymbol{B}_{02}^T\boldsymbol{W}_\infty^{21}+\frac{1}{\gamma^2}\boldsymbol{W}_\infty^{11}\boldsymbol{B}_{01}\boldsymbol{B}_{01}^T\boldsymbol{W}_\infty^{11}=0.\end{aligned} \tag{6.18}$$

Let $v \in \mathrm{Ker}\boldsymbol{W}_\infty^{11}$. Pre-multiplying (6.18) by v^T and post-multiplying by v yields

$$v^T\boldsymbol{C}_{01}^T\boldsymbol{C}_{01}v+\frac{1}{\gamma^2}v^T((\boldsymbol{W}_\infty^{21})^T\boldsymbol{B}_{02}\boldsymbol{B}_{02}^T\boldsymbol{W}_\infty^{21})v=0. \tag{6.19}$$

Equation (6.19) implies that $\boldsymbol{C}_{01}v = 0$ and $\boldsymbol{B}_{02}^T\boldsymbol{W}_\infty^{21}v = 0$. Therefore $\mathrm{Ker}\boldsymbol{W}_\infty^{11} \subset \mathrm{Ker}\boldsymbol{C}_{01}$ and $\mathrm{Ker}\boldsymbol{W}_\infty^{11} \subset \mathrm{Ker}\boldsymbol{B}_{02}^T\boldsymbol{W}_\infty^{21}$. Next, post-multiply (6.18) by v to get

$$\boldsymbol{W}_\infty^{11}(\boldsymbol{A}_1+\frac{1}{\gamma^2}\boldsymbol{B}_{01}\boldsymbol{B}_{02}^T\boldsymbol{W}_\infty^{21})v=\boldsymbol{W}_\infty^{11}\boldsymbol{A}_1v=0,$$

which in turn implies that $\mathrm{Ker}\boldsymbol{W}_\infty^{11}$ is $\boldsymbol{A}_1$-invariant; hence that $\Im m\boldsymbol{W}_\infty^{11}$ is $\boldsymbol{A}_1^T$-invariant or, put in other way, $\hat{\boldsymbol{V}}$ is $\boldsymbol{A}_1^T$-invariant. Because we can assume that $\{\boldsymbol{E},\boldsymbol{A}_0\}$ was in the form (6.16), it is straightforward to see that the following identities hold.

$$\boldsymbol{V}_1^T\boldsymbol{Q}^T\boldsymbol{E}^T\boldsymbol{V}=\begin{bmatrix}\boldsymbol{I}_{\hat{r}} & 0\\ 0 & 0\end{bmatrix},\quad \boldsymbol{V}_1^T\boldsymbol{Q}^T\boldsymbol{A}_0^T\boldsymbol{V}=\begin{bmatrix}\hat{\boldsymbol{V}}^T\boldsymbol{A}_1\hat{\boldsymbol{V}} & *\\ 0 & \boldsymbol{I}\end{bmatrix},$$

where the asterisk sign denotes irrelevent terms. The next thing to see is that $\mathrm{Ker}\boldsymbol{W}_\infty \subset \mathrm{Ker}\boldsymbol{F}_\infty$, which can be easily derived from GARE $Ric_3(\boldsymbol{W}_\infty)=0$; then, accordingly, there exists a matric $\boldsymbol{H}$ such that

$$\boldsymbol{H}\boldsymbol{V}_1^T\boldsymbol{Q}^{-1}=\boldsymbol{F}_\infty. \tag{6.20}$$

Because $\hat{\boldsymbol{A}}_0 = (\boldsymbol{A}_0 + \boldsymbol{B}_2\boldsymbol{F}_\infty)$, it is then easy to see that

$$\boldsymbol{E}_{\hat{r}}\boldsymbol{V}_1^T\boldsymbol{Q}^{-1}=\boldsymbol{V}_1^T\boldsymbol{E},\ \text{and}\ \boldsymbol{F}\boldsymbol{V}_1^T\boldsymbol{Q}^{-1}=\boldsymbol{V}_1^T\hat{\boldsymbol{A}}_0$$

hold for some matrix $\boldsymbol{F}$. This completes the proof. **Q.E.D.**

The following theorem which is the main result of this section characterizes the reduced-order controller by the algebraic equations stated above.

Theorem 6.10. *Consider (4.24). Suppose that Conditions (i) and (ii) of Theorem 4.7 hold. Suppose also that $\{\boldsymbol{E},\boldsymbol{W}_\infty\}$ is regular and impulse-free. Let matrices $\boldsymbol{V}_1$, $\boldsymbol{F}$, $\boldsymbol{H}$ and $\boldsymbol{W}_2$ be given as above and have compatible dimensions. Then the $n-r+\hat{r}$-th order controller*

$$\begin{aligned}\boldsymbol{E}_{\hat{r}}\dot{\boldsymbol{\xi}}&=\boldsymbol{F}\boldsymbol{\xi}+\boldsymbol{G}\boldsymbol{y},\\ \boldsymbol{u}&=\boldsymbol{H}\boldsymbol{\xi},\end{aligned} \tag{6.21}$$

is internally stabilizing for Σ and renders $\|\boldsymbol{T}_{zw}\| < \gamma$, where in (6.21) the matrices $\boldsymbol{E}_{\hat{r}}$, $\boldsymbol{F}$ and H satisfy Equations (6.17) and (6.20), and the matrix $\boldsymbol{G}$ satisfies

$$\boldsymbol{G} = \boldsymbol{V}_1^T \hat{\boldsymbol{B}}_0. \tag{6.22}$$

□

Proof. The proof of the theorem is essentially based on the Bounded Real Lemma(See [116]). Let us first write down the state equations of the closed-loop system (4.24)-(6.21) in the following.

$$\begin{aligned} \boldsymbol{E}_c \dot{\boldsymbol{x}}_c &= \boldsymbol{A}_c \boldsymbol{x}_c + \boldsymbol{B}_c \boldsymbol{w}, \\ \boldsymbol{z} &= \boldsymbol{C}_c \boldsymbol{x}_c, \end{aligned}$$

where

$$\begin{aligned} \boldsymbol{x}_c &= col(\boldsymbol{x}, \boldsymbol{\xi}), \\ \boldsymbol{E}_c &= \begin{bmatrix} \boldsymbol{E} & 0 \\ 0 & \boldsymbol{E}_{\hat{r}} \end{bmatrix}, \\ \boldsymbol{A}_c &= \begin{bmatrix} \boldsymbol{A} & \boldsymbol{B}_2 \boldsymbol{H} \\ \boldsymbol{G}\boldsymbol{C}_2 & \boldsymbol{F} \end{bmatrix}, \\ \boldsymbol{B}_c &= \begin{bmatrix} \boldsymbol{B}_1 \\ \boldsymbol{G}\boldsymbol{D}_{21} \end{bmatrix}, \\ \boldsymbol{C}_c &= \begin{bmatrix} \boldsymbol{C}_1 & \boldsymbol{D}_{12}\boldsymbol{H} \end{bmatrix}. \end{aligned}$$

Define an r.s.e.(restrict system equivalent) transformation pair given by

$$\{\boldsymbol{T}_1, \boldsymbol{T}_2\} = \left\{ \begin{bmatrix} \boldsymbol{I} & 0 \\ \boldsymbol{V}_1^T & \boldsymbol{I} \end{bmatrix}, \begin{bmatrix} \boldsymbol{I} & 0 \\ -\boldsymbol{V}_1^T \boldsymbol{Q}^{-1} & \boldsymbol{I} \end{bmatrix} \right\}.$$

With this transformation matrix in hand, the closed-loop system can now be rewritten as:

$$\begin{aligned} \hat{\boldsymbol{E}}_c \dot{\hat{\boldsymbol{x}}}_c &= \hat{\boldsymbol{A}}_c \hat{\boldsymbol{x}}_c + \hat{\boldsymbol{B}}_c \boldsymbol{w}, \\ \boldsymbol{z} &= \hat{\boldsymbol{C}}_c \hat{\boldsymbol{x}}_c, \end{aligned}$$

where $\boldsymbol{E}_c = \hat{\boldsymbol{E}}_c$, and

$$\begin{aligned} \hat{\boldsymbol{A}}_c &\triangleq \begin{bmatrix} \hat{\boldsymbol{A}}_{11} & \hat{\boldsymbol{A}}_{12} \\ \hat{\boldsymbol{A}}_{21} & \hat{\boldsymbol{A}}_{22} \end{bmatrix} \\ &= \begin{bmatrix} \boldsymbol{A} + \boldsymbol{B}_2 \boldsymbol{H} \boldsymbol{V}_1^T \boldsymbol{Q}^{-1} & \boldsymbol{B}_2 \boldsymbol{H} \\ \boldsymbol{M} & \boldsymbol{F} - \boldsymbol{V}_1^T \boldsymbol{B}_2 \boldsymbol{H} \end{bmatrix}, \\ \hat{\boldsymbol{B}}_c &\triangleq \begin{bmatrix} \hat{\boldsymbol{B}}_1 \\ \hat{\boldsymbol{B}}_2 \end{bmatrix} = \begin{bmatrix} \boldsymbol{B}_1 \\ -\boldsymbol{V}_1^T \boldsymbol{B}_1 + \boldsymbol{G}\boldsymbol{D}_{21} \end{bmatrix}, \\ \hat{\boldsymbol{C}}_c &\triangleq \begin{bmatrix} \hat{\boldsymbol{C}}_1 & \hat{\boldsymbol{C}}_2 \end{bmatrix} = \begin{bmatrix} \boldsymbol{C}_1 + \boldsymbol{D}_{12} \boldsymbol{H} \boldsymbol{V}_1^T \boldsymbol{Q}^{-1} & \boldsymbol{D}_{12} \boldsymbol{H} \end{bmatrix} \end{aligned}$$

with $\boldsymbol{M} \triangleq -\boldsymbol{V}_1^T \boldsymbol{A} + \boldsymbol{G}\boldsymbol{C}_2 - \boldsymbol{V}_1^T \boldsymbol{B}_2 \boldsymbol{H} \boldsymbol{V}_1^T \boldsymbol{Q}^{-1} + \boldsymbol{F}\boldsymbol{V}_1^T$. Set

$$\boldsymbol{X}_c \triangleq \begin{bmatrix} \boldsymbol{X}_\infty & 0 \\ 0 & \boldsymbol{W}_2 \end{bmatrix}.$$

Then, $\boldsymbol{X}_c$ satisfies $\boldsymbol{E}_c^T\boldsymbol{X}_c = \boldsymbol{X}_c^T\boldsymbol{E}_c \geq 0$. It can be shown that the matrix $\boldsymbol{X}_c$ thus constructed is an admissible solution to the following GARE.

$$\begin{aligned}\mathrm{R}ic_S(\boldsymbol{X}) &:= \hat{\boldsymbol{A}}_c^T\boldsymbol{X}_c + \boldsymbol{X}_c^T\hat{\boldsymbol{A}}_c + \hat{\boldsymbol{C}}_c^T\hat{\boldsymbol{C}}_c + \frac{1}{\gamma^2}\boldsymbol{X}_c^T\hat{\boldsymbol{B}}_c\hat{\boldsymbol{B}}_c^T\boldsymbol{X}_c \\ &:= \begin{bmatrix}\boldsymbol{S}_{11} & \boldsymbol{S}_{12} \\ \boldsymbol{S}_{12}^T & \boldsymbol{S}_{22}\end{bmatrix} = 0, \\ \boldsymbol{E}_c^T\boldsymbol{X}_c &= \boldsymbol{X}_c^T\boldsymbol{E}_c \geq 0.\end{aligned}$$

To see this, observe first that $\boldsymbol{S}_{11}$ can be written as

$$\boldsymbol{S}_{11} = \hat{\boldsymbol{A}}_{11}^T\boldsymbol{X}_\infty + \boldsymbol{X}_\infty^T\hat{\boldsymbol{A}}_{11} + \hat{\boldsymbol{C}}_1^T\hat{\boldsymbol{C}}_1 + \frac{1}{\gamma^2}\boldsymbol{X}_\infty\hat{\boldsymbol{B}}_1\hat{\boldsymbol{B}}_1^T\boldsymbol{X}_\infty^T.$$

By using (6.20), it is straightforward to show that $\boldsymbol{S}_{11} = Ric_1(\boldsymbol{X}_\infty) = 0$. Next, as a direct consequence of (6.17) and (6.22), we have

$$\boldsymbol{S}_{12}^T = \boldsymbol{W}_2(-\boldsymbol{V}_1^T\tilde{\boldsymbol{A}} - \boldsymbol{V}_1^T\boldsymbol{B}_2\boldsymbol{F}_\infty + \boldsymbol{G}\tilde{\boldsymbol{C}}_2 + \boldsymbol{F}\boldsymbol{V}_1^T\boldsymbol{Q}^{-1}) = 0.$$

Finally, by some plain calculations, it is shown that

$$\boldsymbol{S}_{22} = \boldsymbol{V}_1^T\boldsymbol{Q}^T \times Ric_3(\boldsymbol{W}_\infty) \times \boldsymbol{Q}\boldsymbol{V}_1 = 0.$$

With the above discussions, it is concluded that $\boldsymbol{X}_c$ is a solution to GARE $Ric_S(\boldsymbol{X}_c) = 0$. It remains to claim that $\boldsymbol{X}_c$ is indeed admissible. Compute the following matrix

$$\hat{\boldsymbol{A}}_c + \frac{1}{\gamma^2}\hat{\boldsymbol{B}}_c\hat{\boldsymbol{B}}_c^T\boldsymbol{X}_c \triangleq \begin{bmatrix}\boldsymbol{H}_{11} & \boldsymbol{H}_{12} \\ \boldsymbol{H}_{21} & \boldsymbol{H}_{22}\end{bmatrix}.$$

Note that $\boldsymbol{H}_{21} = 0$ by Equations (6.17), (6.20) and (6.22). Therefore $\{\boldsymbol{E}_c, \hat{\boldsymbol{A}}_c + \frac{1}{\gamma^2}\hat{\boldsymbol{B}}_c\hat{\boldsymbol{B}}_c^T\boldsymbol{X}_c\}$ is admissible if and only if $\{\boldsymbol{E}, \boldsymbol{H}_{11}\}$ and $\{\boldsymbol{E}_{\hat{r}}, \boldsymbol{H}_{22}\}$ are admissible. The matrix $\boldsymbol{H}_{11}$ has the following form

$$\boldsymbol{H}_{11} = (\boldsymbol{A} - \boldsymbol{B}_2\boldsymbol{R}_1^{-1}\boldsymbol{D}_{12}^T\boldsymbol{C}_1) + (\frac{1}{\gamma^2}\boldsymbol{B}_1\boldsymbol{B}_1^T - \boldsymbol{B}_2\boldsymbol{R}_1^{-1}\boldsymbol{B}_2^T)\boldsymbol{X}_\infty.$$

In view of the above identity, we conclude that $\{\boldsymbol{E}, \boldsymbol{H}_{11}\}$ is admissible because $\boldsymbol{X}_\infty$ is an admissible solution to the GARE $Ric_1(\boldsymbol{X}) = 0$. Now consider the matrix $\boldsymbol{H}_{22}$. It can be shown by a routine calculation that

$$(\boldsymbol{A}_0 + \frac{1}{\gamma^2}\boldsymbol{B}_0\boldsymbol{B}_0^T\boldsymbol{W}_\infty)^T\boldsymbol{V}_1 = \boldsymbol{Q}^{-T}\boldsymbol{V}_1\boldsymbol{H}_{22}^T.$$

This together with the identity $\boldsymbol{E}_{\hat{r}}\boldsymbol{V}_1^T\boldsymbol{Q}^{-1} = \boldsymbol{V}_1^T\boldsymbol{E}$ yields

$$(s\boldsymbol{E} - (\boldsymbol{A}_0 + \frac{1}{\gamma^2}\boldsymbol{B}_0\boldsymbol{B}_0^T\boldsymbol{W}_\infty))^T\boldsymbol{V}_1 = \boldsymbol{Q}^{-T}\boldsymbol{V}_1(s\boldsymbol{E}_{\hat{r}} - \boldsymbol{H}_{22})^T,$$

which simply says that the set of the characteristic polynomials (finte and infinite) of the pencil $s\boldsymbol{E}_{\hat{r}} - \boldsymbol{H}_{22}$ is a subset of the pencil $s\boldsymbol{E} - (\boldsymbol{A}_0 + \frac{1}{\gamma^2}\boldsymbol{B}_0\boldsymbol{B}_0^T\boldsymbol{W}_\infty)$'s. Because the pair $\{\boldsymbol{E}, \boldsymbol{A}_0 + \frac{1}{\gamma^2}\boldsymbol{B}_0\boldsymbol{B}_0^T\boldsymbol{W}_\infty\}$ is admissibe, this has already claimed that $\{\boldsymbol{E}_{\hat{r}}, \boldsymbol{H}_{22}\}$ is also admissible.

So far it has been shown that $\boldsymbol{X}_c$ is an admissible solution to the GARE $Ric_S(\boldsymbol{X}_c) = 0$. By Bounded Real Lemma[116], it is concluded that the reduced-order controller (6.21) is indeed internally stabilizing for Σ and makes $\|\boldsymbol{T}_{zw}\| < \gamma$. This completes the proof. **Q.E.D.**

The algebraic method just proposed certainly provides a way to construct a controller in the conventional state-space form; namely, a controller of the following form

$$\begin{aligned} \dot{\boldsymbol{\xi}} &= \boldsymbol{F}_n\boldsymbol{\xi} + \boldsymbol{G}_n\boldsymbol{y}, \\ \boldsymbol{u} &= \boldsymbol{H}_n\boldsymbol{\xi}, \end{aligned} \tag{6.23}$$

Recall that $\boldsymbol{V}_n \triangleq [\hat{\boldsymbol{V}}^T \ 0]^T$ is an $\hat{r}$-dimensional deflating subspace of $\{\boldsymbol{E}^T, \hat{\boldsymbol{A}}_0^T\}$ with finite spectrum. Then the following identity

$$\hat{\boldsymbol{A}}_0^T\boldsymbol{V}_n = \boldsymbol{E}^T\boldsymbol{V}_n\boldsymbol{F}_n^T \tag{6.24}$$

admits a solution $\boldsymbol{F}_n$. Unfortunately, in the present situation, the assumption that $\{\boldsymbol{E}, \boldsymbol{W}_\infty\}$ is impulse-free is not fulfilled to guarantee the existence of the normal-form reduced-order controller. Right now let us assume that the following identity

$$\boldsymbol{H}_n\boldsymbol{V}_n^T\boldsymbol{E} = \boldsymbol{F}_\infty \tag{6.25}$$

has a solution $\boldsymbol{H}_n$. The following corollary describes a way to examine the existence of the normal controller of the form (6.23).

Corollary 6.11. *Consider (4.24). Suppose that Conditions (i) and (ii) of Theorem 4.7 hold. Let matrices $\boldsymbol{V}_n$, $\boldsymbol{F}_n$ and $\boldsymbol{H}_n$ be given as above and have compatible dimensions. Then the reduced-order normal controller (6.23) is internally stabilizing for Σ and renders $\|\boldsymbol{T}_{zw}\| < \gamma$, where in (6.23) the matrices $\boldsymbol{F}_n$ and $\boldsymbol{H}_n$ satisfy equation (6.24) and (6.25), and the matrix $\boldsymbol{G}_n$ satisfies*

$$\boldsymbol{G}_n = \boldsymbol{V}_n^T\hat{\boldsymbol{B}}_0. \tag{6.26}$$

□

6.2 Mixed $\mathcal{H}_2$ and $\mathcal{H}_\infty$ Control Problems

The $\mathcal{H}_\infty$ control theory has become one of the most important results in the systems and control theory, both in terms of potential practical and theoretical applications. However, the solution to $\mathcal{H}_\infty$ control problem is by no mean unique. In fact, the solutions are parameterized by a free parameter, provided that the parameter belongs to a certain half open ball in the complex plane. With this observation in mind, a question arises naturally: What can we do effectively with the remaining degree of freedom? This is the general motivation for the study of mixed $\mathcal{H}_2/\mathcal{H}_\infty$ control problems. The aim of this section is to study the so-called mixed $\mathcal{H}_2$ and $\mathcal{H}_\infty$ control problem. We will study a mixed $\mathcal{H}_2$ and $\mathcal{H}_\infty$ control problems with internal stability for descriptor systems by means of static feedback law. The problem consists of constructing the robust $\mathcal{H}_\infty$ controllers, while providing the performance requirement that the system output energy measured in $\mathcal{L}_2$-norm is minimized, such that the overall system is asymptotically stable with index one. It is shown that under appropriate assumptions, the probelm for the nonlinear plants is solvable if a pair of cross-coupled Hamilton–Jacobi equations allows smooth solutions. On the other hand, the problem for the linear descriptor systems is solvable if and only if a pair of cross-coupled generalized algebraic Riccati equations have admissible solutions. For simplicity of presentation, the focus will be put on the linear descriptor systems case. Recently, Limebeer *et al.* [69] have developed a design methodology for mixed $\mathcal{H}_2$ and $\mathcal{H}_\infty$ control problems, based on the existing theory of two-player non-zero sum differential games. It is well known that two-player non-zero sum games have two payoff criteria (See, for example, Basar and Olsder[7], and Bryson and Ho[15]). The central idea behind the work of Limebeer *et al.*[69] is to use two performance indices associated with a two-player Nash game to represent the $\mathcal{H}_2$ and $\mathcal{H}_\infty$ criteria, respectively. More specifically, one performance is used to reflect an $\mathcal{H}_\infty$ constraint, while the second is used to reflect $\mathcal{H}_2$ optimality requirements. The output is then that, under some appropriate assumptions, the solution to a mixed $\mathcal{H}_2$ and $\mathcal{H}_\infty$ control problem via static feedback laws is determined by the solution of a pair of cross-coupled Riccati equations. Our aproach to the mixed $\mathcal{H}_2$ and $\mathcal{H}_\infty$ control problems is drawn from the identical concept but, however, in the descriptor systems content, together with a simple completion of squares argument.

6.2.1 Problem Formulation

The problem considered here is described by the standard block diagram shown in Fig. 6.1, where $\boldsymbol{G}$ is the plant and $\boldsymbol{K}$ is the controller. Suppose $\boldsymbol{G}$ is a linear time-invariant descriptor system described by

$$\begin{aligned} \boldsymbol{E}\dot{\boldsymbol{x}} &= \boldsymbol{A}\boldsymbol{x} + \boldsymbol{B}_1\boldsymbol{w} + \boldsymbol{B}_2\boldsymbol{u}, \\ \boldsymbol{z} &= \boldsymbol{C}_1\boldsymbol{x} + \boldsymbol{D}_{12}\boldsymbol{u}, \\ \boldsymbol{y} &= \boldsymbol{C}_2\boldsymbol{x} + \boldsymbol{D}_{21}\boldsymbol{w}, \qquad \boldsymbol{E}\boldsymbol{x}(0) \text{ given}, \end{aligned} \tag{6.27}$$

where $\boldsymbol{x} \in \mathbb{R}^n$ is the state, and $\boldsymbol{w} \in \mathbb{R}^m$ represents a set of exogenous inputs which includes disturbances to be rejected and/or reference commands to be tracked. $\boldsymbol{z} \in \mathbb{R}^p$ is the output to be controlled and $y \in \mathbb{R}^q$ is the measured

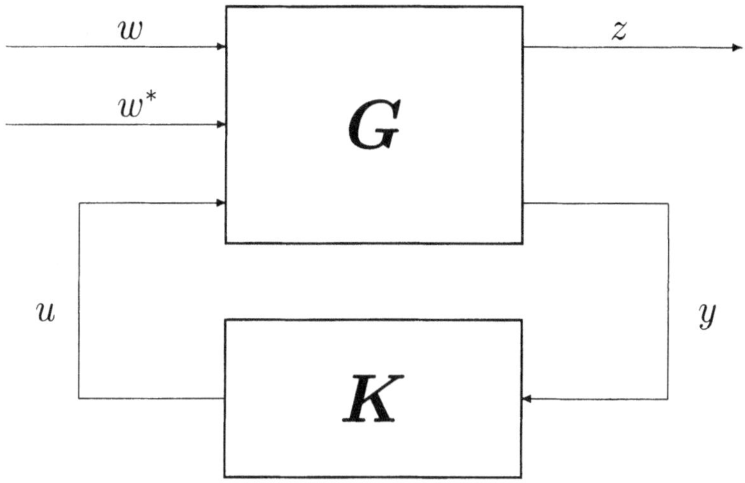

Fig. 6.1. Standard block diagram for mixed $\mathcal{H}_2/\mathcal{H}_\infty$ control problems

output. $\boldsymbol{u} \in \mathbb{R}^l$ is the control input. $\boldsymbol{A}, \boldsymbol{B}_1, \boldsymbol{B}_2, \boldsymbol{C}_1, \boldsymbol{C}_2$, and $\boldsymbol{D}_{12}$, are constant matrices with compatible dimensions. $\boldsymbol{E} \in \mathbb{R}^{n\times n}$ and rank$\boldsymbol{E} = r < n$. The interpretation of $\boldsymbol{w}^*$ will be clear in the sequel. We only consider static state feedback control laws, i.e., $\boldsymbol{u} = \boldsymbol{K}\boldsymbol{y}$. The purpose of the control is threefold.

(1) The resulting closed-loop system is internally stable. Here closed-loop internal stability means that the closed-loop system is regular and impulse-free, and that the states of $\boldsymbol{G}$ and $\boldsymbol{K}$ go to zero from all initial values when $\boldsymbol{w} = 0$.
(2) The closed-loop transfer function $T_{\boldsymbol{zw}}$ has an $\mathcal{L}_2$ gain strict less than γ.
(3) The optimal control $\boldsymbol{u}^*(t)$ is required to regulate the state $\boldsymbol{x}(t)$ as well as to minimize the output energy when the worst-case disturbance $\boldsymbol{w}^*(t)$ is applied to the system.

Condition (3) can be treated as an alternative to the minimization of an auxiliary cost associated with the to-be-controlled system. This is illustrated in Fig. 6.1 (See also [69]).

This problem can be formulated as a two-person nonzero sum game. The two cost functions we will adopt are

$$J_1(\boldsymbol{u}, \boldsymbol{w}) = \int_0^\infty (\|\boldsymbol{z}(t)\|^2 - \gamma^2\|\boldsymbol{w}(t)\|^2)\mathrm{d}t, \tag{6.28}$$

and

$$J_2(\boldsymbol{u}, \boldsymbol{w}) = \int_0^\infty \|\boldsymbol{z}\|^2\mathrm{d}t, \tag{6.29}$$

which are similar to [69]. The first is associated with an $\mathcal{H}_\infty$ criterion, while the second is used for the $\mathcal{H}_2$ optimization part of the problem. In this section, we seek to obtain two control strategies, $\boldsymbol{u}^*$ and $\boldsymbol{w}^*$ that satisy the Nash equilibria defined by

$$J_1(\boldsymbol{u}^*, \boldsymbol{w}^*) \leq J_1(\boldsymbol{u}^*, \boldsymbol{w}), \tag{6.30}$$

$$J_2(\boldsymbol{u}^*, \boldsymbol{w}^*) \leq J_2(\boldsymbol{u}, \boldsymbol{w}^*), \tag{6.31}$$

from which we can, then, deduce that the mixed $\mathcal{H}_2/\mathcal{H}_\infty$ control problem is solvable [69]. Equations (6.30), (6.31), in fact, build the cornerstone for our problem.

The following assumptions are made for the rest of this section.

Assumptions:
(A1) $\{\boldsymbol{E}, \boldsymbol{A}\}$ is regular.
(A2) $\{\boldsymbol{E}, \boldsymbol{A}, \boldsymbol{B}_2\}$ is finite dynamic stabilizable and impulse controllable.
(A3) $\boldsymbol{R}_1 \triangleq \boldsymbol{D}_{12}^T \boldsymbol{D}_{12} > 0$.

Assumptions (A1)-(A2) are due to the existence and stability requirements. Assumption (A3) is made for a technical reason since it provides a sufficient condition for the existence of a Nash equilibrium. Clearly, this assumption can be relaxed in some sense. Nevertheless, we'll not pursuit this point any further here.

6.2.2 Solution via State Feedback

In this subsection, we'll state our first main result. We give necessary and sufficient conditions for the existence of state feedback (i.e., $\boldsymbol{y} = \boldsymbol{x}$) strategies that solve an $\mathcal{H}_2/\mathcal{H}_\infty$ control problem for descriptor systems. The main theorem reads as follows.

Theorem 6.12. *Consider System (6.27) with $\boldsymbol{y} = \boldsymbol{x}$. Assume that Assumptions (A1)-(A3) hold. Suppose that there exist state feedback strategies*

$$\boldsymbol{u}^*(t, \boldsymbol{x}) = \boldsymbol{K}_2 \boldsymbol{x}$$
$$\boldsymbol{w}^*(t, \boldsymbol{x}) = \boldsymbol{K}_1 \boldsymbol{x}$$

such that

i) $T_{zw} \in \mathcal{RH}^\infty$ *with* $\|T_{zw}\|_\infty < \gamma$, *when* $\boldsymbol{u}(t) = \boldsymbol{u}^*(t, \boldsymbol{x})$.
ii) $J_2(\boldsymbol{u}^*, \boldsymbol{w}^*) \leq J_2(\boldsymbol{u}, \boldsymbol{w}^*)$.
iii) $\aleph(\boldsymbol{E}) \cap (\boldsymbol{A} + \boldsymbol{B}_1\boldsymbol{K}_1)^{-1}(\Im m\, \boldsymbol{E}) \cap \aleph(\boldsymbol{C}_1) = \{0\}$.

then the following two coupled GAREs have admissible solution.

$$\begin{aligned}
&(\boldsymbol{A} - \boldsymbol{B}_2\boldsymbol{R}_1^{-1}(\boldsymbol{B}_2^T\boldsymbol{X}_2 + \boldsymbol{D}_{12}^T\boldsymbol{C}_1))^T\boldsymbol{X}_1\\
&+ \boldsymbol{X}_1^T(\boldsymbol{A} - \boldsymbol{B}_2\boldsymbol{R}_1^{-1}(\boldsymbol{B}_2^T\boldsymbol{X}_2 + \boldsymbol{D}_{12}^T\boldsymbol{C}_1))\\
&+ (\boldsymbol{C}_1 - \boldsymbol{D}_{12}\boldsymbol{R}_1^{-1}(\boldsymbol{B}_2^T\boldsymbol{X}_2 + \boldsymbol{D}_{12}^T\boldsymbol{C}_1))^T(\boldsymbol{C}_1 - \boldsymbol{D}_{12}\boldsymbol{R}_1^{-1}(\boldsymbol{B}_2^T\boldsymbol{X}_2 + \boldsymbol{D}_{12}^T\boldsymbol{C}_1))\\
&+ \tfrac{1}{\gamma^2}\boldsymbol{X}_1^T\boldsymbol{B}_1\boldsymbol{B}_1^T\boldsymbol{X}_1 = 0,\\
&\boldsymbol{E}^T\boldsymbol{X}_1 = \boldsymbol{X}_1^T\boldsymbol{E},
\end{aligned} \tag{6.32}$$

$$\begin{aligned}
&(\boldsymbol{A} + \tfrac{1}{\gamma^2}\boldsymbol{B}_1\boldsymbol{B}_1^T\boldsymbol{X}_1)^T\boldsymbol{X}_2 + \boldsymbol{X}_2^T(\boldsymbol{A} + \tfrac{1}{\gamma^2}\boldsymbol{B}_1\boldsymbol{B}_1^T\boldsymbol{X}_1)\\
&+ \boldsymbol{C}_1^T\boldsymbol{C}_1 - (\boldsymbol{X}_2^T\boldsymbol{B}_1 + \boldsymbol{C}_1^T\boldsymbol{D}_{12})\boldsymbol{R}_1^{-1}(\boldsymbol{X}_2^T\boldsymbol{B}_2 + \boldsymbol{C}_1\boldsymbol{D}_{12})^T = 0,\\
&\boldsymbol{E}^T\boldsymbol{X}_2 = \boldsymbol{X}_2^T\boldsymbol{E},
\end{aligned} \tag{6.33}$$

with $\boldsymbol{E}^T\boldsymbol{X}_1,\ \boldsymbol{E}^T\boldsymbol{X}_2 \geq 0$. *Moreover, when these conditions are satisfied, the admissible strategies are given by*

$$\boldsymbol{u}^*(t,\,\boldsymbol{x}) = -\boldsymbol{R}_1^{-1}(\boldsymbol{B}_2^T\boldsymbol{X}_2 + \boldsymbol{D}_{12}^T\boldsymbol{C}_1)\boldsymbol{x}(t). \tag{6.34}$$

$$\boldsymbol{w}^*(t,\,\boldsymbol{x}) = \frac{1}{\gamma^2}\boldsymbol{B}_1^T\boldsymbol{X}_1\boldsymbol{x}(t). \tag{6.35}$$

On the contrary, suppose (6.32) and (6.33) have admissible solutions $\boldsymbol{X}_1, \boldsymbol{X}_2$ *with* $\boldsymbol{E}^T\boldsymbol{X}_1,\ \boldsymbol{E}^T\boldsymbol{X}_2 \geq 0$. *In addition, suppose*

$$\aleph(\boldsymbol{E}) \cap (\boldsymbol{A} + \frac{1}{\gamma^2}\boldsymbol{B}_1\boldsymbol{B}_1^T\boldsymbol{X}_1)^{-1}(\Im m\,\boldsymbol{E}) \cap \aleph(\boldsymbol{C}_1) = \{0\}.$$

Then, Conditions i), ii) are achieved by the admissible strategies (6.34), (6.35); i.e., the $\mathcal{H}_2/\mathcal{H}_\infty$ *is solved via state feedbacks.* □

Proof. Suppose that $\boldsymbol{u}^*(t,\,\boldsymbol{x}) = \boldsymbol{K}_2\boldsymbol{x}$, $\boldsymbol{w}^*(t,\,\boldsymbol{x}) = \boldsymbol{K}_1\boldsymbol{x}$ are the optimal state feedback stragegies. Firstly, with $\boldsymbol{u}^*(t,\,\boldsymbol{x}) = \boldsymbol{K}_2\boldsymbol{x}$, we get the following closed-loop system,

$$\begin{aligned}
\boldsymbol{E}\dot{\boldsymbol{x}}(t) &= (\boldsymbol{A} + \boldsymbol{B}_2\boldsymbol{K}_2)\boldsymbol{x}(t) + \boldsymbol{B}_1\boldsymbol{w}(t),\\
\boldsymbol{z}(t) &= (\boldsymbol{C}_1 + \boldsymbol{D}_{12}\boldsymbol{K}_2)\boldsymbol{x}(t).
\end{aligned}$$

The above system satisfies that $\{\boldsymbol{E}, \boldsymbol{A}+\boldsymbol{B}_2\boldsymbol{K}_2\}$ admissible and $\|T_{\boldsymbol{zw}}\|_\infty < \gamma$, by hypothesis. From Lemma 2.26, it follows that there exists an admissible solution $\boldsymbol{X}_1$ of the GARE

$$\begin{aligned}
&(\boldsymbol{A} + \boldsymbol{B}_2\boldsymbol{K}_2)^T\boldsymbol{X}_1 + \boldsymbol{X}_1^T(\boldsymbol{A} + \boldsymbol{B}_2\boldsymbol{K}_2) + (\boldsymbol{C}_1 + \boldsymbol{D}_{12}\boldsymbol{K}_2)^T(\boldsymbol{C}_1 + \boldsymbol{D}_{12}\boldsymbol{K}_2)\\
&+ \tfrac{1}{\gamma^2}\boldsymbol{X}_1^T\boldsymbol{B}_1\boldsymbol{B}_1^T\boldsymbol{X}_1 = 0,
\end{aligned} \tag{6.36}$$

which satisfies $\boldsymbol{E}^T\boldsymbol{X}_1 = \boldsymbol{X}_1^T\boldsymbol{E} \geq 0$, and is such that $\{\boldsymbol{E}, \boldsymbol{A} + \boldsymbol{B}_2\boldsymbol{K}_2 + \frac{1}{\gamma^2}\boldsymbol{B}_1\boldsymbol{B}_1^T\boldsymbol{X}_1\}$ admissible. This, in turn, implies that

$$\Im m\,\boldsymbol{E} + (\boldsymbol{A} - \frac{1}{\gamma^2}\boldsymbol{B}_1\boldsymbol{B}_1^T\boldsymbol{X}_1)(\aleph(\boldsymbol{E})) + \Im m\,\boldsymbol{B}_2 = \mathbb{R}^n.$$

Next, completing the square yields

$$\int_0^\infty (\gamma^2\|\boldsymbol{w}\|^2 - \|\boldsymbol{z}\|^2)\mathrm{d}t = \int_0^\infty \|\boldsymbol{w} + \frac{1}{\gamma^2}\boldsymbol{B}_1^T\boldsymbol{X}_1\boldsymbol{x}\|^2\mathrm{d}t,$$

which results in

$$\boldsymbol{w}^*(t, \boldsymbol{x}) = \frac{1}{\gamma^2}\boldsymbol{B}_1^T\boldsymbol{X}_1\boldsymbol{x}(t).$$

So $\boldsymbol{K}_1 = -\frac{1}{\gamma^2}\boldsymbol{B}_1^T\boldsymbol{X}_1$. Note that the above integral exists, provided that $\{\boldsymbol{E},\ \boldsymbol{A} + \boldsymbol{B}_2\boldsymbol{K}_2\}$ is admissible. Substituting $\boldsymbol{w}^*(t, \boldsymbol{x}) = \frac{1}{\gamma^2}\boldsymbol{B}_1^T\boldsymbol{X}_1\boldsymbol{x}(t)$ into (6.27) to get

$$\begin{aligned} \boldsymbol{E}\dot{\boldsymbol{x}}(t) &= (\boldsymbol{A} + \tfrac{1}{\gamma^2}\boldsymbol{B}_1\boldsymbol{B}_1^T\boldsymbol{X}_1)\boldsymbol{x}(t) + \boldsymbol{B}_2\boldsymbol{u}, \\ \boldsymbol{z}(t) &= \boldsymbol{C}_1\boldsymbol{x} + \boldsymbol{D}_{12}\boldsymbol{u}, \end{aligned}$$

with

$$\Im m\,\boldsymbol{E} + (\boldsymbol{A} + \frac{1}{\gamma^2}\boldsymbol{B}_1\boldsymbol{B}_1^T\boldsymbol{X}_1)(\aleph(\boldsymbol{E})) + \Im m\,\boldsymbol{B}_2 = \mathbb{R}^n,$$

and

$$\aleph(\boldsymbol{E}) \cap (\boldsymbol{A} + \frac{1}{\gamma^2}\boldsymbol{B}_1\boldsymbol{B}_1^T\boldsymbol{X}_1)^{-1}(\Im m\,\boldsymbol{E}) \cap \aleph(\boldsymbol{C}_1) = \{0\}.$$

As a consequence, from Lemma A.5, the optimal control problem

$$\min_{\boldsymbol{u}} J_2(\boldsymbol{u},\ \boldsymbol{w}^*) = \int_0^\infty \|\boldsymbol{z}\|^2\mathrm{d}t$$

is solved via the state feedback (6.34), where $\boldsymbol{X}_2$ with $\boldsymbol{E}^T\boldsymbol{X}_2 = \boldsymbol{X}_2^T\boldsymbol{E} \geq 0$ is the admissible solution of GARE (6.33) such that $\{\boldsymbol{E}, \boldsymbol{A} + \frac{1}{\gamma^2}\boldsymbol{B}_1\boldsymbol{B}_1^T\boldsymbol{X}_1 - \boldsymbol{B}_2\boldsymbol{R}_1^{-1}(\boldsymbol{B}_2^T\boldsymbol{X}_2 + \boldsymbol{D}_{12}^T\boldsymbol{C}_1)\}$ is admissible. Substituting $\boldsymbol{K}_2 = -\boldsymbol{R}_1^{-1}(\boldsymbol{B}_2^T\boldsymbol{X}_2 + \boldsymbol{D}_{12}^T\boldsymbol{C}_1)$ in (6.36) gives the desired GARE (6.32).

We now establish the opposite implications. Set $\boldsymbol{u}(t) = \boldsymbol{u}^*(t, \boldsymbol{x})$ in (6.27), we have

$$\begin{aligned} \boldsymbol{E}\dot{\boldsymbol{x}}(t) &= (\boldsymbol{A} - \boldsymbol{B}_2\boldsymbol{R}_1^{-1}(\boldsymbol{B}_2^T\boldsymbol{X}_2 + \boldsymbol{D}_{12}^T\boldsymbol{C}_1))\boldsymbol{x}(t) + \boldsymbol{B}_1\boldsymbol{w}(t), \\ \boldsymbol{z}(t) &= (\boldsymbol{C}_1 - \boldsymbol{D}_{12}\boldsymbol{R}_1^{-1}(\boldsymbol{B}_2^T\boldsymbol{X}_2 + \boldsymbol{D}_{12}^T\boldsymbol{C}_1))\boldsymbol{x}(t). \end{aligned}$$

Since the GARE (6.32), by hypothesis, has an admissible solution, it follows, from Lemma 2.26, that $T_{\boldsymbol{zw}} \in \mathcal{RH}^\infty$ and $\|T_{\boldsymbol{zw}}\|_\infty < \gamma$. In particular, $\{\boldsymbol{E}, \boldsymbol{A} + \frac{1}{\gamma^2}\boldsymbol{B}_1\boldsymbol{B}_1^T\boldsymbol{X}_1 - \boldsymbol{B}_2\boldsymbol{R}_1^{-1}(\boldsymbol{B}_2^T\boldsymbol{X}_2 + \boldsymbol{D}_{12}^T\boldsymbol{C}_1)\}$ is admissible, so that $\Im m\,\boldsymbol{E} + (\boldsymbol{A} + \frac{1}{\gamma^2}\boldsymbol{B}_1\boldsymbol{B}_1^T\boldsymbol{X}_1)(\aleph(\boldsymbol{E})) + \Im m\,\boldsymbol{B}_2 = \mathbb{R}^n$.

Set $\boldsymbol{w}(t) = \boldsymbol{w}^*(t,\ \boldsymbol{x})$ in (6.27), we get

$$\begin{aligned} \boldsymbol{E}\dot{\boldsymbol{x}}(t) &= (\boldsymbol{A} + \tfrac{1}{\gamma^2}\boldsymbol{B}_1\boldsymbol{B}_1^T\boldsymbol{X}_1)\boldsymbol{x}(t) + \boldsymbol{B}_2\boldsymbol{u}, \\ \boldsymbol{z}(t) &= \boldsymbol{C}_1\boldsymbol{x} + \boldsymbol{D}_{12}\boldsymbol{u}. \end{aligned}$$

Now we have $\Im m\, \boldsymbol{E} + (\boldsymbol{A} + \frac{1}{\gamma^2}\boldsymbol{B}_1\boldsymbol{B}_1^T\boldsymbol{X}_1)(\aleph(\boldsymbol{E})) + \Im m\, \boldsymbol{B}_2 = \mathbb{R}^n$, from the above discussion, and $\aleph(\boldsymbol{E}) \cap (\boldsymbol{A} + \frac{1}{\gamma^2}\boldsymbol{B}_1\boldsymbol{B}_1^T\boldsymbol{X}_1)^{-1}(\Im m\, \boldsymbol{E}) \cap \aleph(\boldsymbol{C}_1) = \{0\}$, by hypothesis. It ,then, follows from Lemma A.5 that the optimal control problem

$$\min_{\boldsymbol{u}} J_2(\boldsymbol{u}, \boldsymbol{w}^*) = \int_0^\infty \|\boldsymbol{z}\|^2 \mathrm{d}t$$

has a solution $\boldsymbol{u}^*(t, \boldsymbol{x}) = -\boldsymbol{R}_1^{-1}(\boldsymbol{B}_2^T\boldsymbol{X}_2 + \boldsymbol{D}_{12}^T\boldsymbol{C}_1)\boldsymbol{x}(t)$, provided that $\boldsymbol{X}_2$ with $\boldsymbol{E}^T X_2 = \boldsymbol{X}_2^T\boldsymbol{E} \geq 0$ is an admissible solution of the GARE (6.33). This completes the proof. **Q.E.D.**

6.2.3 Solution via Static Output Feedback

This subsection is concerned with the mixed $\mathcal{H}_2$ and $\mathcal{H}_\infty$ control problem for descriptor systems via static output feedback strategies. The main result is summarized in the following statements.

Theorem 6.13. *Consider System (6.27). Assume that Assumptions (A1)-(A3) hold. Suppose there exist matrix K, and matrices $\boldsymbol{Q}_1$ and $\boldsymbol{Q}_2$ with $\boldsymbol{E}^T\boldsymbol{Q}_1 = \boldsymbol{Q}_1^T\boldsymbol{E} \geq 0$ and $\boldsymbol{E}^T\boldsymbol{Q}_2 = \boldsymbol{Q}_2^T\boldsymbol{E} \geq 0$ such that:*

1) $\boldsymbol{R} \triangleq \gamma^2\boldsymbol{I} - \boldsymbol{D}_k'\boldsymbol{D}_k$ is positive definite.
2) $\boldsymbol{Q}_1$ is an admissible solution of the following GARE.

$$\begin{aligned}(\boldsymbol{A}_k + \boldsymbol{B}_{1k}\boldsymbol{R}^{-1}\boldsymbol{D}_k^T\boldsymbol{C}_{1k})^T\boldsymbol{Q}_1 + \boldsymbol{Q}_1^T(\boldsymbol{A}_k + \boldsymbol{B}_{1k}\boldsymbol{R}^{-1}\boldsymbol{D}_k^T\boldsymbol{C}_{1k})\\ +\boldsymbol{Q}_1^T\boldsymbol{B}_{1k}\boldsymbol{R}^{-1}\boldsymbol{B}_{1k}^T\boldsymbol{Q}_1 + \boldsymbol{C}_{1k}^T(\boldsymbol{I} + \boldsymbol{D}_k\boldsymbol{R}^{-1}\boldsymbol{D}_k^T)\boldsymbol{C}_{1k} = 0\end{aligned} \tag{6.37}$$

3) $\boldsymbol{Q}_2$ is an admissible solution of the following GARE.

$$\begin{aligned}&[\boldsymbol{A} + \boldsymbol{B}_1\boldsymbol{R}^{-1}(\boldsymbol{D}_k^T\boldsymbol{C}_{1k} + \boldsymbol{B}_{1k}^T\boldsymbol{Q}_1)]^T\boldsymbol{Q}_2\\ &+ \boldsymbol{Q}_2^T[\boldsymbol{A} + \boldsymbol{B}_1\boldsymbol{R}^{-1}(\boldsymbol{D}_k^T\boldsymbol{C}_{1k} + \boldsymbol{B}_{1k}^T\boldsymbol{Q}_1)]\\ &- \boldsymbol{Q}_2^T\boldsymbol{B}_2\boldsymbol{B}_2^T\boldsymbol{Q}_2 + \boldsymbol{C}_1^T\boldsymbol{C}_1 = 0\end{aligned} \tag{6.38}$$

4)

$$\boldsymbol{K}\boldsymbol{C}_2 + \boldsymbol{K}\boldsymbol{D}_{21}\boldsymbol{R}^{-1}(\boldsymbol{D}_k^T\boldsymbol{C}_{1k} + \boldsymbol{B}_{1k}^T\boldsymbol{Q}_1) + \boldsymbol{B}_2^T\boldsymbol{Q}_2 = 0 \tag{6.39}$$

where

$$\begin{aligned}\boldsymbol{A}_k &= \boldsymbol{A} + \boldsymbol{B}_2\boldsymbol{K}\boldsymbol{C}_2,\\ \boldsymbol{B}_{1k} &= \boldsymbol{B}_1 + \boldsymbol{B}_2\boldsymbol{K}\boldsymbol{D}_{21},\\ \boldsymbol{C}_{1k} &= \boldsymbol{C}_1 + \boldsymbol{D}_{12}\boldsymbol{K}\boldsymbol{C}_2,\\ \boldsymbol{D}_k &= \boldsymbol{D}_{12}\boldsymbol{K}\boldsymbol{D}_{21}.\end{aligned}$$

Then:
If the systems $\boldsymbol{E}\dot{\boldsymbol{x}}(t) = \boldsymbol{A}_k\boldsymbol{x}(t)$ and $\boldsymbol{E}\dot{\boldsymbol{x}}(t) = (\boldsymbol{A} + \boldsymbol{B}_1\boldsymbol{R}^{-1}(\boldsymbol{D}_k^T\boldsymbol{C}_{1k} + \boldsymbol{B}_{1k}^T\boldsymbol{Q}_1))\boldsymbol{x}(t)$ are admissible, the output feedback control $\boldsymbol{u}^*(t) = \boldsymbol{K}\boldsymbol{y}(t)$ and strategy $\boldsymbol{w}^*(t) = \boldsymbol{R}^{-1}(\boldsymbol{D}_k^T\boldsymbol{C}_{1k} + \boldsymbol{B}_{1k}^T\boldsymbol{Q}_1)\boldsymbol{x}(t)$ will result in:

i) $\|T_{zw}\|_\infty < \gamma$ *or* $0 = J_1(u^*, w^*) \leq J_1(u^*, w)$, *and*
ii) $J_2(u^*, w^*) \leq J_2(u, w^*)$

Conversely, if there exist output feedback control $u^*(t) = Ky(t)$ *and strategy* $w^*(t) = K_1 x$ *such that:*

a) The system $E\dot{x}(t) = (A + B_2 K C_2)x(t)$ *is admissible;*
b) $\|T_{zw}\|_\infty < \gamma$ *or,* $0 = J_1(u^*, w^*) \leq J_1(u^*, w)$;
c) $J_2(u^*, w^*) \leq J_2(u, w^*)$.

then $R := \gamma^2 I - D_{21}^T K^T K D_{21} > 0$, *and there exists an admissible solution* Q_1 *of (6.37) with* $E^T Q_1 = Q_1^T E \geq 0$. *In addition,* $K_1 = R^{-1}(D_k^T C_{1k} + B_{1k}^T Q_1)$. *Furthermore, if the system*

$$E\dot{x}(t) = (A + B_1 R^{-1}(D_k^T C_{1k} + B_{1k}^T(t) Q_1))x(t)$$

is admissible, then there exists an admissible solution Q_2 *of (6.38) with* $E^T Q_2 = Q_2^T E \geq 0$ *and satisfying (6.39).* □

Proof.
Sufficiency.(i) Set $u(t) = u^*(t) = Ky(t) = KC_2 x(t) + KD_{21} w(t)$ to get the closed-loop system as follows

$$\begin{aligned} E\dot{x}(t) &= A_k x(t) + B_{1k} w(t) \\ z(t) &= C_{1k} x(t) + D_k w(t) \end{aligned}$$

Since $x(0) = 0$ and the homogeneous system $E\dot{x}(t) = A_k x(t)$ is stable and impulse-free, completing the square yields

$$\begin{aligned} J_1(u^*, w) &= \int_0^\infty (\gamma^2 w^T(t) w(t) - z^T(t) z(t)) \mathrm{d}t \\ &= \int_0^\infty [\gamma^2 w^T(t) w(t) - z^T(t) z(t) - \frac{\mathrm{d}}{\mathrm{d}t} x^T(t) E^T Q_1 x(t)] \mathrm{d}t \\ &= \int_0^\infty (w(t) - w^*(t))^T R (w(t) - w^*(t)) \mathrm{d}t \\ &= \int_0^\infty \|R^{\frac{1}{2}}(w(t) - w^*(t))\|^2 \mathrm{d}t \geq 0 \end{aligned}$$

When $w(t) = w^*(t)$, we have $J_1(u^*, w^*) = 0$. Thus

$$J_1(u^*, w) \geq J_1(u^*, w^*) = 0.$$

Defining the invertible operator $\mathbf{L}$ from $w(t)$ to $R^{\frac{1}{2}}(w(t) - w^*(t))$ by

$$\begin{aligned} E\dot{x}(t) &= A_k x(t) + B_{1k} w(t) \\ R^{\frac{1}{2}}(w(t) - w^*(t)) &= -R^{-\frac{1}{2}}(D_k^T C_{1k} + B_{1k} Q_1) x(t) + R^{\frac{1}{2}} w(t) \end{aligned}$$

gives

$$\begin{aligned}\gamma^2\|\boldsymbol{w}(t)\|_2^2 - \|\boldsymbol{z}(t)\|_2^2 &= \|R^{\frac{1}{2}}(\boldsymbol{w}(t) - \boldsymbol{w}^*(t))\|_2^2 \\ &= \|\mathbf{L}\boldsymbol{w}(t)\|_2^2 \\ &\geq \epsilon\|\boldsymbol{w}(t)\|_2^2\end{aligned}$$

for some positive ϵ. Consequently, $\|T_{\boldsymbol{zw}}\|_\infty < \gamma$.

(ii) Setting $\boldsymbol{w}(t) = \boldsymbol{w}^*(t)$ gives

$$\begin{aligned}\boldsymbol{E}\dot{\boldsymbol{x}}(t) &= (\boldsymbol{A} + \boldsymbol{B}_1\boldsymbol{R}^{-1}(\boldsymbol{D}_k^T\boldsymbol{C}_{1k} + \boldsymbol{B}_{1k}^T\boldsymbol{Q}_1)\boldsymbol{x}(t) + \boldsymbol{B}_2\boldsymbol{u}(t) \\ \boldsymbol{z}(t) &= \boldsymbol{C}_1\boldsymbol{x}(t) + \boldsymbol{D}_{12}\boldsymbol{u}(t)\end{aligned}$$

Observing that the minimization problem

$$\min_{\boldsymbol{u}}\{J_2(\boldsymbol{u}, \boldsymbol{w}^*) = \int_0^\infty \boldsymbol{z}^T(t)\boldsymbol{z}(t)\mathrm{d}t\}$$

is a standard optimal control problem with the homogeneous system $\boldsymbol{E}\dot{\boldsymbol{x}}(t) = (\boldsymbol{A} + \boldsymbol{B}_1\boldsymbol{R}^{-1}(\boldsymbol{D}_k^T\boldsymbol{C}_{1k} + \boldsymbol{B}_{1k}^T\boldsymbol{Q}_1))\boldsymbol{x}(t)$ being exponentially stable gives

$$\begin{aligned}\boldsymbol{u}^*(t) &= -\boldsymbol{B}_2^T\boldsymbol{Q}_2\boldsymbol{x}(t) \\ &= \boldsymbol{K}\boldsymbol{y}^*(t)\end{aligned}$$

in which $\boldsymbol{y}^*(t) = \boldsymbol{C}_2\boldsymbol{x}(t) + \boldsymbol{D}_{21}\boldsymbol{w}^*(t)$ is the measured output when $\boldsymbol{w}(t) = \boldsymbol{w}^*(t)$. Here we have used (6.39). This results in $J_2(\boldsymbol{u}^*, \boldsymbol{w}^*) \leq J_2(\boldsymbol{u}, \boldsymbol{w}^*)$ as required.

Necessity. We now move on to establish the opposite implication.
(i) Implementing $\boldsymbol{u}^*(t) = \boldsymbol{K}\boldsymbol{y}(t)$ gives

$$\begin{aligned}\boldsymbol{E}\dot{\boldsymbol{x}}(t) &= (\boldsymbol{A} + \boldsymbol{B}_2\boldsymbol{K}\boldsymbol{C}_2)\boldsymbol{x}(t) + (\boldsymbol{B}_1 + \boldsymbol{B}_2\boldsymbol{K}\boldsymbol{D}_{21})\boldsymbol{w}(t) \\ &\triangleq \boldsymbol{A}_k\boldsymbol{x}(t) + \boldsymbol{B}_{1k}\boldsymbol{w}(t) \\ \boldsymbol{z}(t) &= (\boldsymbol{C}_1 + \boldsymbol{D}_{12}\boldsymbol{K}\boldsymbol{C}_2)\boldsymbol{x}(t) + \boldsymbol{D}_{12}\boldsymbol{K}\boldsymbol{D}_{21}\boldsymbol{w}(t) \\ &\triangleq \boldsymbol{C}_{1k}\boldsymbol{x}(t) + \boldsymbol{D}_k\boldsymbol{w}(t)\end{aligned}$$

where $\boldsymbol{E}\dot{\boldsymbol{x}}(t) = (\boldsymbol{A} + \boldsymbol{B}_2\boldsymbol{K}\boldsymbol{C}_2)\boldsymbol{x}(t)$ is stable and impulse-free, and $\|T_{\boldsymbol{zw}}\|_\infty < \gamma$ by assumption. It now follows from Lemma 2.26 that there exists a matrix $\boldsymbol{Q}_1$ satisfying

$$\begin{aligned}0 = {}&\boldsymbol{A}_k^T\boldsymbol{Q}_1 + \boldsymbol{Q}_1^T\boldsymbol{A}_k + \boldsymbol{C}_{1k}^T\boldsymbol{C}_{1k} \\ &+(\boldsymbol{Q}_1^T\boldsymbol{B}_{1k}(t) + \boldsymbol{C}_{1k}^T\boldsymbol{D}_k(t))\boldsymbol{R}^{-1}(\boldsymbol{B}_{1k}^T\boldsymbol{Q}_1 + \boldsymbol{D}_k^T\boldsymbol{C}_{1k})\end{aligned}$$

such that the system

$$\boldsymbol{E}\dot{\boldsymbol{x}}(t) = (\boldsymbol{A}_k + \boldsymbol{B}_{1k}(\gamma^2\boldsymbol{I} - \boldsymbol{D}_k^T\boldsymbol{D}_k)^{-1}(\boldsymbol{B}_{1k}^T\boldsymbol{Q}_1 + \boldsymbol{D}_k^T\boldsymbol{C}_{1k}))\boldsymbol{x}(t)$$

is stable and impulse-free. This is exactly the GARE given in (6.37). A similar argument as in the sufficiency proof gives

$$\begin{aligned}
&J_1(\boldsymbol{w}, \boldsymbol{u}^*) \\
&= \int_0^\infty [\gamma^2 \boldsymbol{w}^T(t)\boldsymbol{w}(t) - \boldsymbol{z}^T(t)\boldsymbol{z}(t) - \frac{\mathrm{d}}{\mathrm{d}t}(\boldsymbol{x}^T(t)\boldsymbol{E}^T\boldsymbol{Q}_1\boldsymbol{x}(t))]\mathrm{d}t \\
&= \int_0^\infty [\boldsymbol{w}(t) - \boldsymbol{R}^{-1}(\boldsymbol{D}_k^T\boldsymbol{C}_{1k} + \boldsymbol{B}_{1k}^T\boldsymbol{Q}_1)\boldsymbol{x}(t)]^T\boldsymbol{R} \\
&\quad \cdot[\boldsymbol{w}(t) - \boldsymbol{R}^{-1}(\boldsymbol{D}_k^T\boldsymbol{C}_{1k} + \boldsymbol{B}_{1k}^T\boldsymbol{Q}_1)\boldsymbol{x}(t)]\mathrm{d}t
\end{aligned}$$

Since $0 = J_1(\boldsymbol{u}^*, \boldsymbol{w}^*) \leq J_1(\boldsymbol{u}^*, \boldsymbol{w})$, we get

$$\begin{aligned}
&\boldsymbol{w}^*(t) = \boldsymbol{R}^{-1}(\boldsymbol{D}_k^T\boldsymbol{C}_{1k} + \boldsymbol{B}_{1k}^T\boldsymbol{Q}_1)\boldsymbol{x}(t) \\
\Rightarrow\ &\boldsymbol{K}_1 = \boldsymbol{R}^{-1}(\boldsymbol{D}_k^T\boldsymbol{C}_{1k} + \boldsymbol{B}_{1k}^T\boldsymbol{Q}_1)
\end{aligned}$$

(ii) Implementing $\boldsymbol{w}^*(t) = \boldsymbol{R}^{-1}(\boldsymbol{D}_k^T\boldsymbol{C}_{1k} + \boldsymbol{B}_{1k}^T\boldsymbol{Q}_1)\boldsymbol{x}(t)$ gives

$$\begin{aligned}
\boldsymbol{E}\dot{\boldsymbol{x}}(t) &= [\boldsymbol{A} + \boldsymbol{B}_1\boldsymbol{R}^{-1}(\boldsymbol{D}_k^T\boldsymbol{C}_{1k} + \boldsymbol{B}_{1k}^T\boldsymbol{Q}_1)]\boldsymbol{x}(t) + \boldsymbol{B}_2\boldsymbol{u}(t) \\
\boldsymbol{z}(t) &= \boldsymbol{C}_1\boldsymbol{x}(t) + \boldsymbol{D}_{12}\boldsymbol{u}(t) \\
\boldsymbol{y}(t) &= (\boldsymbol{C}_2 + \boldsymbol{D}_{21}\boldsymbol{R}^{-1}(\boldsymbol{D}_k^T\boldsymbol{C}_{1k} + \boldsymbol{B}_{1k}^T\boldsymbol{Q}_1))\boldsymbol{x}(t)
\end{aligned}$$

Suppose the system $\boldsymbol{E}\dot{\boldsymbol{x}}(t) = (\boldsymbol{A} + \boldsymbol{B}_1\boldsymbol{R}^{-1}(\boldsymbol{D}_k^T\boldsymbol{C}_{1k} + \boldsymbol{B}_{1k}^T\boldsymbol{Q}_1))\boldsymbol{x}(t)$ is stable and impulse-free. Since $J_2(\boldsymbol{u}^*, \boldsymbol{w}^*) \leq J_2(\boldsymbol{u}, \boldsymbol{w}^*)$, it follows that $\boldsymbol{u} = \boldsymbol{u}^*(t)$ is the unique solution to the standard optimal control problem

$$\min_{\boldsymbol{u}}\{J_2(\boldsymbol{u}, \boldsymbol{w}^*) = \int_0^\infty \boldsymbol{z}^T(t)\boldsymbol{z}(t)\mathrm{d}t\}$$

As a result, $\boldsymbol{u}^*(t) = -\boldsymbol{B}_2^T\boldsymbol{Q}_2\boldsymbol{x}(t)$, where $\boldsymbol{Q}_2$ is an admissible solution of the following GARE

$$\begin{aligned}
0 = &[\boldsymbol{A} + \boldsymbol{B}_1\boldsymbol{R}^{-1}(\boldsymbol{D}_k^T\boldsymbol{C}_{1k} + \boldsymbol{B}_{1k}^T\boldsymbol{Q}_1)]^T\boldsymbol{Q}_2 \\
&+\boldsymbol{Q}_2^T[\boldsymbol{A} + \boldsymbol{B}_1\boldsymbol{R}^{-1}(\boldsymbol{D}_k^T\boldsymbol{C}_{1k} + \boldsymbol{B}_{1k}^T\boldsymbol{Q}_1)] \\
&-\boldsymbol{Q}_2^T\boldsymbol{B}_2\boldsymbol{B}_2^T\boldsymbol{Q}_2 + \boldsymbol{C}_1^T\boldsymbol{C}_1,
\end{aligned} \tag{6.40}$$

with $\boldsymbol{E}^T\boldsymbol{Q}_2 = \boldsymbol{Q}_2^T\boldsymbol{E} \geq 0$. Finally,

$$\begin{aligned}
\boldsymbol{u}^*(t) &= \boldsymbol{K}\boldsymbol{y}(t) \\
&= [\boldsymbol{K}\boldsymbol{C}_2 + \boldsymbol{K}\boldsymbol{D}_{21}\boldsymbol{R}^{-1}(\boldsymbol{D}_k^T\boldsymbol{C}_{1k} + \boldsymbol{B}_{1k}^T\boldsymbol{Q}_1)]\boldsymbol{x}(t) \\
&= -\boldsymbol{B}_2^T\boldsymbol{Q}_2\boldsymbol{x}(t) \\
\Rightarrow\quad &\boldsymbol{K}\boldsymbol{C}_2 + \boldsymbol{K}\boldsymbol{D}_{21}\boldsymbol{R}^{-1}(\boldsymbol{D}_k^T\boldsymbol{C}_{1k} + \boldsymbol{B}_{1k}^T\boldsymbol{Q}_1) \\
&+\boldsymbol{B}_2^T\boldsymbol{Q}_2 = 0
\end{aligned} \tag{6.41}$$

This completes the proof. **Q.E.D.**

6.2.4 Generalization to Nonlinear Systems

In this subsection, we consider a general nonlinear descriptor system governed by a set of differential-algebraic equations given by

$$\boldsymbol{\Sigma} \triangleq \begin{cases} \boldsymbol{E}\dot{\boldsymbol{x}} = \boldsymbol{F}(\boldsymbol{x},\, \boldsymbol{w},\, \boldsymbol{u}), \ \boldsymbol{w} \in \mathcal{W} \subset \mathbb{R}^l, \ \ \boldsymbol{u} \in \mathcal{U} \subset \mathbb{R}^m, \\ \quad \boldsymbol{z} = \boldsymbol{Z}(\boldsymbol{x},\, \boldsymbol{w},\, \boldsymbol{u}), \ \boldsymbol{z} \in \mathcal{Z} \subset \mathbb{R}^s, \end{cases} \tag{6.42}$$

associated with the performance criteria

$$I_1(\boldsymbol{u}, \boldsymbol{w}) = \int_0^\infty (\|\boldsymbol{z}(t)\|^2 - \gamma^2 \|\boldsymbol{w}(t)\|^2)\mathrm{d}t, \tag{6.43}$$

and

$$I_2(\boldsymbol{u}, \boldsymbol{w}) = \int_0^\infty \|\boldsymbol{z}\|^2 \mathrm{d}t, \tag{6.44}$$

where $\boldsymbol{x} \in \mathcal{X}$. Here $\boldsymbol{u}$ stands for the vector of control inputs, $\boldsymbol{w}$ is the exogenous input, and $\boldsymbol{z}$ denotes the penalty function, which is required to achieve the prescribed performance specifications. It is assumed throughout that $\boldsymbol{F}(0,\, 0,\, 0) = 0$ and $\boldsymbol{Z}(0,\, 0,\, 0) = 0$. For technical reason, we assume the following standing assumptions.

(NA1) The matrix $\boldsymbol{D}_{12}$ has rank m and the matrix $\boldsymbol{D}_{11}^T\boldsymbol{D}_{11} - \gamma^2\boldsymbol{I}$ is negative definite, where $\boldsymbol{D}_{12} = (\dfrac{\partial \boldsymbol{Z}}{\partial \boldsymbol{u}})_{(\boldsymbol{x},\boldsymbol{w},\boldsymbol{u})=(0,0,0)}$ and $\boldsymbol{D}_{11} = (\dfrac{\partial \boldsymbol{Z}}{\partial \boldsymbol{w}})_{(\boldsymbol{x},\boldsymbol{w},\boldsymbol{u})=(0,0,0)}$.

(NA2) Any bounded trajectory $\boldsymbol{x}(t)$ of the system $\boldsymbol{E}\dot{\boldsymbol{x}}(t) = F(\boldsymbol{x}(t),\, 0,\, \boldsymbol{u}(t))$ satisfying $\boldsymbol{Z}(\boldsymbol{x}(t),\, 0,\, \boldsymbol{u}(t)) = 0$ for all $t \geq 0$ is such that $\lim_{t\to\infty} \boldsymbol{x}(t) = 0$.

(NA3) The matrix pencil

$$\begin{bmatrix} \boldsymbol{A} - \mathrm{j}\omega\boldsymbol{E} & \boldsymbol{B}_2 \\ \boldsymbol{C}_1 & \boldsymbol{D}_{12} \end{bmatrix}$$

has full column rank for all $\omega \in \mathbb{R} \cup \{\infty\}$, where $\boldsymbol{A} = (\dfrac{\partial \boldsymbol{F}}{\partial \boldsymbol{x}})_{(\boldsymbol{x},\boldsymbol{w},\boldsymbol{u})=(0,0,0)}$, $\boldsymbol{B}_2 = (\frac{\partial \boldsymbol{F}}{\partial \boldsymbol{u}})_{(\boldsymbol{x},\boldsymbol{w},\boldsymbol{u})=(0,0,0)}$, and $\boldsymbol{C}_1 = (\dfrac{\partial \boldsymbol{Z}}{\partial \boldsymbol{x}})_{(\boldsymbol{x},\boldsymbol{w},\boldsymbol{u})=(0,0,0)}$.

To describe the main result of this subsection, we introduce the following two Hamilton–Jacobi equations.

$$\mathfrak{H}_V(\boldsymbol{x}, \boldsymbol{w}, \boldsymbol{u}) \triangleq \boldsymbol{V}^T\boldsymbol{F}(\boldsymbol{x}, \boldsymbol{w}, \boldsymbol{u}) + \|\boldsymbol{Z}(\boldsymbol{x}, \boldsymbol{w}, \boldsymbol{u})\|^2 - \gamma^2\|\boldsymbol{w}\|^2, \tag{6.45}$$

$$\mathfrak{H}_U(\boldsymbol{x}, \boldsymbol{w}, \boldsymbol{u}) \triangleq \boldsymbol{U}^T\boldsymbol{F}(\boldsymbol{x}, \boldsymbol{w}, iu) + \|\boldsymbol{Z}(\boldsymbol{x}, \boldsymbol{w}, \boldsymbol{u})\|^2. \tag{6.46}$$

With the previous two equations in hand, it is easy to derive the following identity by some routine algebraic calculations.

$$\begin{bmatrix} \dfrac{\partial^2 \mathfrak{H}_U}{\partial \boldsymbol{u}^2}(0,0,0) & \dfrac{\partial^2 \mathfrak{H}_U}{\partial \boldsymbol{w} \partial \boldsymbol{u}}(0,0,0) \\ \dfrac{\partial^2 \mathfrak{H}_V}{\partial \boldsymbol{u} \partial \boldsymbol{w}}(0,0,0) & \dfrac{\partial^2 \mathfrak{H}_V}{\partial \boldsymbol{w}^2}(0,0,0) \end{bmatrix} = \begin{bmatrix} 2\boldsymbol{D}_{12}^T\boldsymbol{D}12 & 0 \\ 0 & 2(\gamma^2\boldsymbol{I} - \boldsymbol{D}_{11}^T\boldsymbol{D}_{11}) \end{bmatrix}$$

The righthand side of the above identity is nonsingular by Assumption (NA1). Hence, by the Implicit Function Theorem, there exist unique smooth functions $\boldsymbol{u}^*(\boldsymbol{x})$ and $\boldsymbol{w}^*(\boldsymbol{x})$, locally defined on a neighborhood of the origin, such that

$$\frac{\partial \mathfrak{H}_U}{\partial \boldsymbol{u}}(\boldsymbol{x}, \boldsymbol{w}^*(\boldsymbol{x}), \boldsymbol{u}^*(\boldsymbol{x})) = 0, \quad \text{and} \quad \frac{\partial \mathfrak{H}_V}{\partial \boldsymbol{w}}(\boldsymbol{x}, \boldsymbol{w}^*(\boldsymbol{x}), \boldsymbol{u}^*(\boldsymbol{x})) = 0 \tag{6.47}$$

with

$$\boldsymbol{u}^*(0) = 0, \quad \text{and} \quad \boldsymbol{w}^*(0) = 0.$$

Suppose that the functions $\boldsymbol{u}^*(\bullet)$ and $\boldsymbol{w}^*(\bullet)$ are available. Then the following result is readily obtained, which extends Theorem 6.12 to nonlinear descriptor systems of the form (6.42). The proof of the theorem can be accomplished by means of the similar arguments as those used in the previous theorem.

Theorem 6.14. *Consider the nonlinear descriptor system (6.42). Suppose that the Assumptions (NA1) to (NA3) hold. Suppose also that the following hypotheses is satisfied.*

(NH1) *There exists a smooth function* $\mathsf{V}(\bullet) : \mathcal{X} \to \mathbb{R}^n$ *vanishing at* $\boldsymbol{E}\boldsymbol{x} = 0$ *and positive elsewhere, locally defined on a neighborhood of the equilibrium point 0 in* $\mathcal{X}$, *such that*

$$\boldsymbol{V}^T(\boldsymbol{x})\boldsymbol{F}(\boldsymbol{x}, \boldsymbol{w}^*(\boldsymbol{x}), \boldsymbol{u}^*(\boldsymbol{x})) + \|\boldsymbol{Z}(\boldsymbol{x}, \boldsymbol{w}^*(\boldsymbol{x}), \boldsymbol{u}^*(\boldsymbol{x}))\|^2 - \gamma^2 \boldsymbol{w}^*(\boldsymbol{x})\|^2 = 0, \tag{6.48}$$

where $\boldsymbol{w}^*(\boldsymbol{x})$ *and* $\boldsymbol{u}^*(\boldsymbol{x})$ *are local solutions of the equation (6.47). The function* $\boldsymbol{V}(\boldsymbol{x})$ *is defined by*

$$\frac{\partial \mathsf{V}(\boldsymbol{x})}{\partial \boldsymbol{x}} = \boldsymbol{V}^T(\boldsymbol{x})\boldsymbol{E},$$

with $\boldsymbol{E}^T\boldsymbol{V}(\boldsymbol{x})_{\boldsymbol{x}} = \boldsymbol{V}(\boldsymbol{x})_{\boldsymbol{x}}^T\boldsymbol{E} \geq 0$.

(NH2) *There exists a smooth function* $\mathsf{U}(\bullet) : \mathcal{X} \to \mathbb{R}^n$ *vanishing at* $\boldsymbol{E}\boldsymbol{x} = 0$ *and positive elsewhere, locally defined on a neighborhood of the equilibrium point 0 in* $\mathcal{X}$, *such that*

$$\boldsymbol{U}^T(\boldsymbol{x})\boldsymbol{F}(\boldsymbol{x}, \boldsymbol{w}^*(\boldsymbol{x}), \boldsymbol{u}^*(\boldsymbol{x})) + \|\boldsymbol{Z}(\boldsymbol{x}, \boldsymbol{w}^*(\boldsymbol{x}), \boldsymbol{u}^*(\boldsymbol{x}))\|^2 = 0, \tag{6.49}$$

where $\boldsymbol{w}^*(\boldsymbol{x})$ *and* $\boldsymbol{u}^*(\boldsymbol{x})$ *are local solutions of the equation (6.47). The function* $\boldsymbol{U}(\boldsymbol{x})$ *is defined by*

$$\frac{\partial \mathsf{U}(\boldsymbol{x})}{\partial \boldsymbol{x}} = \boldsymbol{U}^T(\boldsymbol{x})\boldsymbol{E},$$

with $\boldsymbol{E}^T\boldsymbol{U}(\boldsymbol{x})_{\boldsymbol{x}} = \boldsymbol{U}(\boldsymbol{x})_{\boldsymbol{x}}^T\boldsymbol{E} \geq 0$.

Then the closed-loop system

$$\begin{aligned} \boldsymbol{E}\dot{\boldsymbol{x}}(t) &= \boldsymbol{F}(\boldsymbol{x}(t),\, \boldsymbol{w}(t),\, \boldsymbol{u}^*(\boldsymbol{x}(t))) \\ \boldsymbol{z} &= \boldsymbol{Z}(\boldsymbol{x}(t),\, \boldsymbol{w}(t),\, \boldsymbol{u}^*(\boldsymbol{x}(t))) \end{aligned}$$

is locally asymptotically stable with index one at the equilibrium point $\boldsymbol{x} = 0$. *Furthermore, the following conditions are satisfied.*

(a) *The trajectory of the closed loop system* $\dot{x} = F(x, w, u^*)$ *with* $Ex_0 = 0$ *satisfies*

$$\int_0^\infty \|Z(x, w^*(x), u^*(x))\|^2 \mathrm{d}t \leq \int_0^\infty \gamma^2 w^*(x)\|^2 \mathrm{d}t, \quad \forall w \in \mathcal{L}_2[0, \infty).$$

(b)

$$\begin{aligned}\min_{u\in\Omega} I_2(w^*(x), u) &= \min_{\substack{u\in\Omega\\ w=w^*(x)}} \int_0^\infty \|Z(x, w^*(x), u^*(x))\|^2 \mathrm{d}t \\ &= I_2(w^*(x), u^*(x)) = \mathtt{U}(x_0)\end{aligned}$$

where Ω *is the feasibility control set defined by*

$$\Omega \triangleq \{u | E\dot{x} = F(x, w^*(x), u) \text{ is locally asymptotically stable with index one at } x = 0\}.$$

□

Notes and References

1. Recently, there have been a great many contributions devoted to the reduced-order $\mathcal{H}_\infty$ controller design for linear systems, see, for example, DeShetler and Ridgley[22], Gahinet and Apkarian[31], Gu *et al.*[38], Haddad and Bernstein [39], Hsu *et al.*[44], Hyland and Bernstein[45], Iwasaki and Skelton[50] [51], Juang *et al.*[54], Li and Chang[68], Stoorvogel *et al.*[108] Sweriduk and Calise[110], Xin *et al.*[124].
2. The present treatment on the reduced-order controllers design for descriptor systems given in this chapter is essentially based on the technique proposed by Yung [133], [129].
3. The mixed $\mathcal{H}_2$ and $\mathcal{H}_\infty$ control problems have been investigate by several authors for linear state-variable systems, by means of different approaches. The reader may be refered to, for example, Doyle *et al.*[25], Bernstein and Haddad[11], Khargonekar and Rotea[58], Limebeer *et al.*[69], Mustafa[87], Yeh *et al.*[126], Zhou *et al.*[141].
4. One approach to mixed $\mathcal{H}_2/\mathcal{H}_\infty$ problems is introduced by Bernstein and Haddad [11]. Their approach is to minimize an $\mathcal{H}_2$ type cost, while provided that an $\mathcal{H}_\infty$ performance bound is achieved. Another mixed $\mathcal{H}_2/\mathcal{H}_\infty$ problem is addressed in [25][141]. They consider a standard system configuration with auxiliary input signal: one is assumed to be white noise, while the other one is a signal of bounded power. It was shown later by Yeh *et al.*[126] that the solution of Bernstein and Haddad[11] and the solution presented by Doyle *et al.*[25] and Zhou *et al.*[141] are actually dual to each other, although they look quite different for the solutions proposed by Bernstein and Haddad[11] are characterized in terms of three modified

Riccati equations, while the mixed $\mathcal{H}_2/\mathcal{H}_\infty$ controller derived in [25] and [141] is parameterized in terms of a pair of cross-coupled Riccati equations and a standard $\mathcal{H}_\infty$ Riccati equation. Khargonekar and Rotea [58] have examined multiple-objective control problems, which include mixed $\mathcal{H}_2/\mathcal{H}_\infty$ problem as a special case. Their work is quite the contrary to the other work in this area. In fact, they give an algorithmic solution based on convex optimization. Most recently, Limebeer *et al.*[69] developed a design methodology for the mixed $\mathcal{H}_2$ and $\mathcal{H}_\infty$ control problem via the solution of a two-player nonzero-sum Nash game. It is well known that two-player nonzero-sum differential games have two performance criteria[7], and that the central idea behind the work in [69] is to use the two performance indices to reflect the $\mathcal{H}_2$ optimization and $\mathcal{H}_\infty$ constraint separately. For the finite horizon linear time-varying sytem case and the infinite horizon linear time-invariant system case, it turns out the, under some appropriate assumptions, the solution to the mixed $\mathcal{H}_2$ and $\mathcal{H}_\infty$ control problem via state feedback is determined by the solution of a pair of cross-coupled Riccati equations.

5. For nonlinear systems, the mixed $\mathcal{H}_2$ and $\mathcal{H}_\infty$ control problem was fully addressed in Lin[70] for both the continuous and discrete time cases. He used a method that is basically motivated by Limebeer *et al.*[69].
6. The technique used in this chapter to solve the mixed $\mathcal{H}_2$ and $\mathcal{H}_\infty$ control problem for descriptor systems is essentially the same as Limebeer *et al.*[69] and Yung and Shaw[131].

7

Conclusions

7.1 Concluding Remarks

In this monograph, on the one hand, we have studied the $\mathcal{H}_\infty$ control problems for dynamical systems governed by a set of differential-algebraic equations, in an inclusive manner. It has been shown that most of the existing results for conventional state-variable systems can be extended, under appropriate assumptions, to descriptor systems. On the other hand, we have examined certain properties of the DAEs, which include the solvability, stability, controllability, observability, and dissipativeness. Some useful preliminary results have been first developed in Chapter 2. We have derived the Lyapunov stability theorem and LaSalle's Invariant Principle for stability test of descriptor systems. Furthermore, the dissipation property has been investigated in the descriptor systems context. It has been shown that the dissipation property can be characterized by the dissipation inequality, which turn out a Hamilton–Jacobi inequality for the nonlinear case or a generalized algebraic Riccati inequality for the linear case. Most of the results in the chapter as well as the Youla parameterization for descriptor systems elaborated in Chapter 3 construct a cornerstone to the solutions of the standard $\mathcal{H}_\infty$ control problem. In Chapter 4, For the nonlinear case, we have given a family of output feedbak controllers solving the $\mathcal{H}_\infty$ control problem for nonilnear differential-algebraic equations. The resulting closed-loop system has index one which guarantees that the system has no bifurcation induced by singularity. Both necessary and sufficient conditions have been also provided in that chapter. For the linear case, we have derived all solutions to the $\mathcal{H}_\infty$ output feedback control problem. The set of all $\mathcal{H}_\infty$ controllers solving the problem has been given in terms of a linear fractional transformation of the form $\mathcal{LFT}(\boldsymbol{M}_{OF}, \boldsymbol{Q})$ with thc frcc parameter $\boldsymbol{Q}$ satisfying $\|\boldsymbol{Q}\|_\infty < \gamma$.

In Chapter 5, we have studied the balanced realization for descriptor systems, with applications to model reduction and network synthesis. The conclusions drawn in that chapter generalize some well known results in the literature.

Finally, in Chapter 6, we have proposed two further topics in the area of the $\mathcal{H}_\infty$ control for descriptor systems. They are the reduced-order controller design for the full-order generalized plant and the mixed $\mathcal{H}_2$ and $\mathcal{H}_\infty$ control problem. The main results in that chapter are the extensions of the results for conventional state-variable systems. For the reduced-order controllers design, it has been shown that the existenc of a reduced-order controller is guaranteed by the solvability of the standard $\mathcal{H}_\infty$ problem together with that three algebraic equations have solutions. The controller thus obtained is in a conventional state-space. Thereafter, the controller can be implemented by the classical controller synthesis technique. For the mixed $\mathcal{H}_2$ and $\mathcal{H}_\infty$ control problem, we have solved the problem via a Nash game approach. Compared to the varieties of other approaches, the present one has the advantage that it provides a more intuitive interpretation of the intrinsic nature of the mixed $\mathcal{H}_2$ and $\mathcal{H}_\infty$ control problem. Furthermore, the control strategy thus obtained has a explicit state feedback form, which is easily implemented. However, it can be recognized that the control law provided in this monograph has a bit of practical limitation.

Some topics that are worth further studies are as follows.

Minimum Entropy Control Consider a linear dynamical system $\boldsymbol{G}(s)$. Suppose that $\boldsymbol{G}(s) \in \mathcal{RH}^\infty$ with $\|\boldsymbol{G}(s)\|_\infty < \gamma$. Then the entropy of $\boldsymbol{G}(s)$ is defined by

$$I(\boldsymbol{G}, \gamma) \triangleq -\frac{\gamma^2}{2\pi} \int_{-\infty}^{\infty} \ln |\det \left(\boldsymbol{I} - \gamma^{-2} \boldsymbol{G}^*(\mathrm{j}\omega) \boldsymbol{G}(\mathrm{j}\omega) \right)| \mathrm{d}\omega. \tag{7.1}$$

It has been shown by Glover and Mustafa [35] that the central controller given by Doyle *et al.*[24] expressed in terms of the solutions of two algebraic Riccati equations(ARE) minimizes the entropy integral (7.1) and, thus, the central controller would be the preferred controller to use in practical applications. It is expected that the controller obtained in Chapter 4, which is expressed in terms of two generalized algebraic equations, has a similar minimum-entropy property as well.

$\mathcal{H}_\infty$-Balanced Truncation A major drawback of the original balancing method as described in Chapter 5 is that it only applies to stable systems. Furthermore, the method emphasizes the open-loop input-output characteristics of the system, while it is *a prior* not clear if it yields good approximations in closed-loop configurations. This suggests the $\mathcal{H}_\infty$ balancing technique. $\mathcal{H}_\infty$ balancing was introduced with the aim of finding a model reduction method for a system(not necessarily stable) together with its corresponding $\mathcal{H}_\infty$ compensator. For conventional linear state-variable systems, the $\mathcal{H}_\infty$ balancing technique has been introduced by Mustafa and Glover[89][90]. Similar procedure was extended to nonlinear systems by Scherpen[102][103]. Nevertheless, the $\mathcal{H}_\infty$-balanced method for descriptor systems has not been discussed yet in the literature. It is believed that

the same technique can be applied to truncate not only the less important finite dynamic modes of descriptor systems, but also the nondynamic modes. Meanwhile, the $\mathcal{H}_\infty$ robust performance can be guaranteed.

7.2 An Open Problem for DAEs[12]

In this monograph, we have considered a more general class of dynamical systems, namely the DAE systems. However, this class of systems is necessarily linear in the derivative term. Hence, the results presented here cannot be applied to the general fully implicit DAEs.

Problem 7.1. *Consider a state-space systems described by implicit differential-algebraic equations*

$$\boldsymbol{F}(\boldsymbol{x}(t),\, \dot{\boldsymbol{x}}(t),\, \boldsymbol{w}) = 0, \tag{7.2}$$

where $\boldsymbol{x}$ are local coordinates for some n-dimensional state-space manifold $\mathcal{X}$, and $\boldsymbol{w} \in \mathbb{W} = \mathbb{R}^s$ are the external variables(inputs and outputs). The system is dissipative with respect to a supply rate $s : \mathbb{W} \to \mathbb{R}$ if there exists a function $S : \mathcal{X} \to \mathbb{R}^+$, such that

$$S(\boldsymbol{x}(t_1)) \leq S(\boldsymbol{x}(t_0)) + \int_{t_0}^{t_1} s(\boldsymbol{w}(t))\mathrm{d}t,$$

for all trajectories $\boldsymbol{x}(\bullet)$, $\boldsymbol{w}(\bullet)$ satisfying (7.2). The remaining problem is how we can construct the function $S(\bullet)$ such that the system (7.2) is dissipative and in the mean time the internal stability is guaranteed.

Recently, there has been much effort working on dynamical systems represented by differential-algebraic equations of the form

$$\boldsymbol{R}\left(\frac{\mathrm{d}}{\mathrm{d}t}\right)\boldsymbol{w} = 0, \tag{7.3}$$

where $\boldsymbol{R}(\bullet)$ is a matrix polynomial. The dynamic variable $\boldsymbol{w}$, in this case, is a collection of the time trajectories satisfying (7.3), which is called the *behavior* of the system. Control theory based on the behavioral approach has now been well developed, and our result, to some extent, can be viewed as a special case of it, but is quite different in nature. The main difficulty of the behavioral approach is that its nonlinear systems counterpart is miscellaneous. In particular, the stability theroy is still an open problem to the control public. To be precise, let's consider the nonlinear system of differential equations

$$\boldsymbol{f}\left(\boldsymbol{w}, \frac{\mathrm{d}}{\mathrm{d}t}\boldsymbol{w}, \frac{\mathrm{d}^2}{\mathrm{d}t^2}\boldsymbol{w}, \ldots, \frac{\mathrm{d}^n}{\mathrm{d}t^n}\boldsymbol{w}\right) = 0, \tag{7.4}$$

where $\boldsymbol{f} : (\mathrm{I\!R}^q)^{n+1} \rightarrow \mathrm{I\!R}^\bullet$ is a polynomial map. In the language of Willems[122], Equation (7.4) defines the dynamical system

$$\Sigma_{\boldsymbol{f}} \triangleq (\mathrm{I\!R}, \mathrm{I\!R}^q, \mathcal{B}_{\boldsymbol{f}})$$

with the behavior

$$\mathcal{B}_{\boldsymbol{f}} \triangleq \left\{ \boldsymbol{w} : \mathrm{I\!R} \rightarrow \mathrm{I\!R}^q \mid \boldsymbol{f}\left(\boldsymbol{w}, \frac{\mathrm{d}}{\mathrm{d}t}\boldsymbol{w}, \ldots, \frac{\mathrm{d}^n}{\mathrm{d}t^n}\boldsymbol{w}\right) = 0 \right\}.$$

An important question is put forward: What is the Lyapunov stability theorem for the system (7.3) looking like? The answer to this question is in fact a key factor to Problem 7.1, since, as we have all experienced, that the stroage functions usually consist of appropriate Lyapunov functions so that the asymptotical stability of the system can be guaranteed in the same time. It is conjectured that the aformentioned statement is also true for the general fully implicit DAE systems (7.2). A full exploration in this direction is left for future research.

A

Generalized Algebraic Riccati Equations

In the first part of this appendix, we study the coupled generalized algebraic Riccati equation

$$\begin{cases} \boldsymbol{A}^T\boldsymbol{X} + \boldsymbol{X}^T\boldsymbol{A} + \boldsymbol{Q} + \boldsymbol{X}^T\boldsymbol{R}\boldsymbol{X} = 0, \\ \boldsymbol{E}^T\boldsymbol{X} = \boldsymbol{X}^T\boldsymbol{E}, \end{cases} \tag{A.1}$$

and the generalized algebraic Riccati inequality (GARI)

$$\begin{cases} \boldsymbol{A}^T\boldsymbol{P} + \boldsymbol{P}^T\boldsymbol{A} + \boldsymbol{Q} + \boldsymbol{P}^T\boldsymbol{R}\boldsymbol{P} < 0, \\ \boldsymbol{E}^T\boldsymbol{P} = \boldsymbol{P}^T\boldsymbol{E}, \end{cases} \tag{A.2}$$

together with the associated matrix pencil, called *Hamiltonian pencil*,

$$s\begin{bmatrix} \boldsymbol{E} & 0 \\ 0 & \boldsymbol{E}^T \end{bmatrix} - \begin{bmatrix} \boldsymbol{A} & \boldsymbol{R} \\ -\boldsymbol{Q} & -\boldsymbol{A}^T \end{bmatrix} \tag{A.3}$$

where $\boldsymbol{A}$, $\boldsymbol{Q}$, $\boldsymbol{R} \in \mathbb{R}^{n\times n}$ with $\boldsymbol{Q}$, $\boldsymbol{R}$ symmetric. This pencil was named *Hamiltonian* since the matrix

$$\bar{\boldsymbol{H}} \triangleq \begin{bmatrix} \boldsymbol{A} & \boldsymbol{R} \\ -\boldsymbol{Q} & -\boldsymbol{A}^T \end{bmatrix}$$

is a Hamiltonian matrix.

Definition A.1. *A solution $\boldsymbol{X}$ to GARE (A.1) is said to be admissible if the pair $\{\boldsymbol{E},\ \boldsymbol{A}+\boldsymbol{R}\boldsymbol{X}\}$ is admissible.* □

The following two lemmas summarize some properties relevent to the GARE, GARI, and the Hamiltonian pencil.

Lemma A.2. *Suppose that $\{\boldsymbol{E},\ A\}$ is impulse-free. Suppose that $\boldsymbol{Q} = \boldsymbol{Q}^T$, $\boldsymbol{R} \geq 0$ and the pencil $\boldsymbol{P}_2(s) \triangleq s\begin{bmatrix} -\boldsymbol{E} & 0 \end{bmatrix} + \begin{bmatrix} \boldsymbol{A} & \boldsymbol{R} \end{bmatrix}$ has full row rank on the imaginary axis. Furthermore, suppose that GARI*

$$\begin{cases} \boldsymbol{A}^T\boldsymbol{P} + \boldsymbol{P}^T\boldsymbol{A} + \boldsymbol{Q} + \boldsymbol{P}^T\boldsymbol{R}\boldsymbol{P} < 0, \\ \boldsymbol{E}^T\boldsymbol{P} = \boldsymbol{P}^T\boldsymbol{E}, \end{cases} \tag{A.4}$$

has a solution $\boldsymbol{P}$ with $\boldsymbol{E}^T\boldsymbol{P} = \boldsymbol{P}^T\boldsymbol{E} \geq 0$. Under these conditions, the Hamiltonian pencil

$$s\begin{bmatrix} \boldsymbol{E} & 0 \\ 0 & \boldsymbol{E}^T \end{bmatrix} - \begin{bmatrix} \boldsymbol{A} & \boldsymbol{R} \\ -\boldsymbol{Q} & -\boldsymbol{A}^T \end{bmatrix} \triangleq s\bar{\boldsymbol{E}} - \bar{\boldsymbol{H}} \tag{A.5}$$

has no pure imaginary zeros and $\{\bar{\boldsymbol{E}}, \bar{\boldsymbol{H}}\}$ is impulse-free. □

Proof. Without loss of generality, we take $\{\boldsymbol{E}, \boldsymbol{A}\}$ in the following form,

$$\{\boldsymbol{E}, \boldsymbol{A}\} = \left\{ \begin{bmatrix} \boldsymbol{I} & 0 \\ 0 & 0 \end{bmatrix}, \begin{bmatrix} \boldsymbol{A}_1 & 0 \\ 0 & \boldsymbol{I} \end{bmatrix} \right\}, \tag{A.6}$$

and

$$\boldsymbol{P} = \begin{bmatrix} \boldsymbol{P}_{11} & 0 \\ \boldsymbol{P}_{21} & -\boldsymbol{P}_{22} \end{bmatrix} \quad \text{with}\, \boldsymbol{P}_{11} > 0, \boldsymbol{P}_{22} > 0. \tag{A.7}$$

See also [82]. Set $\boldsymbol{Q} = \begin{bmatrix} \boldsymbol{Q}_{11} & \boldsymbol{Q}_{12} \\ \boldsymbol{Q}_{21} & \boldsymbol{Q}_{22} \end{bmatrix}$, $\boldsymbol{B} = \begin{bmatrix} \boldsymbol{B}_1 \\ \boldsymbol{B}_2 \end{bmatrix}$, and $\boldsymbol{R} = \boldsymbol{B}\boldsymbol{B}^T = \begin{bmatrix} \boldsymbol{B}_1\boldsymbol{B}_1^T & \boldsymbol{B}_1\boldsymbol{B}_2^T \\ \boldsymbol{B}_2\boldsymbol{B}_1^T & \boldsymbol{B}_2\boldsymbol{B}_2^T \end{bmatrix}$, where the partition is compatible with (A.6), and set

$$\boldsymbol{S} = \boldsymbol{A}^T\boldsymbol{P} + \boldsymbol{P}^T\boldsymbol{A} + \boldsymbol{Q} + \boldsymbol{P}^T\boldsymbol{B}\boldsymbol{B}^T\boldsymbol{P} = \begin{bmatrix} \boldsymbol{S}_{11} & \boldsymbol{S}_{12} \\ \boldsymbol{S}_{21} & \boldsymbol{S}_{22} \end{bmatrix}.$$

Then $\boldsymbol{S} < 0$ by hypothesis. We first show that Hamiltonian pencil is column-reduced. This is equivalent to showing that

$$\boldsymbol{H}_{22} \triangleq \begin{bmatrix} -\boldsymbol{I} & \boldsymbol{B}_2\boldsymbol{B}_2^T \\ -\boldsymbol{Q}_{22} & I \end{bmatrix} \tag{A.8}$$

is nonsingular. Observe now that

$$\begin{aligned} \boldsymbol{S}_{22} &= -\boldsymbol{P}_{22} - \boldsymbol{P}_{22} + \boldsymbol{Q}_{22} + \boldsymbol{P}_{22}\boldsymbol{B}_2\boldsymbol{B}_2^T\boldsymbol{P}_{22} \\ &= (-\boldsymbol{I})^T\boldsymbol{P}_{22} + \boldsymbol{P}_{22}(-\boldsymbol{I}) + \boldsymbol{Q}_{22} + \boldsymbol{P}_{22}\boldsymbol{B}_2\boldsymbol{B}_2^T\boldsymbol{P}_{22} < 0 \end{aligned}$$

and clearly, $[\boldsymbol{I} - \mathrm{j}\omega\boldsymbol{I} \quad \boldsymbol{R}_{22}]$ has full row rank for all $\omega \in \mathbb{R}$. Then, by standard results of algebraic Riccati inequailty(ARI), this implies that (A.8) has no eigenvalues on $\mathrm{j}\omega$-axis (see [60], Lemma A.2.4), i.e., $\boldsymbol{H}_{22}$ is nonsingular.

We now show that Hamiltonian pencil has no pure imaginary zeros. We assume, for convenience, that all signals may be complex (i.e., $\mathcal{C}^n$) at this time. Observe the following identity

$$\begin{aligned} &\boldsymbol{x}^*(\boldsymbol{A}^T\boldsymbol{P} + \boldsymbol{P}^T\boldsymbol{A})\boldsymbol{x} + \boldsymbol{x}^*\boldsymbol{Q}\boldsymbol{x} - \boldsymbol{u}^*\boldsymbol{u} + \boldsymbol{x}^*\boldsymbol{P}^T\boldsymbol{B}\boldsymbol{u} + \boldsymbol{u}^*\boldsymbol{B}^T\boldsymbol{P}\boldsymbol{x} \\ &= \boldsymbol{x}^*(\boldsymbol{A}^T\boldsymbol{P} + \boldsymbol{P}^T\boldsymbol{A} + \boldsymbol{Q} + \boldsymbol{P}^T\boldsymbol{B}\boldsymbol{B}^T\boldsymbol{P})\boldsymbol{x} - (\boldsymbol{u} - \boldsymbol{B}^T\boldsymbol{P})^*(\boldsymbol{u} - \boldsymbol{B}^T\boldsymbol{P}) \\ &= \boldsymbol{x}^*\boldsymbol{S}\boldsymbol{x} - (\boldsymbol{u} - \boldsymbol{B}^T\boldsymbol{P})^*(\boldsymbol{u} - \boldsymbol{B}^T\boldsymbol{P}) \leq \boldsymbol{x}^*\boldsymbol{S}\boldsymbol{x}. \end{aligned} \tag{A.9}$$

Consider the descriptor system

$$\boldsymbol{E}\dot{\boldsymbol{x}} = \boldsymbol{A}\boldsymbol{x} + \boldsymbol{B}\boldsymbol{u} \tag{A.10}$$

Choose an input $\boldsymbol{u}(\bullet)$, an initial condition $\boldsymbol{E}\boldsymbol{x}(0) \in \mathcal{C}^n$ and let $\boldsymbol{x}(\bullet)$ denote the corresponding solution. Observing that

$$\frac{\mathrm{d}(\boldsymbol{x}^*(t)\boldsymbol{P}^T\boldsymbol{E}\boldsymbol{x}(t))}{\mathrm{d}t} = \boldsymbol{x}^*(\boldsymbol{A}^T\boldsymbol{P} + \boldsymbol{P}^T\boldsymbol{A})\boldsymbol{x} + \boldsymbol{x}^*\boldsymbol{P}^T\boldsymbol{B}\boldsymbol{u} + \boldsymbol{u}^*\boldsymbol{B}^T\boldsymbol{P}\boldsymbol{x}$$

and using (A.9) yields

$$\frac{\mathrm{d}(\boldsymbol{x}^*(t)\boldsymbol{P}^T\boldsymbol{E}\boldsymbol{x}(t))}{\mathrm{d}t} + \boldsymbol{x}^*\boldsymbol{Q}\boldsymbol{x} - \boldsymbol{u}^*\boldsymbol{u} \le \boldsymbol{x}^*\boldsymbol{S}\boldsymbol{x}. \tag{A.11}$$

Suppose, by contradiction, the pencil has a zero on the imaginary axis. By definition, there exist vectors $\boldsymbol{x}_0 \in \mathcal{C}^n$, $\boldsymbol{p}_0 \in \mathcal{C}^n$ and a number $\omega_0 \in \mathbb{R}$ such that

$$\begin{bmatrix} \boldsymbol{A} - \mathrm{j}\omega_0\,\boldsymbol{E} & \boldsymbol{R} \\ -\boldsymbol{Q} & -\boldsymbol{A}^T - \mathrm{j}\omega_0\,\boldsymbol{E}^T \end{bmatrix} \begin{bmatrix} \boldsymbol{x}_0 \\ \boldsymbol{p}_0 \end{bmatrix} = \begin{bmatrix} 0 \\ 0 \end{bmatrix}. \tag{A.12}$$

Note that $\boldsymbol{x}_0 \neq 0$. Otherwise,

$$\begin{aligned} \boldsymbol{R}\boldsymbol{p}_0 &= 0 \\ (\boldsymbol{A}^T + \mathrm{j}\omega_0\,\boldsymbol{E}^T)\boldsymbol{p}_0 &= 0 \end{aligned}$$

This leads to a contradiction, because, by hypothesis, $\boldsymbol{P}_2(s)$ has full row rank. From (A.12), we have

$$\begin{aligned} \boldsymbol{p}_0^*(\boldsymbol{A} - \mathrm{j}\omega_0\,\boldsymbol{E})\boldsymbol{x}_0 + \boldsymbol{p}_0^*\boldsymbol{R}\boldsymbol{p}_0 &= 0, \\ \boldsymbol{x}_0^*\boldsymbol{Q}\boldsymbol{x}_0 + \boldsymbol{x}_0^*(\boldsymbol{A} - \mathrm{j}\omega_0\,\boldsymbol{E})^*\boldsymbol{p}_0 &= 0, \end{aligned}$$

ans therefore

$$\boldsymbol{x}_0^*\boldsymbol{Q}\boldsymbol{x}_0 - \boldsymbol{p}_0^*\boldsymbol{R}\boldsymbol{p}_0 = 0.$$

Set $\boldsymbol{u}(t) = \boldsymbol{B}^T\boldsymbol{p}_0 e^{j\omega_0 t}$ and note that $\boldsymbol{E}\boldsymbol{x}(t) = \boldsymbol{E}\boldsymbol{x}_0 e^{j\omega_0 t}$ is the solution of (A.10) satisfying $\boldsymbol{E}\boldsymbol{x}(0) = \boldsymbol{E}\boldsymbol{x}_0$. Then

$$\begin{aligned} \boldsymbol{x}^*(t)\boldsymbol{P}^T\boldsymbol{E}\boldsymbol{x}(t) &= \boldsymbol{x}_0^*\boldsymbol{P}^T\boldsymbol{E}\boldsymbol{x}_0, \\ \boldsymbol{x}^*(t)\boldsymbol{Q}\boldsymbol{x}(t) - \boldsymbol{u}^*(t)\boldsymbol{u}(t) &= \boldsymbol{x}_0^*\boldsymbol{Q}\boldsymbol{x}_0 - \boldsymbol{p}_0^*\boldsymbol{R}\boldsymbol{p}_0 = 0, \\ \boldsymbol{x}^*(t)\boldsymbol{S}\boldsymbol{x}(t) &= \boldsymbol{x}_0^*\boldsymbol{S}\boldsymbol{x}_0. \end{aligned}$$

Inequality (A.11) yields $\boldsymbol{x}_0^*\boldsymbol{S}\boldsymbol{x}_0 \ge 0$ which is a contradiction, because $\boldsymbol{S}$ is negative definite and $\boldsymbol{x}_0 \neq 0$. **Q.E.D.**

Lemma A.3. *Suppose that $\{\boldsymbol{E}, \boldsymbol{A}\}$ is impulse-free. Suppose that $\boldsymbol{Q} \ge 0$, $\boldsymbol{R} = \boldsymbol{R}^T$ and the pencil $\boldsymbol{P}_1(s) \triangleq s\begin{bmatrix} -\boldsymbol{E} \\ 0 \end{bmatrix} + \begin{bmatrix} \boldsymbol{A} \\ \boldsymbol{Q} \end{bmatrix}$ has full column rank on the imaginary axis. Furthermore, suppose that GARI (A.4) has a solution $\boldsymbol{P}$ with*

$\boldsymbol{E}^T\boldsymbol{P} = \boldsymbol{P}^T\boldsymbol{E} \geq 0$. *Under these conditions, there exists an admissible solution* $\boldsymbol{X}_E$ *to the GARE*

$$\begin{cases} \boldsymbol{A}^T\boldsymbol{X} + \boldsymbol{X}^T\boldsymbol{A} + \boldsymbol{Q} + \boldsymbol{X}^T\boldsymbol{R}\boldsymbol{X} = 0, \\ \boldsymbol{E}^T\boldsymbol{X} = \boldsymbol{X}^T\boldsymbol{E}, \end{cases} \tag{A.13}$$

with $\boldsymbol{E}^T\boldsymbol{X}_E = \boldsymbol{X}_E^T\boldsymbol{E} \geq 0$, *and having the property that* $\{\boldsymbol{E}, \boldsymbol{A} + \boldsymbol{R}\boldsymbol{X}_E\}$ *is admissible. Moreover,* $\rho(\boldsymbol{P}) > \rho(\boldsymbol{X}_E)$. □

Proof. The hypothesis on GARI (A.4) implies that

$$s\begin{bmatrix} \boldsymbol{E}^T & 0 \\ 0 & \boldsymbol{E} \end{bmatrix} - \begin{bmatrix} \boldsymbol{A}^T & \boldsymbol{Q} \\ -\boldsymbol{R} & -\boldsymbol{A} \end{bmatrix} = 0$$

has no zeros on jω-axis and is column-reduced. It follows that

$$s\begin{bmatrix} \boldsymbol{E} & 0 \\ 0 & \boldsymbol{E}^T \end{bmatrix} - \begin{bmatrix} \boldsymbol{A} & \boldsymbol{R} \\ -\boldsymbol{Q} & -\boldsymbol{A}^T \end{bmatrix} = 0 \tag{A.14}$$

has no zeros on jω-axis and is column-reduced. Set

$$\boldsymbol{Q} = \begin{bmatrix} \boldsymbol{Q}_{11} & \boldsymbol{Q}_{12} \\ \boldsymbol{Q}_{12}^T & \boldsymbol{Q}_{22} \end{bmatrix} \quad \text{and} \quad \boldsymbol{R} = \begin{bmatrix} \boldsymbol{R}_{11} & \boldsymbol{R}_{12} \\ \boldsymbol{R}_{12}^T & \boldsymbol{R}_{22} \end{bmatrix}$$

where the partition is compatible with (A.6). From (A.4), there exists a positive definite matrix $\boldsymbol{P}_{22}$ such that

$$(-\boldsymbol{I})\boldsymbol{P}_{22} + \boldsymbol{P}_{22}(-\boldsymbol{I}) + \boldsymbol{Q}_{22} + \boldsymbol{P}_{22}^T\boldsymbol{R}_{22}\boldsymbol{P}_{22} < 0. \tag{A.15}$$

Clearly $[\boldsymbol{I} - \mathrm{j}\omega\boldsymbol{I} \quad \boldsymbol{Q}_{22}]^T$ has full column rank for all $\omega \in \mathbb{R}$. This, again from the standard results of ARI (see [60], Lemma A.2.5), implies that there exists a matrix $\boldsymbol{X}_{22} = \boldsymbol{X}_{22}^T \geq 0$ satisfying the ARE

$$-\boldsymbol{X}_{22} - \boldsymbol{X}_{22} + \boldsymbol{Q}_{22} + \boldsymbol{X}_{22}\boldsymbol{R}_{22}\boldsymbol{X}_{22} = 0$$

with $\rho(\boldsymbol{P}_{22}) > \rho(\boldsymbol{X}_{22})$ [60][93]. Rewrite Pencil (A.14) as

$$s\left[\begin{array}{cc|cc} \boldsymbol{I} & 0 & 0 & 0 \\ 0 & \boldsymbol{I} & 0 & 0 \\ \hline 0 & 0 & 0 & 0 \\ 0 & 0 & 0 & 0 \end{array}\right] - \left[\begin{array}{cc|cc} \boldsymbol{A}_1 & \boldsymbol{R}_{11} & \boldsymbol{R}_{12} & 0 \\ -\boldsymbol{Q}_{11} & -\boldsymbol{A}_1^T & 0 & -\boldsymbol{Q}_{12} \\ \hline 0 & \boldsymbol{R}_{12}^T & \boldsymbol{R}_{22} & \boldsymbol{I} \\ \boldsymbol{Q}_{12}^T & 0 & \boldsymbol{I} & \boldsymbol{Q}_{22} \end{array}\right]$$

$$\triangleq s\begin{bmatrix} \mathcal{I} & 0 \\ 0 & 0 \end{bmatrix} - \begin{bmatrix} \mathcal{T}_1 & \mathcal{T}_2 \\ \mathcal{T}_3 & \mathcal{T}_4 \end{bmatrix}$$

From ARI (A.15), the above pencil can be simplified as

$$s\mathcal{I} - [\mathcal{T}_1 - \mathcal{T}_2\mathcal{T}_4^{-1}\mathcal{T}_3] \triangleq s\mathcal{I} - \begin{bmatrix} \mathcal{A}_0 & \mathcal{R}_0 \\ -\mathcal{Q}_0 & -\mathcal{A}_0^T \end{bmatrix}.$$

The existence of $\boldsymbol{P}_{22}$ to ARI (A.15) implies that $\mathcal{Q}_0 \geq 0$ (See Willems[120], Lemma 1). Moreover, GARI (A.4) and ARI (A.15) together imply that there exists a positive definite matrix $\boldsymbol{P}_0\,(=\boldsymbol{P}_{11})$ satisfying

$$\boldsymbol{S}(\boldsymbol{P}_0) \triangleq \mathcal{A}_0^T \boldsymbol{P}_0 + \boldsymbol{P}_0 \mathcal{A}_0 + \mathcal{Q}_0 + \boldsymbol{P}_0 \mathcal{R}_0 \boldsymbol{P}_0 < 0. \tag{A.16}$$

This, together with the hypothesis on pencil $\boldsymbol{P}_1(s)$, implies that there exists a stabilizing solution $X_0 \geq 0$ satisfying the ARE $\boldsymbol{S}(\boldsymbol{X}_0) = 0$ since one can deduce that $[\mathcal{A}_0^T - \mathrm{j}\omega I \quad \mathcal{Q}_0]^T$ has full column rank for all $\omega \in \mathbb{R}$ from the assumption that $[\boldsymbol{A}^T - \mathrm{j}\omega \boldsymbol{E}^T \quad \boldsymbol{Q}]^T$ has full column rank on the jω-axis. Note that $\rho(\boldsymbol{P}_0) > \rho(\boldsymbol{X}_0)$. Set

$$\boldsymbol{X}_{21} = \boldsymbol{L}_2 - \boldsymbol{X}_{22}\boldsymbol{L}_1 \quad \text{and} \quad \boldsymbol{X}_E = \begin{bmatrix} \boldsymbol{X}_0 & 0 \\ \boldsymbol{X}_{21} & \boldsymbol{X}_{22} \end{bmatrix},$$

where

$$\begin{bmatrix} \boldsymbol{L}_1 \\ \boldsymbol{L}_2 \end{bmatrix} \triangleq -\mathcal{T}_4^{-1}\mathcal{T}_3 \begin{bmatrix} \boldsymbol{I} \\ \boldsymbol{X}_0 \end{bmatrix}.$$

It is easy to see that $\boldsymbol{X}_E$ satisfys the GARE (A.13) with $\{\boldsymbol{E},\, \boldsymbol{A} + \boldsymbol{R}\boldsymbol{X}_E\}$ admissible and $\rho(\boldsymbol{P}) > \rho(\boldsymbol{X}_E)$. This completes the proof. **Q.E.D.**

The next two propositions are devoted to the relationship between least square optimization problems and generalized algebraic Riccati equations.

Proposition A.4. *Consider the following linear time-invariant descriptor system*

$$\boldsymbol{\Sigma}: \begin{cases} \boldsymbol{E}\dot{\boldsymbol{x}} = \boldsymbol{A}\boldsymbol{x} + \boldsymbol{B}\boldsymbol{u}, \\ \boldsymbol{y} = \boldsymbol{C}\boldsymbol{x} \end{cases}$$

together with its Weierstrass form representation

$$\begin{aligned} \begin{bmatrix} \boldsymbol{I} & 0 \\ 0 & \boldsymbol{N} \end{bmatrix} \begin{bmatrix} \dot{\hat{\boldsymbol{x}}}_1 \\ \dot{\hat{\boldsymbol{x}}}_2 \end{bmatrix} &= \begin{bmatrix} \boldsymbol{A}_1 & 0 \\ 0 & \boldsymbol{I} \end{bmatrix} \begin{bmatrix} \hat{\boldsymbol{x}}_1 \\ \hat{\boldsymbol{x}}_2 \end{bmatrix} + \begin{bmatrix} \boldsymbol{B}_1 \\ \boldsymbol{B}_2 \end{bmatrix} \begin{bmatrix} \boldsymbol{u}_1 \\ \boldsymbol{u}_2 \end{bmatrix}, \\ \boldsymbol{y} &= \begin{bmatrix} \boldsymbol{C}_1 & \boldsymbol{C}_2 \end{bmatrix} \begin{bmatrix} \hat{\boldsymbol{x}}_1 \\ \hat{\boldsymbol{x}}_2 \end{bmatrix}. \end{aligned} \tag{A.17}$$

Suppose that $\{\boldsymbol{E},\, \boldsymbol{A}\}$ is regular, impulse-free and that $\{\boldsymbol{E}, \boldsymbol{A}, \boldsymbol{B}\}$ is finite dynamics stabilizable and impulse controllable. Suppose also that $\boldsymbol{I} - \boldsymbol{B}_2^T \boldsymbol{C}_2^T \boldsymbol{C}_2 \boldsymbol{B}_2 > 0$. Furthermore, assume that the Hamiltonian system

$$\begin{bmatrix} \boldsymbol{E} & 0 \\ 0 & \boldsymbol{E}^T \end{bmatrix} \begin{bmatrix} \dot{\boldsymbol{x}} \\ \dot{\boldsymbol{\lambda}} \end{bmatrix} = \begin{bmatrix} \boldsymbol{A} & \boldsymbol{B}\boldsymbol{B}^T \\ -\boldsymbol{C}^T\boldsymbol{C} & -\boldsymbol{A}^T \end{bmatrix} \begin{bmatrix} \boldsymbol{x} \\ \boldsymbol{\lambda} \end{bmatrix} \tag{A.18}$$

is regular, impulse-free and has no finite dynamic modes on the imaginary axis. Then there exists an admissible solution $\boldsymbol{X}$ to the GARE

$$\begin{cases} \boldsymbol{A}^T\boldsymbol{X} + \boldsymbol{X}^T\boldsymbol{A} + \boldsymbol{C}^T\boldsymbol{C} + \boldsymbol{X}^T\boldsymbol{B}\boldsymbol{B}^T\boldsymbol{X} = 0, \\ \boldsymbol{E}^T\boldsymbol{X} = \boldsymbol{X}^T\boldsymbol{E}. \end{cases}$$

□

Remark. Here a solution $\boldsymbol{X}$ to the GARE is called an admissible solution if $\{\boldsymbol{E}, \boldsymbol{A}+\boldsymbol{B}\boldsymbol{B}^T\boldsymbol{X}\}$ is admissible. It is noted that $\boldsymbol{X}$ might not be unique, but $\boldsymbol{E}^T\boldsymbol{X} = \boldsymbol{X}^T\boldsymbol{E}$ is unique. For details, see, e.g., [56], [111], [116] and the references quoted therein.

Proof. Consider a linear dynamical system of the form,

$$\begin{aligned} \dot{\boldsymbol{x}} &= -\boldsymbol{x} + \boldsymbol{B}_2\boldsymbol{u} \\ \boldsymbol{y} &= \boldsymbol{C}_2\boldsymbol{x} \end{aligned}$$

It is easy to verify that $\boldsymbol{I}-\boldsymbol{B}_2^T\boldsymbol{C}_2^T\boldsymbol{C}_2\boldsymbol{B}_2 > 0$ if and only if the system is strictly bounded real with an upper bound 1. Then, using a standard result on algebraic Riccati equations(ARE)([60], Theorem 2.3.1), there exists a stabilizing solution $\boldsymbol{X}_{22} = \boldsymbol{X}_{22}^T \geq 0$ satisfying the ARE

$$(-\boldsymbol{X}) + (-\boldsymbol{X}) + \boldsymbol{C}_2^T\boldsymbol{C}_2 + \boldsymbol{X}\boldsymbol{B}_2\boldsymbol{B}_2^T\boldsymbol{X} = 0.$$

(Note that a solution to ARE $\boldsymbol{A}^T\boldsymbol{X} + \boldsymbol{X}\boldsymbol{A} + \boldsymbol{Q} + \boldsymbol{X}\boldsymbol{R}\boldsymbol{X} = 0$ is said to be stabilizing if the matrix $(\boldsymbol{A}+\boldsymbol{R}\boldsymbol{X})$ is stable.) Set $\boldsymbol{x} = \begin{bmatrix} \boldsymbol{x}_1 \\ \boldsymbol{x}_2 \end{bmatrix}$ and $\lambda = \begin{bmatrix} \boldsymbol{\lambda}_1 \\ \boldsymbol{\lambda}_2 \end{bmatrix}$ compatible with (A.17). Then Equation (A.18) can be rearranged in the following form.

$$\left[\begin{array}{cc|cc} \boldsymbol{I} & 0 & 0 & 0 \\ 0 & \boldsymbol{I} & 0 & 0 \\ \hline 0 & 0 & 0 & 0 \\ 0 & 0 & 0 & 0 \end{array}\right] \begin{bmatrix} \dot{\boldsymbol{x}}_1 \\ \dot{\boldsymbol{\lambda}}_1 \\ \dot{\boldsymbol{x}}_2 \\ \dot{\boldsymbol{\lambda}}_2 \end{bmatrix} = \left[\begin{array}{cc|cc} \boldsymbol{A}_1 & \boldsymbol{B}_1\boldsymbol{B}_1^T & \boldsymbol{B}_1\boldsymbol{B}_2^T & 0 \\ -\boldsymbol{C}_1^T\boldsymbol{C}_1 & -\boldsymbol{A}_1^T & 0 & -\boldsymbol{C}_1^T\boldsymbol{C}_2 \\ \hline 0 & \boldsymbol{B}_2\boldsymbol{B}_1^T & \boldsymbol{B}_2\boldsymbol{B}_2^T & \boldsymbol{I} \\ \boldsymbol{C}_2^T\boldsymbol{C}_1 & 0 & \boldsymbol{I} & \boldsymbol{C}_2^T\boldsymbol{C}_2 \end{array}\right] \begin{bmatrix} \boldsymbol{x}_1 \\ \boldsymbol{\lambda}_1 \\ \boldsymbol{x}_2 \\ \boldsymbol{\lambda}_2 \end{bmatrix}$$

$$\triangleq \begin{bmatrix} \mathcal{I} & 0 \\ 0 & 0 \end{bmatrix} \begin{bmatrix} \dot{\boldsymbol{\xi}} \\ \dot{\boldsymbol{\eta}} \end{bmatrix} = \begin{bmatrix} \mathcal{T}_1 & \mathcal{T}_2 \\ \mathcal{T}_3 & \mathcal{T}_4 \end{bmatrix} \begin{bmatrix} \boldsymbol{\xi} \\ \boldsymbol{\eta} \end{bmatrix}$$

Since, by hypothesis, $\boldsymbol{I}-\boldsymbol{B}_2^T\boldsymbol{C}_2^T\boldsymbol{C}_2\boldsymbol{B}_2 > 0$, this implies that $\mathcal{T}_4$ is nonsingular. Hence, the above system can be simplified as

$$\dot{\boldsymbol{\xi}} = [\mathcal{T}_1 - \mathcal{T}_2\mathcal{T}_4^{-1}\mathcal{T}_3]\boldsymbol{\xi} \quad \triangleq \quad \dot{\boldsymbol{\xi}} = \begin{bmatrix} \mathcal{A}_0 & \mathcal{R}_0 \\ -\mathcal{Q}_0 & -\mathcal{A}_0^T \end{bmatrix} \boldsymbol{\xi}. \tag{A.19}$$

It is easy to verify that $\mathcal{R}_0 \geq 0$ (since $\boldsymbol{I}-\boldsymbol{B}_2^T\boldsymbol{C}_2^T\boldsymbol{C}_2\boldsymbol{B}_2 > 0$) and $\{\mathcal{A}_0, \mathcal{R}_0\}$ is stabilizable (since $\{\boldsymbol{E}, \boldsymbol{A}, \boldsymbol{B}\}$ is finite dynamic stabilizable). Then, by a standard result on ARE ([60], Lemma A.2.3), this implies that there exists a stabilizing solution $\boldsymbol{X}_0 = \boldsymbol{X}_0^T \geq 0$ to the ARE $\boldsymbol{S}(\boldsymbol{X}_0) \triangleq \mathcal{A}_0^T\boldsymbol{X}_0 + \boldsymbol{X}_0\mathcal{A}_0 + \mathcal{Q}_0 + \boldsymbol{X}_0\mathcal{R}_0\boldsymbol{X}_0 = 0$. Set

$$\boldsymbol{X}_{21} = \boldsymbol{L}_2 - \boldsymbol{X}_{22}\boldsymbol{L}_1 \quad \text{and} \quad \boldsymbol{X}_E = \begin{bmatrix} \boldsymbol{X}_0 & 0 \\ \boldsymbol{X}_{21} & \boldsymbol{X}_{22} \end{bmatrix},$$

where

$$\begin{bmatrix} L_1 \\ L_2 \end{bmatrix} \triangleq -\mathcal{T}_4^{-1}\mathcal{T}_3 \begin{bmatrix} I \\ X_0 \end{bmatrix}.$$

It now follows that the GARE $A^T X_E + X_E^T A + C^T C + X_E^T B B^T X_E = 0$, $E^T X_E = X_E^T E \geq 0$ with $\{E,\ A + BB^T X_E\}$ admissible. This completes the proof. **Q.E.D.**

Consider the following constrained minimization problem,

$$\min_{u} \int_0^{\infty} (x^T C^T C x + 2x^T C^T D u + u^T R u)\mathrm{d}t \tag{A.20}$$

subject to

$$E\dot{x} = Ax + Bu, \quad Ex(0) = 0, \tag{A.21}$$

in which $R \triangleq D^T D > 0$. The following proposition is readily obtained [119].

Proposition A.5. *Consider (A.20) and (A.21). Suppose that $\{E,\ A,\ B\}$ is finite dynamic stabilizable and detectable, and $\Im m\, E + A(\aleph(E)) + \Im m\, B = \mathbb{R}^n$ and $\aleph(E) \cap A^{-1}(\Im m\, E) \cap \aleph(C) = \{0\}$, then there exists a unique optimal control for (A.20), which can be implemented by*

$$u^*(t) = -B^T X x = Kx, \tag{A.22}$$

where

$$X = \begin{bmatrix} X_{11} & 0 \\ X_{21} & X_{22} \end{bmatrix}$$

satisfyng the GARE

$$\begin{cases} A^T X + X^T A + C^T C - (X^T B + C^T D) R^{-1} (B^T X + C^T D)^T = 0, \\ E^T X = X^T E \geq 0. \end{cases} \tag{A.23}$$

Moreover, the optimal closed-loop system

$$E\dot{x} = (A - BR^{-1}(B^T X + D^T C))\, x$$

is admissible. □

Remark. It was pointed out in [119] that X_{22} can be chosen almost arbitrarily. In this paper, we only consider the X_{22} matrix which satisfying the GARE (A.23) since it, in addition, provides the existence of a certain spectral factor. However, it can be shown that such X_{22} matrix always exists under the assumptions that that $\Im m\, E + A(\aleph(E)) + \Im m\, B = \mathbb{R}^n$ and $\aleph(E) \cap A^{-1}(\Im m\, E) \cap \aleph(C) = \{0\}$. We first perform an SVD on A_{22} such that $A_{22} = U \Sigma_{22} V$ where U, V are orthogonal and Σ_{22} a diagonal matrix with diagonal elements either greater than or equal to zero. Next, define two nonsingular matrices as

$$\boldsymbol{M} = \begin{bmatrix} \boldsymbol{I} & 0 \\ 0 & -\boldsymbol{U}^* \end{bmatrix}, \qquad \boldsymbol{N} = \begin{bmatrix} \boldsymbol{I} & 0 \\ 0 & \boldsymbol{V}^* \end{bmatrix}.$$

Then the resulting transformed system,

$$\tilde{\boldsymbol{E}} = \boldsymbol{MEN}, \quad \tilde{\boldsymbol{A}} = \boldsymbol{MAN}, \quad \tilde{\boldsymbol{B}} = \boldsymbol{MB}, \quad \tilde{\boldsymbol{C}} = \boldsymbol{CN},$$

has the property that $\{\tilde{\boldsymbol{A}}_{22}, \tilde{\boldsymbol{B}}_2, \tilde{\boldsymbol{C}}_2\}$ is stabilizable and detectable. Rename these matrices as $\{\boldsymbol{A}_{22}, \boldsymbol{B}_2, \boldsymbol{C}_2\}$. It is clear that there exists a matrix $\boldsymbol{X}_{22}$ satisfying

$$\boldsymbol{A}_{22}^T \boldsymbol{X}_{22} + \boldsymbol{X}_{22} \boldsymbol{A} + \boldsymbol{C}_2^T \boldsymbol{C}_2 - (\boldsymbol{X}_{22} \boldsymbol{B}_2 + \boldsymbol{C}_2^T \boldsymbol{D}) \boldsymbol{R}^{-1} (\boldsymbol{B}_2^T \boldsymbol{X}_{22} + \boldsymbol{D}^T \boldsymbol{C}_2) = 0,$$

i.e., the condition in Proposition A.5 is satisfied.

Lemma A.6. *Let $\boldsymbol{X}^-$ (respectively, $\boldsymbol{X}^+$) be a solution of the generalized algebraic Riccati equation (A.1) having the property that the pair $\{\boldsymbol{E}, \boldsymbol{A} + \boldsymbol{R}\boldsymbol{X}^-\}$ is admissible (respectively, $\{-\boldsymbol{E}, -(\boldsymbol{A} + \boldsymbol{R}\boldsymbol{X}^+)\}$ is admissible). Then, there exist a real number ϵ_0, a family of symmetric matrices $\boldsymbol{S}_\epsilon < 0$, and a family of matrices $\boldsymbol{P}_\epsilon$, defined for $\epsilon \in [0, \epsilon_0]$ and continuously depending on ϵ, satisfying*

$$\begin{cases} \boldsymbol{A}^T \boldsymbol{P}_\epsilon + \boldsymbol{P}_\epsilon^T \boldsymbol{A} + \boldsymbol{Q} + \boldsymbol{P}_\epsilon^T \boldsymbol{R} \boldsymbol{P}_\epsilon = \boldsymbol{S}_\epsilon, \\ \boldsymbol{E}^T \boldsymbol{P}_\epsilon = \boldsymbol{P}_\epsilon^T \boldsymbol{E} \geq 0 \end{cases}$$

and such that $\boldsymbol{x}^T \boldsymbol{E}^T (\boldsymbol{X}^- - \boldsymbol{P}_\epsilon) \boldsymbol{x} < 0$ (respectively, $\boldsymbol{x}^T \boldsymbol{E}^T (\boldsymbol{X}^+ - \boldsymbol{P}_\epsilon) \boldsymbol{x} > 0$) for all $\boldsymbol{x} \in \Im m\, \boldsymbol{E}$. □

B

Center Manifold Theory

The purpose of the second part of the appendix is to review some elements of center manifold theory and extend certain existing results for conventional systems to descriptor systems. The proposed development is rather straightforward but is useful in this monograph. Consider a nonlinear descriptor system described by the following set of DAEs.

$$\begin{aligned} \dot{\boldsymbol{x}} &= \boldsymbol{f}_1(\boldsymbol{x}, \boldsymbol{y}, \boldsymbol{z}), \\ \dot{\boldsymbol{y}} &= \boldsymbol{f}_2(\boldsymbol{x}, \boldsymbol{y}, \boldsymbol{z}), \quad (\boldsymbol{x}, \boldsymbol{y}, \boldsymbol{z}) \in \mathcal{X} \subset \mathbb{R}^n, \\ 0 &= \boldsymbol{f}_2(\boldsymbol{x}, \boldsymbol{y}, \boldsymbol{z}). \end{aligned} \tag{B.1}$$

or in a more compact form

$$\boldsymbol{E}\dot{\boldsymbol{x}} = \boldsymbol{F}(\boldsymbol{x}) \tag{B.2}$$

where $\boldsymbol{f}_1$, $\boldsymbol{f}_2$ and $\boldsymbol{f}_3$ are C^r functions defined on an open subset $\mathcal{X}$ of $\mathbb{R}^n$ with $\boldsymbol{f}_1(0,0,0) = 0$, $\boldsymbol{f}_2(0,0,0) = 0$ and $\boldsymbol{f}_3(0,0,0) = 0$. For a conventional nonlinear system

$$\dot{\boldsymbol{x}} = \boldsymbol{f}(\boldsymbol{x}), \ \boldsymbol{f}(0) = 0 \tag{B.3}$$

It is well known that the local asymptotical stability of the equilibrium point (in this case, $\boldsymbol{x} = 0$) can be determined, to some extent by the behavior of the linear approximation of f at $\boldsymbol{x} = 0$. This important result is commonly known as the *Principle of Stability in the First Approximation.* As a matter of fact, this is also true for the nonlinear descriptor system (B.1). For, let

$$\mathcal{F} = \left[\frac{\partial \boldsymbol{F}}{\partial \boldsymbol{x}}\right]_{\boldsymbol{x}=0}$$

denote the Jacobian matrix of $\boldsymbol{F}$ at $\boldsymbol{x} = 0$. Then

1. if the pair $\{\boldsymbol{E}, \mathcal{F}\}$ is stable and impulse-free, then the system (B.1) (or equivalently, the system (B.2)) is asymptotically stable with index one.
2. if one or more eigenvalues of the pair $\{\boldsymbol{E}, \mathcal{F}\}$ are in the right-half complex plane, then $\boldsymbol{x} = 0$ is an unstable equilibrium of (B.1).

If the system (B.1) has index one, then locally the solution trajectory $(\boldsymbol{x}(t), \boldsymbol{y}(t))$ is constrained to the surface $\boldsymbol{f}_2(\boldsymbol{x}, \boldsymbol{y}, \boldsymbol{k}(\boldsymbol{x}, \boldsymbol{y})) = 0$ for some function $\boldsymbol{k}(\bullet, \bullet)$ with $\boldsymbol{k}(0,0) = 0$. Moreover, the original DAEs reduces to ordinary differential equations of the form

$$\begin{aligned} \dot{\boldsymbol{x}} &= \boldsymbol{f}_1(\boldsymbol{x}, \boldsymbol{y}, \boldsymbol{k}(\boldsymbol{x}, \boldsymbol{y})), \\ \dot{\boldsymbol{y}} &= \boldsymbol{f}_2(\boldsymbol{x}, \boldsymbol{y}, \boldsymbol{k}(\boldsymbol{x}, \boldsymbol{y})). \end{aligned} \tag{B.4}$$

Then, locally on the constrained surface, the results of the center manifold theory applies equally well to the descriptor systems (B.1). However, some details still need to be worked out. Define

$$\mathcal{S} := \{(\boldsymbol{x}, \boldsymbol{y}, \boldsymbol{z}) \in \mathcal{X} \mid \boldsymbol{z} = (\boldsymbol{x}, \boldsymbol{y})\}$$

It is easily verified that $\mathcal{S}$ is a C^r submanifold of $\mathcal{X}$. We call it *constrained manifold.*

Definition B.1. *A C^r constrained manifold $\mathcal{S}$ of $\mathcal{U}$ is said to be locally invariant for (B.1) if for each $\boldsymbol{x}^0 \in \mathcal{S}$, there exist $t_1 < 0 < t_2$ such that the integral curve $\boldsymbol{x}(t)$ of (B.1) satisfying $\boldsymbol{x}(0) = \boldsymbol{x}^0$ is such that $\boldsymbol{x}(t) \in \mathcal{S}$ for all $t \in (t_1, t_2)$.* □

In the following, we present some interesting results about the asymptotic properties of certain nonlinear descriptor systems, that are useful. The proofs can be made with the help of the center manifold theory and the previous definition.

Lemma B.2. *Consider a system*

$$\begin{aligned} \boldsymbol{E}_z \dot{\boldsymbol{z}} &= \boldsymbol{f}(\boldsymbol{z}, \boldsymbol{y}), \\ \boldsymbol{E}_y \dot{\boldsymbol{y}} &= \boldsymbol{A}\boldsymbol{y} + \boldsymbol{p}(\boldsymbol{z}, \boldsymbol{y}). \end{aligned} \tag{B.5}$$

Suppose that $\mathbf{p}(\boldsymbol{z}, 0) = 0$ for all $\boldsymbol{z}$ near 0 and

$$\frac{\partial \boldsymbol{p}}{\partial \boldsymbol{y}}(0,0) = 0.$$

If $\boldsymbol{E}_z \dot{\boldsymbol{z}} = \boldsymbol{f}(\boldsymbol{z}, 0)$ has an asymptotically stable equilibrium at $\boldsymbol{z} = 0$, which is of index one, and the pair $\{\boldsymbol{E}_y, \boldsymbol{A}\}$ is stable and impulse-free, then the system (B.5) has an asymptotically stable equilibrium at $(\boldsymbol{z}, \boldsymbol{y}) = (0,0)$. Moreover, it is of index one. □

Lemma B.3. *Consider the system*

$$\begin{aligned} \boldsymbol{E}_z \dot{\boldsymbol{z}} &= \boldsymbol{q}(\boldsymbol{z}, \boldsymbol{y}) \\ \boldsymbol{E}_y \dot{\boldsymbol{y}} &= \boldsymbol{g}(\boldsymbol{z}, \boldsymbol{y}). \end{aligned} \tag{B.6}$$

Suppose $(\boldsymbol{z}, \boldsymbol{y}) = (0,0)$ is an equilibrium of (B.6), the equilibrium $\boldsymbol{z} = 0$ of $\boldsymbol{E}_z \dot{\boldsymbol{z}} = \boldsymbol{q}(\boldsymbol{z}, 0)$ is asymptotically stable with index one, and the equilibrium $\boldsymbol{y} = 0$ of $\dot{\boldsymbol{y}} = \boldsymbol{g}(\boldsymbol{y})$ is stable with index one. Then the equilibrium

$(\boldsymbol{z}, \boldsymbol{y}) = (0, 0)$ *of the system (B.6) is stable with index one. Furthermore, if the equilibrium* $\boldsymbol{y} = 0$ *of* $\dot{\boldsymbol{y}} = \boldsymbol{g}(\boldsymbol{y})$ *is asymptotically stable with index one, then the equilibrium* $(\boldsymbol{z}, \boldsymbol{y}) = (0, 0)$ *of the system (B.6) is asymptotically stable with index one.* □

References

1. A. Ailon, *Controllability of Generalized Linear Time-Invariant Systems*, IEEE Transactions on Automatic Control, Vol. **AC**-32, No. 5, pp. 429-432, 1987.
2. B. D. O. Anderson and S. Vongpanitlerd, *Network Analysis and Synthesis - A Modern Systems Theory Approach*, Prentice-Hall, Englewood Cliffs, N. J., 1973.
3. V. A. Armentano, *The pencil $(sE - A)$ and Controllability-Observability for Generalized Linear Systems: a Geometric Approach*, SIAM J. Control and Optimization, Vol. 24, No. 4, 616-638, 1986.
4. U. M. Ascher and L. R. Petzold, *Computer Methods for Ordinary Differential Equations and Differential-Algebraic Equations*, Society for Industrial and Applied Mathematics, Philadelphia, 1998.
5. K. J. Astrom, *Theory and applications of adaptive control–A survey*, Automatica, Vol. 19, No. 1, pp. 471-481, 1983.
6. J. A. Ball, J. W. Helton, and M. L. Walker, *$\mathcal{H}_\infty$ Control for Nonlinear Systems with Output Feedback*, IEEE Transactions on Automatic Control, Vol. **AC**-38, No. 4, pp. 546-559, 1993.
7. T. Basar and G. J. Olsder, *Dynamic Non Cooperative Game Theory*, New York: Academic Press, 1982.
8. T. Basar and P. Bernhard, *$\mathcal{H}_\infty$ Optimal Control and Related Minimax Design Problems*, Birkhauser, 1990.
9. V. Belevitch. *Theory of 2n-Terminal Networks with Applications to Conference Telephony*, Electrical Communication, Vol. 27, No. 3, pp. 231-244, 1950.
10. J. Bender and A. J. Laub, *The Linear-Quadratic Optimal Regulator for Descriptor System*, IEEE Transactions on Automatic Control, Vol. **AC**-23, No. 1, 672-688, 1987.
11. D. S. Bernstein and W. M. Haddad, *LQG Control with an $\mathcal{H}_\infty$ Performance Bound: A Riccati Equation Approach*, IEEE Transactions on Automatic Control, Vol. **AC**-34, No. 3, pp. 295-305, 1989.
12. V. D. Blondel, E. D. Sontag, M. Vidyasagar, and J. C. Willems(Eds), *Open Problems in Mathematical Systems and Control Theory*, Springer-Verlag, London, 1999.
13. M. Boutayeb and M. Darouach, *Observers Design for Nonlinear Descriptor Systems*, Proceedings of the 34th IEEE Conference on Decision and Control, New Orleans, LA, USA, pp. 2369-2374, 1995

14. K. E. Brenan, S. L. Campbell, and L. R. Petzold, *Numerical Solution of Initial-Value Problems in Differential-Algebraic Equations*, SIAM's Classics in Applied Mathematics series, 1996.
15. A. E. Bryson and Y. C. Ho, *Applied Optimal Control*, Revised Printing, Taylor & Francis, U. S. A., 1969.
16. S. L. Campbell, *Singular Systems of Differential Equations* I, Pitman, New York, 1980.
17. S. L. Campbell, *Singular Systems of Differential Equations* II, Pitman, New York, 1982.
18. C. T. Chen, *Linear System Theory and Design*, Holt, Rinehart and Winston, Inc., New York, 1984.
19. D. J. Cobb, *Descriptor Variable Systems and Optimal State Regulation*, IEEE Transactions on Automatic Control, Vol. **AC**-28, No. 5, pp. 601-611, 1983.
20. D. J. Cobb, *Controllability, Observability, and Duality in Singular Systems*, IEEE Transactions on Automatic Control, **AC**-29, No. 12, pp. 1076-1082, 1984.
21. B. R. Copeland and M. G. Safonov, it A Generalized Eigenproblem Solution for Singular $\mathcal{H}_2$ and $\mathcal{H}_\infty$ Problems, Control and Dynamic Systems, Vol. 50, pp. 331-394, 1992.
22. D. R. DeShelter and D. B. Ridgely, *Reduced-Order $\mathcal{H}_2$ and $\mathcal{H}_\infty$ Compensation via Gradient Techniques*, Proceedings of the 31th IEEE Conference on Decision and Control, Tuscon, Arizona, pp. 2262-2267, 1992.
23. J. C. Doyle, *Lecture Notes in Advances in Multivariable Control*, ONR/Honeywell Workshop Minneapolis, 1984.
24. J. C. Doyle, K. Glover, P. P. Khargonekar, and B. A. Francis, *State-Space Solutions to Standard H_2 and H_∞ Control Problems*, IEEE Transactions on Automatic Control, Vol. **AC**-34, No. 8, pp. 831-846, 1989.
25. J. C. Doyle, K. Zhou, K. Glover, and B. Bodenheimer, *Mixed $\mathcal{H}_2$ and $\mathcal{H}_\infty$ Performance Objectives II: Optimal Control*, IEEE Transactions on Automatic Control, Vol. **AC**-39, No. 8, pp. 1575-1587, 1994.
26. B. Dziurla and R. W. Newcomb, *The Drazin Inverse and Semi-State Equations*, Proceedings of the International Symposium on Mathematical Theory of Networks and Systems, Delft, pp. 283-289, 1979.
27. D. F. Enns, *Model Reduction with Balanced Realizations: An Error Bound and Frequency Weighted Generalization*, Proceedings of the 23rd IEEE Conference on Decision and Control, Las Vegas, pp. 127-132, 1984.
28. F. Fagnani and J. C. Willems, *Representations of Symmetric Linear Dynamical Systems*, SIAM Jounnal on Control and Optimization, Vol. 31, No. 5, pp. 1267-1293, 1993.
29. C. H. Fang and F. R. Chang, *Analysis of Stability Robustness for Generalized State-Space Systems with Structured Perturbations*, Systems and Control Letters, Vol. 21, No. 2, pp. 109-114, 1993.
30. B. A. Francis, *A Course in $\mathcal{H}_\infty$ Control Theory*, Lecture Notes in Control and Information Sciences, 88, Springer-Verlag, Berlin, 1987.
31. P. Gahinet and P. Apkarian, *An LMI-based Parameterization of All $\mathcal{H}_\infty$ Controllers with Application*, Proceedings of the 32rd IEEE Conference on Decison and Control, San Antonio, TX, pp. 656-661, 1993.
32. F. R. Gantmacher, *The Theory of Matrices*, Vol. I and II, Chelsea, New York, 1959.

33. C. W. Gear, B. Leimkuhler, and G. K. Gupta, *Automatic Integration of Euler-Lagrange Equations with Constraints*, Journal of Computational and Applied Mathematics, Vol. 12, No.1, pp. 77-90, 1985.
34. K. Glover, *All Optimal Hankel-norm Approximations of Linear Multivariable Systems and Their $\mathcal{L}_\infty$-error Bounds*, International Journal of Control, Vol. 39, No. 6, pp. 1115-1193, 1984.
35. K. Glover and D. Mustafa, *Derivation of the Maximum Entropy $\mathcal{H}_\infty$ Controller and a State Space Formula for Its Entropy*, International Journal of Control, Vol. 50, No. 3, pp. 899-916, 1989.
36. K. Glover, D. J. M. Limebeer, J. C. Doyle, E. M. Kasenally, and M. G. Safonov, *A Characterization of All Solutions to the Four Block General Distance Problem*, SIAM Journal on Control and Optimization, Vol. 29, No. 2, pp. 283-324, 1991.
37. M. Green and D. J. N. Limebeer, *Linear Robust Control*, Information and System Sciences Series, Prentice Hall, Englewood Cliffs, NJ, 1995.
38. D. W. Gu, B. W. Choi, and I. Postlethwaite, *Low-Order $\mathcal{H}_\infty$ Suboptimal Controllers*, Proceedings of the 12th World Congress of IFAC, Sydney, Australia, pp. 347-350, 1993.
39. W. M. Haddad and D. S. Bernstein, *Generalized Riccati Equations for the Full and Reduced-Order Mixed-Norm $\mathcal{H}_2/\mathcal{H}_\infty$ Standard Problem*, Systems and Control Letters, Vol. 14, pp. 185-197, 1990.
40. E. Hairer, S. Nörsett, and G. Wanner, *Solving Ordinary Differential Equations* II: *Stiff and Differential-Algebraic Problems*, Springer-Verlag, Berlin, 1991.
41. G. E. Hayton, P. Fretwell, and A. C. Pugh, *Fundamental Equivalence of Generalized State Space Systems*, IEEE Transactions on Automatic Control, Vol. **AC**-31, No. 5, pp. 431-439, 1986.
42. D. J. Hill and P. J. Moylan, *Stability of Nonlinear Dissipative Systems*, IEEE Transactions on Automatic Control, Vol. **AC**-21, No. 3, pp. 708-711, 1976.
43. D. J. Hill and P. J. Moylan, *Dissipative Dynamical Systems: Basic Input-Output and State Properties*, Journal of Franklin Institute, Vol. 309, No. 2, pp. 327-357, 1980.
44. C. S. Hsu, X. Yu, H. H. Yeh, and S. S. Banda, *$\mathcal{H}_\infty$ Compensator Design with Minimal-Order Observers*, IEEE Transactions on Automatic Control, Vol. **AC**-39, No. 9, pp. 1679-1681, 1994.
45. D. C. Hyland and D. S. Bernstein, *The Optimal Porjection Equations for Fixed-Order Dynamic Compensation*, IEEE Transactions on Automatic Control, Vol. **AC**-29, No. 5, pp. 1034-1037, 1984.
46. J. I. Imura and T. Yoshikawa, *Parametrization of All Stabilizing Controllers of Nonlinear Systems*, Systems and Control Letters, Vol. 29, No. 4, pp. 207-213, 1997.
47. A. Isidori, *$\mathcal{H}_\infty$ Control via Measurement Feedback for Affine Nonlinear Systems*, International Journal of Robust and Nonlinear Control, Vol. 4, No. 4, pp. 553-574, 1994.
48. A. Isidori and A. Astolfi, *Disturbance Attenuation and $\mathcal{H}_\infty$ Control via Measurement Feedback in Nonlinear Systems*, IEEE Transactions on Automatic Control, Vol. **AC**-37, No. 7, 1992.
49. A. Isidori and W. Kang, *$\mathcal{H}_\infty$ Control via Measurement Feedback for General Nonlinear Systems*, IEEE Trans. on Automatic Control, Vol. **AC**-40, No. 3, pp. 466-472, 1995.

50. T. Iwasaki and R. E. Skelton, *All Low Order $\mathcal{H}_\infty$ Controllers with Covariance Upper Bound*, Proceedings of the 1993 American Control Conference, San Francisco, California, pp. 2180-2184.
51. T. Iwasaki and R. E. Skelton, *All Controllers for the General $\mathcal{H}_\infty$ Control Problem: LMI Existence Conditions and State Space Formulas*, Automatica, Vol. 30, pp. 1307-1317, 1994.
52. E. A. Jonckheere and L. M. Silverman, *Singular Value Analysis of Deformable Systems*, Journal of Circuits, Systems and Signal Processing, Vol. 1, No. 2, pp. 447-470, 1982.
53. E. A. Jonckheere and L. M. Silverman, *A New Set of Invariants for Linear Systems - Applications to Reduced Order Compensator Design*, IEEE Transactions on Automatic Control, Vol. **AC**-28, No. 4, pp. 953-964, 1983.
54. J. C. Juang, H. H. Yeh, and S. S. Banda, *Observer-Based Compensators for Fix-Order $\mathcal{H}_\infty$ Control*, International Journal of Control, Vol. 64, pp. 441-461, 1996.
55. N. Karcanias, *Regular State-Space Realizations of Singular System Control Problems*, Proceedings of the 28th Conference on Decision and Control, Los Angeles, CA, pp. 1144-1146, 1987.
56. A. Kawamoto and T. Katayama, *The Dissipation Inequality and Generalized Algebraic Riccati Equation for Linear Quadratic Control Problem of Descriptor System*, 1996 IFAC 13th World Congress, San Francisco, USA, 103-108, 1996.
57. A. Kawamoto and T. Katayama, *Standard $\mathcal{H}_\infty$ Control Problem for Descriptor System*, Proceedings of the 36th Conference on Decision and Control, San Diego, California, pp. 4130-4133, 1997.
58. P. P. Khargonekar and M .A. Rotea, *Mixed $\mathcal{H}_2/\mathcal{H}_\infty$ Control: A Convex Optimization Approach*, IEEE Transactions on Automatic Control, Vol. **AC**-36, No. 7, pp. 824-837, 1991.
59. C. F. Klamm, Jr., B. D. O. Anderson, and R. W. Newcomb, *Stability of Passive Time-Variable Circuits*, Proceedings of the IEE, Vol. 114, No. 1, pp. 71-75, 1967.
60. H. W. Knobloch, A. Isidori, and D. Flockerzi, *Topics in Control Theory*, Birkhäuser, Basel, Switzerland, 1993.
61. V. Kucera, *Stationary LQG Control of Singular Systems*, IEEE Transactions on Automatic Control, Vol. **AC**-31, No. 1, pp. 31-39, 1986.
62. M. Kuijper, *First-Order Representations of Linear Systems.* Birkhäuser, Boston, 1994.
63. P. Lancaster and L. Rodman, *Algebraic Riccati Equations*, Oxford Science Publication, 1995.
64. W. W. Leontieff, *Static and Dynamic Theory*, in Studies in the Structure of the American Economy (ed: W. W. Leontieff), Oxford University Press, New York, 1953.
65. F. L. Lewis, *Preliminary Notes on Optimal Control for Singular Systems*, Proc. IEEE 24th CDC, Ft. Lauderdale, pp. 266-272, 1985.
66. F. L. Lewis, *A Survey of Linear Singular Systems*, Circuits, Systems, and Signal Process, Vol. 5, No. 1, pp. 3-36, 1986.
67. F. L. Lewis, *A Tutorial on the Geometric Analysis of Linear Time-invariant Implicit Systems*, Automatica, Vol. 28, No. 1, pp. 119-137, 1992.
68. X. P. Li and B. C. Chang, *A Parameterization Approach to Reduced-Order $\mathcal{H}_\infty$ Controller Design*, Proceedings of the 32nd IEEE Conference on Decision and Control, San Antonio, TX, pp. 2909-2912, 1993.

69. D. J. N. Limebeer, B. D. O. Anderson, and B. Hendel, *A Nash Game Approach to Mixed $\mathcal{H}_2/\mathcal{H}_\infty$ Control*, IEEE Transactions on Automatic Control, Vol. **AC**-39, No. 1, pp. 69-82, 1994.
70. W. Lin, *Mixed $\mathcal{H}_2/\mathcal{H}_\infty$ Control via State Feedback for Nonlinear Systems*, International Journal of Control, Vol. 64, No. 5, pp. 899-922, 1996.
71. W. Lin and C. I. Byrnes, *$\mathcal{H}_\infty$-Control of Discrete-Time Nonlinear Systems*, IEEE Transactions on Automatic Control, Vol. **AC**-41, No. 4, pp. 494-510, 1996.
72. W. Q. Liu, W. Y. Yan, and K. L. Teo, *A Frequency Domain Approach to Control of Singular Systems*, IEEE Transactions on Automatic Control, Vol. **AC**-42, No. 3, pp. 672-688, 1997.
73. P. Lotstedt and L. Petzold, *Numerical Solution of Nonlinear Differential Equations with Algebraic Constraints, I: Convergence Results for Backward Differentiation Formulas*, Mathematics of Computation, Vol. 46, No. 174, pp. 491-516, 1986.
74. W. M. Lu and J. C. Doyle, *$\mathcal{H}_\infty$ Control of Nonlinear Systems via Output Feedback: A Class of Controllers*, Proceedings of the 32nd Conference on Decision and Control, San Antonio, Texas, pp. 166-171, 1993.
75. W. M. Lu and J. C. Doyle, *$\mathcal{H}_\infty$ Control of Nonlinear System via Output Feedback: Controllers Parameterization*, IEEE Transactions on Automatic Control, Vol. **AC**-39, No. 12, pp. 2517-2521, 1994.
76. W. M. Lu, *A State-Space Approach to Parameterization of Stabilizing of Controllers for Nonlinear Systems*, IEEE Transactions on Automatic Control, **AC**-40, No. 12, pp. 1576-1588, 1995.
77. D. G. Luenberger, *Singular Dynamic Leontieff Systems*, Econometrica, Vol. 45, pp. 991-995, 1977.
78. D. G. Luenberger, *Dynamic Equations in Descriptor Form*, IEEE Transactions on Automatic Control, Vol. **AC**-22, No. 3, pp. 312-321, 1977.
79. D. G. Luenberger, *Non-Linear Descriptor Systems*, Journal of Economic Dynamics and Control, Vol. 1, pp. 212-242, 1979.
80. J. W. Manke, B. Dembart, M. A. Epton, A. M. Erisman, P. Lu, R. F. Sincovec, and E. L. Yip, *Solvability of Large Scale Descriptor Systems*, Report, Boeing Computer Services Company, Seattle, WA, 1979.
81. R. März, *On Initial Value Problems in Differential-Algebraic Equations and Their Numerical Treatment*, Computing, Vol. 35, No. 1, pp. 13-37, 1985.
82. I. Masubuchi, Y. Kamitane, A. Ohara, and N. Suda, *$\mathcal{H}_\infty$ Control for Descriptor Systems: A Matrix Inequalities Approach*, Automatica, Vol. 33, No. 4, pp. 669-673, 1997.
83. N. H. McClamroch, *Singular Systems of Differential Equations as Dynamic Models for Constrained Robot Systems*, Proceedings of the IEEE Conference on Robotics and Automations, San Francisco, California, pp. 21-28, 1986.
84. B. C. Moore, *Principle Component Analysis in Linear Systems: Controllability, Observability and Model Reduction*, IEEE Transactions on Automatic Control, Vol. **AC**-26, No. 1, pp. 17-32, 1981.
85. J. B. Moore, K. Glover, and A. Telford, *All Stabilizing Controllers as Frequency Shaped State Estimate Feedback*, IEEE Transactions on Automatic Control, **AC**-35, No.1, pp. 203-208, 1990.
86. C. T. Mullis and R. A. Roberts, *Synthesis of Minimum Roundoff Noise Fixed Point Digital Filter*, IEEE Transactions on Circuits and Systems, Vol. **CAS**-23, No. 2, pp. 551-562, 1976.

87. D. Mustafa, *Relations Between Maximum Entropy $\mathcal{H}_2$ Control and Combined $\mathcal{H}_\infty$/ LQG Control*, Systems and Control Letters, Vol. 12, No. 3, pp. 193-203, 1989.
88. D. Mustafa, *$\mathcal{H}_\infty$ Characteristic Values*, Proceedings of the 28th IEEE Conference on Decision and Control, pp. 1483-1487, 1989.
89. D. Mustafa and K. Glover, *Minimum Entropy $\mathcal{H}_\infty$ Control*, Lecture Notes in Control and Information Science, No. 146, Springer-Verlag, Berlin, 1990.
90. D. Mustafa and K. Glover, *Controller Reduction by $\mathcal{H}_\infty$-Balanced Truncation*, IEEE Transactions on Automatic Control, Vol. **AC**-36, No. 6, pp. 668-682, 1991.
91. R. W. Newcomb and B. Dziurla, *Some Circuits and Systems Applications of Semistate Theory*, Circuits, Systems and Signal Processing, Vol. 8, No. 3, pp. 235-260, 1989.
92. K. Ozcaldiran and F. L. Lewis, *A Geometric Approach to Eigenstructure Assignment for Singular Systems*, IEEE Transactions on Automatic Control, Vol. **AC**-32, No. 2, pp. 629-631, 1987.
93. I. R. Petersen, B. D. O. Anderson, and E. A. Jonckheere, *A First Principles Solution to the Nonsingular H_∞ Control Problem*, International Journal of Robust and Nonlinear Control, Vol. 1, No. 2, pp. 153-170, 1991.
94. L. Petzold, *Numerical Solution of Differential/Algebraic Systems by Implicit Runge-Kutta Methods*, Proceedings of the 27th Midwest Symposium on Circuits and Systems, Morgantown, West Virginia, pp. 678-691, 1984.
95. V. M. Popov, *Hyperstability of Control Systems*, Berlin: Springer-Verlag, 1973.
96. S. Reich, *On a Geometric Interpretation of Differential-Algebraic Equations*, Circuits, Systems and Signal Processing, Vol. 9, No. 3, pp. 369-382, 1990.
97. S. Reich, *On an Existence and Uniqueness Theory for Nonlinear Differential-Algebraic Equations*, Circuits, Systems and Signal Processing, Vol. 10, No. 3, pp. 343-359, 1991.
98. M. G. Safonov, E. A. Jonckheere, M. Verma, and D. J. M. Limebeer, *Synthesis of Positive Real Multivariable Feedback Systems*, International Journal of Control, Vol. 45, No. 3, 817-842, 1987.
99. M. T. Sampei, T. Mita, and M. Nakamichi, *An Algebraic Approach to $\mathcal{H}_\infty$ Output Feedback Control Problem*, Systems and Control Letters, Vol. 14, No. 1, pp. 13-24, 1990.
100. J. M. A. Scherpen, *Balancing for Nonlinear Systems*, Systems and Control Letters, Vol. 21, No. 2, pp. 143-153, 1993.
101. J. M. A. Scherpen, *Balancing for Unstable Nonlinear Systems*, Proceedings of the 32nd IEEE Conference on Decision and Control, San Antonio, Texas, pp. 14-19, 1993.
102. J. M. A. Scherpen, *Balancing for Nonlinear System*, doctoral dissertation, University of Twente, 1994.
103. J. M. A. Scherpen, *$\mathcal{H}_\infty$ Balancing for Nonlinear Systems*, International Journal of Robust and Nonlinear Control, Vol. 6, No. 7, pp. 645-688, 1996.
104. A. J. van der Schaft, *On a State Space Approach to Nonlinear $\mathcal{H}_\infty$ Control*, Systems and Control Letters, Vol. 16, No. 1, pp. 1-8, 1991.
105. A. J. van der Schaft, *$\mathcal{L}_2$-Gain and Passivity Techniques in Nonlinear Control*, Lecture Notes in Control and Information Sciences **2**18, Springer Verlag, Berlin, New York, 1996.

106. D. N. Shields, *Feedback Stabilization of a Class of Singular Nonlinear Systems*, IMA Journal of Mathematical Control and Information, Vol. 10, No. 4, pp. 305-322, 1993.
107. D. N. Shields, *Observer Design and Detection for Nonlinear Descriptor Systems*, International Journal of Control, Vol. 67, No. 2, pp. 153-168, 1997.
108. A. A. Stoorvogel, A. Saberi, and B. M. Chen, *A Reduced-Order Observer-Based Controller Design for $\mathcal{H}_\infty$ Optimization*, Proceedings of the AIAA Guidence, Navigation, and Control Conference, New Orleans, Louisiana, pp. 716-722, 1991.
109. W. Sun, P. P. Khargonekar, and D. Shim, *Solution to the Positive Real Control Problem for Linear Time-Invariant Systems*, IEEE Transactions on Automatic Control, Vol. **AC**-39, No. 10, 2034-2046, 1994.
110. D. Sweriduk and A. J. Calise, *Robust Fixed-Order Dynamic Compensation: A Differential Game Approach*, Proceedings of the IEEE Conference on Aerospace Control Systems, West Lake Village, California, pp. 458-462, 1993.
111. K. Takaba, N. Morihira, and T. Katayama, *H_∞ Control for Descriptor Systems - A J-Spectral Factorization Approach*, Proceedings of 33rd Conference on Decision and Control, Lake Buena Vista, Florida, pp. 2251-2256, 1994.
112. T. Takaba, N. Morihira, and T. Katayama, *A Generalized Lyapunov Theorem for Descriptor System*, Systems and Control Letters, Vol. 4, No. 1, pp. 49-51, 1995.
113. V. Venkatasubramanian, H. Schättler, and J. Zaborszky, *Local Bifurcations and Feasibility Regions in Differential-Algebraic Systems*, IEEE Transactions on Automatic Control, Vol. **AC**-40, No. 12, pp. 1992-2013, 1995.
114. G. Verghese, B. C. Levy, and T. Kailath, *A Generalized State-Space for Singular Systems*, IEEE Transaction on Automatic Control, Vol. **AC**-26, No. 4, pp. 811-831, 1981.
115. M. Vidyasagar, *On the Stabilization of Nonlinear Systems Using State Detection*, IEEE Transactions on Automatic Control, Vol. **AC**-25, No. 4, pp. 773-779, 1980.
116. H. S. Wang, C. F. Yung, and F. R. Chang, *Bounded Real Lemma and $\mathcal{H}_\infty$ Control for Descriptor Systems*, IEE Proceeding D: Control Theory and Its Applications, Vol. 145, No. 3, pp. 316-322, 1998.
117. H. S. Wang, C. F. Yung, and F. R. Chang, *$\mathcal{H}_\infty$ Control for Nonlinear Differential-Algebraic Equations*, IEEE Transactions on Automatic Control, Vol. **AC**-47, No. 11, pp. 1919-1925, 2002.
118. H. S. Wang, C. F. Yung, and Mei-Chi Chen, *Reduced-Order $\mathcal{H}_\infty$ Controller Design for Descriptor System*, Proceedings of the 40th IEEE Conference on Decision and Control, Orlando, Florida, pp. 3710-3715, 2001.
119. Y. Y. Wang, P. M. Frank, and D. J. Clements, *The Robustness Properties of the Linear Quadratic Regulators for Singular Systems*, IEEE Transactions on Automatic Control, Vol. **AC**-38, No. 1, pp. 96-100, 1993.
120. J. C. Willems, *Least Square Stationary Optimal Control and the Algebraic Riccati Equation*, IEEE Transactions on Automatic Control, Vol. **AC**-16, No. 6, pp. 621-634, 1971.
121. J. C. Willems, *Dissipative Dynamical Systems part I: General Theory, part II: Linear Systems with Quadratic Supply Rates*, Archive for Rational Mechanics and Analysis, Vol. **45**, pp. 321-393, 1972.
122. J. C. Willems, *Paradigms and Puzzles in the Theory of Dynamical Systems*, IEEE Transactions on Automatic Control, Vol **AC**-36, No. 2, pp. 259-294, 1991.

123. H. Wu and K. Mizukami, *Stability and Robust Stabilization of Nonlinear Descriptor Systems with Uncertainties*, Proceedings of the 33rd Conference on Decision and Control, Lake Buena Vista, Florida, pp. 2772-2777, 1994.
124. X. Xin, L. Guo, and C. B. Feng, *Reduced-Order Controllers for Continuous and Discrete Time Singualr $\mathcal{H}_\infty$ Control Problems Based on LMI*, Automatica, Vol. 32, pp. 171-185.
125. H. Xu and K. Mizukami, *Hamilton–Jacobi Equation for Descriptor Systems*, Systems and Control Letters, Vol. 21, No. 4, pp. 321-327, 1993.
126. H. H. Yeh, S. S. Banda, and B. C. Chang, *Necessary and Sufficient Conditions for Mixed $\mathcal{H}_2$ and $\mathcal{H}_\infty$ Optimal Control*, IEEE Transactions on Automatic Control, Vol. **AC**-37, No. 3, pp. 355-358, 1992.
127. D. C. Youla, H. A. Jabr and, J. J. Bongiorno, *Modern Wiener-Hopf Design of Optimal Controllers: II. The Multivariable Case*, IEEE Transactions on Automatic Control, Vol. **AC**-21, No. 1, pp. 319-338, 1976.
128. C. F. Yung, *State-Space Characterization of All Solutions to Positive Real Control Problem*, Proceedings of 1997 American Control Conference, Albuquerque, New Mexico, pp. 2587-2591, 1997.
129. C. F. Yung, *Reduced-Order $\mathcal{H}_\infty$ Controller Design - An Algebraic Riccati Equation Approcah*, Automatica, Vol. 36, No. 6, pp. 923-926, 2000.
130. C. F. Yung, Y. P. Lin, and F. B. Yeh, *A Family of Nonlinear $\mathcal{H}_\infty$ Output Feedback Controllers*, IEEE Transactions on Automatic Control, Vol. **AC**-41, No. 2, pp. 232-236, 1996.
131. C. F. Yung and J. W. Shaw, *Mixed $\mathcal{H}_2/\mathcal{H}_\infty$ Control for Linear Time-Varying Systems: Infinite Horizon Case*, Journal of Control Systems and Technology, Vol. 6, No. 4, pp. 269-276, 1998.
132. C. F. Yung, H. S. Wang, and F. R. Chang, *State-Space Characterization of All $\mathcal{H}_\infty$ Controllers for Descriptor Systems: A Youla Parameterization Approach*, submitted for publication. See also Proceedings of 1999 Taiwan Automatic Control Conference, pp. 366-371, 1999.
133. C. F. Yung and H. S. Wang, *$\mathcal{H}_\infty$ Controller Reduction for Nonlinear Systems*, Automatica, Vol. 37, No. 11, pp. 1797-1802, 2001.
134. C. F. Yung, J. L. Wu, and T. T. Lee, *Parameterization of $\mathcal{H}_\infty$ State Feedback Controllers*, Automatica, Vol. 33, No. 8, pp. 1587-1590, 1997.
135. C. F. Yung, J. L. Wu, and T. T. Lee, *$\mathcal{H}_\infty$ Control for More General Nonlinear Systems*, IEEE Transactions on Automatic Control, Vol. **AC**-43, No. 12, pp. 1724-1727, 1998.
136. C. F. Yung, *Reduced-Order $\mathcal{H}_\infty$ Controller Design - An Algebraic Riccati Equation Approach*, Automatica, Vol. 36, pp. 923-926, 2000.
137. G. Zames, *On the Input-Output Stability of Nonlinear Time-Varying Feedback Systems, Part I*, IEEE Transaction on Automatic Control, Vol. **AC**-11, No. 2, pp. 228-238, 1966.
138. G. Zames, *On the Input-Output Stability of Nonlinear Time-Varying Feedback Systems, Part II*, IEEE Transaction on Automatic Control, Vol. **AC**-11, No. 3, pp. 465-476, 1966.
139. K. Zhou, *On the Parameterization of $\mathcal{H}_\infty$ Controllers*, IEEE Transactions on Automatic Control, Vol. **AC**-37, No. 9, pp. 1442-1445, 1992.
140. K. Zhou (with J. C. Doyle and K. Glover), *Robust and Optimal Control*, Prentice Hall, Upper Saddle River, New Jersey, 1996.

141. K. Zhou, K. Glover, B. Bodenheimer, and J. C. Doyle, *Mixed $\mathcal{H}_2$ and $\mathcal{H}_\infty$ Performance Objectives I: Robust Performance Analysis*, IEEE Transactions on Automatic Control, Vol. **AC**-39, No. 8, pp. 1564-1574, 1994.

Index

Lecture Notes in Control and Information Sciences

Edited by M. Thoma and M. Morari

Further volumes of this series can be found on our homepage: springer.com

Vol. 325: Amato, F.
Robust Control of Linear Systems Subject to Uncertain Time-Varying Parameters
180 p. 2006 [3-540-23950-2]

Vol. 324: Christofides, P.; El-Farra, N.
Control of Nonlinear and Hybrid Process Systems
446 p. 2005 [3-540-28456-7]

Vol. 323: Bandyopadhyay, B.; Janardhanan, S.
Discrete-time Sliding Mode Control
147 p. 2005 [3-540-28140-1]

Vol. 322: Meurer, T.; Graichen, K.; Gilles, E.D. (Eds.)
Control and Observer Design for Nonlinear Finite and Infinite Dimensional Systems
422 p. 2005 [3-540-27938-5]

Vol. 321: Dayawansa, W.P.; Lindquist, A.; Zhou, Y. (Eds.)
New Directions and Applications in Control Theory
400 p. 2005 [3-540-23953-7]

Vol. 320: Steffen, T.
Control Reconfiguration of Dynamical Systems
290 p. 2005 [3-540-25730-6]

Vol. 319: Hofbaur, M.W.
Hybrid Estimation of Complex Systems
148 p. 2005 [3-540-25727-6]

Vol. 318: Gershon, E.; Shaked, U.; Yaesh, I.
H_∞ Control and Estimation of State-muliplicative Linear Systems
256 p. 2005 [1-85233-997-7]

Vol. 317: Ma, C.; Wonham, M.
Nonblocking Supervisory Control of State Tree Structures
208 p. 2005 [3-540-25069-7]

Vol. 316: Patel, R.V.; Shadpey, F.
Control of Redundant Robot Manipulators
224 p. 2005 [3-540-25071-9]

Vol. 315: Herbordt, W.
Sound Capture for Human/Machine Interfaces: Practical Aspects of Microphone Array Signal Processing
286 p. 2005 [3-540-23954-5]

Vol. 314: Gil', M.I.
Explicit Stability Conditions for Continuous Systems
193 p. 2005 [3-540-23984-7]

Vol. 313: Li, Z.; Soh, Y.; Wen, C.
Switched and Impulsive Systems
277 p. 2005 [3-540-23952-9]

Vol. 312: Henrion, D.; Garulli, A. (Eds.)
Positive Polynomials in Control
313 p. 2005 [3-540-23948-0]

Vol. 311: Lamnabhi-Lagarrigue, F.; Loría, A.; Panteley, V. (Eds.)
Advanced Topics in Control Systems Theory
294 p. 2005 [1-85233-923-3]

Vol. 310: Janczak, A.
Identification of Nonlinear Systems Using Neural Networks and Polynomial Models
197 p. 2005 [3-540-23185-4]

Vol. 309: Kumar, V.; Leonard, N.; Morse, A.S. (Eds.)
Cooperative Control
301 p. 2005 [3-540-22861-6]

Vol. 308: Tarbouriech, S.; Abdallah, C.T.; Chiasson, J. (Eds.)
Advances in Communication Control Networks
358 p. 2005 [3-540-22819-5]

Vol. 307: Kwon, S.J.; Chung, W.K.
Perturbation Compensator based Robust Tracking Control and State Estimation of Mechanical Systems
158 p. 2004 [3-540-22077-1]

Vol. 306: Bien, Z.Z.; Stefanov, D. (Eds.)
Advances in Rehabilitation
472 p. 2004 [3-540-21986-2]

Vol. 305: Nebylov, A.
Ensuring Control Accuracy
256 p. 2004 [3-540-21876-9]

Vol. 304: Margaris, N.I.
Theory of the Non-linear Analog Phase Locked Loop
303 p. 2004 [3-540-21339-2]

Vol. 303: Mahmoud, M.S.
Resilient Control of Uncertain Dynamical Systems
278 p. 2004 [3-540-21351-1]

Vol. 302: Filatov, N.M.; Unbehauen, H.
Adaptive Dual Control: Theory and Applications
237 p. 2004 [3-540-21373-2]

Vol. 301: de Queiroz, M.; Malisoff, M.; Wolenski, P. (Eds.)
Optimal Control, Stabilization and Nonsmooth Analysis
373 p. 2004 [3-540-21330-9]

Vol. 300: Nakamura, M.; Goto, S.; Kyura, N.; Zhang, T.
Mechatronic Servo System Control
Problems in Industries and their Theoretical Solutions
212 p. 2004 [3-540-21096-2]

Vol. 299: Tarn, T.-J.; Chen, S.-B.; Zhou, C. (Eds.)
Robotic Welding, Intelligence and Automation
214 p. 2004 [3-540-20804-6]

Vol. 298: Choi, Y.; Chung, W.K.
PID Trajectory Tracking Control for Mechanical Systems
127 p. 2004 [3-540-20567-5]

Vol. 297: Damm, T.
Rational Matrix Equations in Stochastic Control
219 p. 2004 [3-540-20516-0]

Vol. 296: Matsuo, T.; Hasegawa, Y.
Realization Theory of Discrete-Time Dynamical Systems
235 p. 2003 [3-540-40675-1]

Vol. 295: Kang, W.; Xiao, M.; Borges, C. (Eds)
New Trends in Nonlinear Dynamics and Control, and their Applications
365 p. 2003 [3-540-10474-0]

Vol. 294: Benvenuti, L.; De Santis, A.; Farina, L. (Eds)
Positive Systems: Theory and Applications (POSTA 2003)
414 p. 2003 [3-540-40342-6]

Vol. 293: Chen, G. and Hill, D.J.
Bifurcation Control
320 p. 2003 [3-540-40341-8]

Vol. 292: Chen, G. and Yu, X.
Chaos Control
380 p. 2003 [3-540-40405-8]

Vol. 291: Xu, J.-X. and Tan, Y.
Linear and Nonlinear Iterative Learning Control
189 p. 2003 [3-540-40173-3]

Vol. 290: Borrelli, F.
Constrained Optimal Control of Linear and Hybrid Systems
237 p. 2003 [3-540-00257-X]

Vol. 289: Giarré, L. and Bamieh, B.
Multidisciplinary Research in Control
237 p. 2003 [3-540-00917-5]

Vol. 288: Taware, A. and Tao, G.
Control of Sandwich Nonlinear Systems
393 p. 2003 [3-540-44115-8]

Vol. 287: Mahmoud, M.M.; Jiang, J.; Zhang, Y.
Active Fault Tolerant Control Systems
239 p. 2003 [3-540-00318-5]

Vol. 286: Rantzer, A. and Byrnes C.I. (Eds)
Directions in Mathematical Systems Theory and Optimization
399 p. 2003 [3-540-00065-8]

Vol. 285: Wang, Q.-G.
Decoupling Control
373 p. 2003 [3-540-44128-X]

Vol. 284: Johansson, M.
Piecewise Linear Control Systems
216 p. 2003 [3-540-44124-7]

Vol. 283: Fielding, Ch. et al. (Eds)
Advanced Techniques for Clearance of Flight Control Laws
480 p. 2003 [3-540-44054-2]

Vol. 282: Schröder, J.
Modelling, State Observation and Diagnosis of Quantised Systems
368 p. 2003 [3-540-44075-5]

Vol. 281: Zinober A.; Owens D. (Eds)
Nonlinear and Adaptive Control
416 p. 2002 [3-540-43240-X]

Vol. 280: Pasik-Duncan, B. (Ed)
Stochastic Theory and Control
564 p. 2002 [3-540-43777-0]

Vol. 279: Engell, S.; Frehse, G.; Schnieder, E. (Eds)
Modelling, Analysis, and Design of Hybrid Systems
516 p. 2002 [3-540-43812-2]

Vol. 278: Chunling D. and Lihua X. (Eds)
H_∞ Control and Filtering of Two-dimensional Systems
161 p. 2002 [3-540-43329-5]

Vol. 277: Sasane, A.
Hankel Norm Approximation for Infinite-Dimensional Systems
150 p. 2002 [3-540-43327-9]

Vol. 276: Bubnicki, Z.
Uncertain Logics, Variables and Systems
142 p. 2002 [3-540-43235-3]

Vol. 275: Ishii, H.; Francis, B.A.
Limited Data Rate in Control Systems with Networks
171 p. 2002 [3-540-43237-X]

Vol. 274: Yu, X.; Xu, J.-X. (Eds)
Variable Structure Systems: Towards the 21st Century
420 p. 2002 [3-540-42965-4]

Vol. 273: Colonius, F.; Grüne, L. (Eds)
Dynamics, Bifurcations, and Control
312 p. 2002 [3-540-42560-9]

Vol. 272: Yang, T.
Impulsive Control Theory
363 p. 2001 [3-540-42296-X]

Vol. 271: Rus, D.; Singh, S.
Experimental Robotics VII
585 p. 2001 [3-540-42104-1]

Vol. 270: Nicosia, S. et al.
RAMSETE
294 p. 2001 [3-540-42090-8]

Vol. 269: Niculescu, S.-I.
Delay Effects on Stability
400 p. 2001 [1-85233-291-316]

Vol. 268: Moheimani, S.O.R. (Ed)
Perspectives in Robust Control
390 p. 2001 [1-85233-452-5]

Vol. 267: Bacciotti, A.; Rosier, L.
Liapunov Functions and Stability in Control Theory
224 p. 2001 [1-85233-419-3]

Vol. 266: Stramigioli, S.
Modeling and IPC Control of Interactive Mechanical Systems – A Coordinate-free Approach
296 p. 2001 [1-85233-395-2]

Vol. 265: Ichikawa, A.; Katayama, H.
Linear Time Varying Systems and Sampled-data Systems
376 p. 2001 [1-85233-439-8]

www.ingramcontent.com/pod-product-compliance
Lightning Source LLC
LaVergne TN
LVHW022042200926
848481LV00003B/748
* 9 7 8 1 8 4 6 2 8 2 8 9 8 *